Methods in Molecular Biology

Series Editor
John M. Walker
School of Life and Medical Sciences
University of Hertfordshire
Hatfield, Hertfordshire, AL10 9AB, UK

For further volumes:
http://www.springer.com/series/7651

The Hippo Pathway

Methods and Protocols

Edited by

Alexander Hergovich

Cancer Institute, University College London, London, UK

Humana Press

Editor
Alexander Hergovich
Cancer Institute
University College London
London, UK

ISSN 1064-3745 ISSN 1940-6029 (electronic)
Methods in Molecular Biology
ISBN 978-1-4939-8909-6 ISBN 978-1-4939-8910-2 (eBook)
https://doi.org/10.1007/978-1-4939-8910-2

Library of Congress Control Number: 2018963058

This Humana Press imprint is published by the registered company Springer Science+Business Media, LLC part of Springer Nature.
The registered company address is: 233 Spring Street, New York, NY 10013, U.S.A.

Preface

Over the past 15 years, the Hippo pathway has emerged as an essential regulator of organ growth and tissue homeostasis. Beginning with Drosophila geneticists, who laid much of the groundwork in Hippo signaling, we have now come to a greater understanding of this conserved and essential pathway across species. The work of many research teams since then has uncovered the crucial roles of the Hippo pathway in cell proliferation, death, differentiation, stemness, and other essential processes. Equally important, combinatorial approaches ranging from structural biology to mouse genetics have helped to establish the Hippo pathway as a vital signal transduction network that integrates numerous intracellular and extracellular signaling cues.

Before presenting the content of this book, I am very fleetingly providing an overview of our current understanding of the core of the Hippo pathway, as a kind of crash course for newcomers to the Hippo field. The intracellular core of the mammalian Hippo pathway is composed of scaffolding proteins (such as NF2/Merlin and MOB1), serine/threonine kinases (such as MST1/2 and LATS1/2; aka Hippo and Warts in flies), and the co-transcriptional regulators YAP/TAZ (aka Yorkie in flies). In the on state of the Hippo pathway, MST1/2 phosphorylate and thereby activate LATS1/2 in complex with MOB1. The activated LATS1/2-MOB1 complex then phosphorylates YAP/TAZ, hence resulting in the inactivation of YAP/TAZ (at least regarding their co-transcriptional activities). In the off state, MST1/2-LATS1/2 signaling is inactive, thus allowing YAP/TAZ to drive pro-survival and pro-growth transcriptional programs in the nucleus. To date, YAP/TAZ are considered the main effectors of the Hippo pathway, and therefore quite a number of chapters in this book are devoted to studies of YAP/TAZ.

This volume of the *Methods in Molecular Biology* series is a collection of expert lab protocols and commentaries for studies in the Hippo pathway. More specifically, this book documents the most common experimental approaches, recording methods ranging from single-molecule analysis to complex genetics and imaging in multicellular organisms. Thus, this book covers numerous expert methodologies to examine Hippo signaling on the structural, molecular, cellular, and organismal level. Notably, all chapters have been written by specialists in their fields who have repeatedly tested and validated these methods in the lab. On the one hand, this book introduces scientists from diverse research fields, such as genetics, biochemistry, structure, molecular, and cell biology, to the Hippo pathway. On the other hand, this book provides experimental templates for newcomers to exciting studies of the Hippo pathway. In this regard, we would like to emphasize that each methods chapter is complemented by a "Notes" section in order to further equip researchers with information on how to tackle any problem or difficulty that might arise when using a given technique.

This book has 26 chapters distributed over four main parts. The first part is focused on fly genetics, containing chapters that present to newcomers how the Hippo pathway was initially discovered and subsequently dissected genetically. Thus, approaches regarding mosaic genetics and immunofluorescence microscopy in flies are presented in the first part, which is comprised of six chapters. The second part addresses procedures that are useful for the molecular and cell biological analysis of the Hippo pathway. More specifically, the 11 chapters of the second part describe state-of-the-art methods including studies of the subcellular localization, co-transcriptional activities, and CRISPR-mediated manipulation of

YAP/TAZ; the regulation of the Hippo pathway by phosphorylation events; and three-dimensional tissue culture models and mechano-sensing. The third part consists of 7 chapters, covering experimental applications to study key components of the Hippo pathway using structural biology and biochemistry. These provide insights into state-of-the-art assays to study protein–protein interactions of Hippo core components both structurally and biochemically, as well as measurements of the phosphorylation and activity status of Hippo core components. The fourth and final part of this book gives information on studies of the Hippo pathway using mouse genetics. Given the present availability of excellent reviews of the Hippo pathway in cancer mouse models, we include here only two chapters that are focused on Hippo components in early embryonic development and regenerative medicine in mice, respectively.

Collectively, I hope that this book will be an essential part of many laboratory libraries to assist newcomers as well as experts, although I would be most pleased if this book would be mainly used on the lab bench rather than being stored away on a bookshelf. I am confident that this book equips newcomers and specialists alike with key methodologies to accurately define the status of Hippo signaling in their experimental settings. I sincerely believe that this book will encourage the scientific community to embrace coherent standards when studying the Hippo pathway.

I extend my thanks and appreciation to all the contributing authors and Prof. John Walker (University of Hertfordshire, UK) for their invaluable efforts during the process of assembling this book. Very special thanks go to Joanna Hergovich Lisztwan for her editing efforts. I also thank University College London and Springer Nature for their support required to produce this book. Last but not least, I would like to thank God for continuously encouraging me in the course of this book journey.

London, UK *Alexander Hergovich*

Contents

Part IV The Hippo Pathway and Mouse Models

Contributors

YOICHI ASAOKA • *Department of Systems Biochemistry in Pathology and Regeneration, Yamaguchi University Graduate School of Medicine, Yamaguchi, Japan*

MARDELLE ATKINS • *Department of Biological Sciences, Sam Houston State University, Huntsville, TX, USA*

ISABELLE BEQUIGNON • *IGF, Univ Montpellier, CNRS, INSERM, Montpellier, France*

MERVE BILGIC • *IGF, Univ Montpellier, CNRS, INSERM, Montpellier, France*

TREVER G. BIVONA • *Department of Medicine, University of California, San Francisco, San Francisco, CA, USA; Department of Cellular and Molecular Pharmacology, University of California, San Francisco, San Francisco, CA, USA; Helen Diller Family Comprehensive Cancer Center, University of California, San Francisco, San Francisco, CA, USA*

XIAOLEI CAO • *Life Sciences Institute and Innovation Center for Cell Signaling Network, Zhejiang University, Hangzhou, Zhejiang, China*

FABIENNE CHARRIER-SAVOURNIN • *Cisbio Bioassays, F-30200, Codolet, France*

MARIA CHATZIFRANGKESKOU • *Department of Oncology, University of Oxford, Oxford, UK*

LANFEN CHEN • *State Key Laboratory of Cellular Stress Biology, Innovation Center for Cell Signaling Network, School of Life Sciences, Xiamen University, Xiamen, Fujian, China*

QINGHUA CHEN • *State Key Laboratory of Cellular Stress Biology, Innovation Center for Cell Signaling Network, School of Life Sciences, Xiamen University, Xiamen, Fujian, China*

RUI CHEN • *Department of Pharmacology and Moores Cancer Center, University of California, La Jolla, CA, USA*

SIRIO DUPONT • *Department of Molecular Medicine, School of Medicine, University of Padova, Padova, Italy*

ELODIE DUPUIS • *Cisbio Bioassays, F-30200, Codolet, France*

RICHARD G. FEHON • *Department of Molecular Genetics and Cell Biology, The University of Chicago, Chicago, IL, USA; Committee on Development, Regeneration and Stem Cell Biology, The University of Chicago, Chicago, IL, USA*

TRISTAN FRUM • *Department of Biochemistry and Molecular Biology, Michigan State University, East Lansing, MI, USA*

ALEXANDER D. FULFORD • *Centre for Tumour Biology, Barts Cancer Institute, Queen Mary University of London, London, UK*

HIROKO FURUMOTO • *Department of Systems Biochemistry in Pathology and Regeneration, Yamaguchi University Graduate School of Medicine, Yamaguchi, Japan*

MAKOTO FURUTANI-SEIKI • *Department of Systems Biochemistry in Pathology and Regeneration, Yamaguchi University Graduate School of Medicine, Yamaguchi, Japan*

NEIL J. GANEM • *Department of Pharmacology and Experimental Therapeutics, The Cancer Center, Boston University School of Medicine, Boston, MA, USA; Division of Hematology and Oncology, Department of Medicine, Boston University School of Medicine, Boston, MA, USA*

REWATEE GOKHALE • *Department of Oncological Sciences, The Icahn School of Medicine at Mount Sinai, New York, NY, USA*

MARTA GOMEZ • *University College London Cancer Institute, London, UK*

KUN-LIANG GUAN • *Department of Pharmacology and Moores Cancer Center, University of California, La Jolla, CA, USA*
FRANZISKA HADERK • *Department of Medicine, University of California, San Francisco, San Francisco, CA, USA; Department of Cellular and Molecular Pharmacology, University of California, San Francisco, San Francisco, CA, USA; Helen Diller Family Comprehensive Cancer Center, University of California, San Francisco, San Francisco, CA, USA*
TOSHIO HAKOSHIMA • *Structural Biology Laboratory, Nara Institute of Science and Technology, Nara, Japan*
CARSTEN G. HANSEN • *University of Edinburgh Centre for Inflammation Research, Queen's Medical Research Institute, Edinburgh, UK*
CARL-PHILIPP HEISENBERG • *IST Austria, Klosterneuburg, Austria*
ALEXANDER HERGOVICH • *Cancer Institute, University College London, London, UK*
JULIA HICKS-BERTHET • *Department of Biochemistry, Boston University School of Medicine, Boston, MA, USA*
LIXIN HONG • *State Key Laboratory of Cellular Stress Biology, Innovation Center for Cell Signaling Network, School of Life Sciences, Xiamen University, Xiamen, Fujian, China*
KENNETH D. IRVINE • *Waksman Institute and Department of Molecular Biology and Biochemistry, Rutgers University, Piscataway, NJ, USA*
SUN-YONG KIM • *Structural Biology Laboratory, Nara Institute of Science and Technology, Nara, Japan*
NATHAN M. KINGSTON • *Department of Biochemistry, Boston University School of Medicine, Boston, MA, USA*
YAVUZ KULABEROGLU • *University College London Cancer Institute, London, UK; Department of Pharmacology, University of Cambridge, Cambridge, UK*
ZHI-CHUN LAI • *Intercollege Graduate Degree Program in Molecular, Cellular and Integrative Biosciences, The Pennsylvania State University, University Park, PA, USA; Department of Biology, The Pennsylvania State University, University Park, PA, USA; Department of Biochemistry and Molecular Biology, The Pennsylvania State University, University Park, PA, USA*
ODILE LECHA • *IGF, Univ Montpellier, CNRS, INSERM, Montpellier, France*
YUXI LI • *State Key Laboratory of Cellular Stress Biology, Innovation Center for Cell Signaling Network, School of Life Sciences, Xiamen University, Xiamen, Fujian, China*
KUI LIN • *State Key Laboratory of Microbial Metabolism, School of Life Sciences and Biotechnology, The Joint International Research Laboratory of Metabolic and Developmental Sciences, Shanghai Jiao Tong University, Shanghai, China*
QINGXU LIU • *State Key Laboratory of Cellular Stress Biology, Innovation Center for Cell Signaling Network, School of Life Sciences, Xiamen University, Xiamen, Fujian, China*
XUELIAN LUO • *Departments of Pharmacology and Biophysics, University of Texas Southwestern Medical Center, Dallas, TX, USA*
DAVID MATALLANAS • *Systems Biology Ireland, University College Dublin, Belfield, Dublin, Ireland; School of Medicine, University College Dublin, Belfield, Dublin, Ireland*
YURIKA MATSUI • *Intercollege Graduate Degree Program in Molecular, Cellular and Integrative Biosciences, The Pennsylvania State University, University Park, PA, USA*
HITOSHI MORITA • *Laboratory for Developmental Biology, Center for Medical Education and Sciences, Graduate School of Medical Sciences, University of Yamanashi, Yamanashi, Japan*
LISHENG NI • *Departments of Pharmacology and Biophysics, University of Texas Southwestern Medical Center, Dallas, TX, USA*

NIAMH A. O'DRISCOLL • *Systems Biology Ireland, University College Dublin, Belfield, Dublin, Ireland; School of Medicine, University College Dublin, Belfield, Dublin, Ireland*
ERIC O'NEILL • *Department of Oncology, University of Oxford, Oxford, UK*
VICTOR OLIVAS • *Department of Medicine, University of California, San Francisco, San Francisco, CA, USA; Department of Cellular and Molecular Pharmacology, University of California, San Francisco, San Francisco, CA, USA; Helen Diller Family Comprehensive Cancer Center, University of California, San Francisco, San Francisco, CA, USA*
JULIE PANNEQUIN • *IGF, Univ Montpellier, CNRS, INSERM, Montpellier, France*
CATHIE M. PFLEGER • *Department of Oncological Sciences, The Icahn School of Medicine at Mount Sinai, New York, NY, USA; The Graduate School of Biomedical Sciences, The Icahn School of Medicine at Mount Sinai, New York, NY, USA; The Tisch Cancer Institute, The Icahn School of Medicine at Mount Sinai, New York, NY, USA*
JEAN-PHILIPPE PIN • *IGF, Univ Montpellier, CNRS, INSERM, Montpellier, France*
STEVEN W. PLOUFFE • *Department of Pharmacology and Moores Cancer Center, University of California, La Jolla, CA, USA*
LAURENT PRÉZEAU • *IGF, Univ Montpellier, CNRS, INSERM, Montpellier, France*
RYAN J. QUINTON • *Department of Pharmacology and Experimental Therapeutics, The Cancer Center, Boston University School of Medicine, Boston, MA, USA*
AMY RALSTON • *Department of Biochemistry and Molecular Biology, Michigan State University, East Lansing, MI, USA*
VALENTINA RAUSCH • *University of Edinburgh Centre for Inflammation Research, Queen's Medical Research Institute, Edinburgh, UK*
CORDELIA RAUSKOLB • *Waksman Institute and Department of Molecular Biology and Biochemistry, Rutgers University, Piscataway, NJ, USA*
NINA REUVEN • *Department of Molecular Genetics, Weizmann Institute of Science, Rehovot, Israel*
PAULO S. RIBEIRO • *Centre for Tumour Biology, Barts Cancer Institute, Queen Mary University of London, London, UK*
MAÏTÉ ROMIER • *Cisbio Bioassays, F-30200, Codolet, France*
THOMAS ROUX • *Cisbio Bioassays, F-30200, Codolet, France*
MATAN SHANZER • *Department of Molecular Genetics, Weizmann Institute of Science, Rehovot, Israel*
YOSEF SHAUL • *Department of Molecular Genetics, Weizmann Institute of Science, Rehovot, Israel*
TING SU • *Department of Molecular Genetics and Cell Biology, The University of Chicago, Chicago, IL, USA*
ANDREW M. TILSTON-LUNEL • *Department of Biochemistry, Boston University School of Medicine, Boston, MA, USA*
SHERZOD A. TOKAMOV • *Department of Molecular Genetics and Cell Biology, The University of Chicago, Chicago, IL, USA; Committee on Development, Regeneration and Stem Cell Biology, The University of Chicago, Chicago, IL, USA*
ERIC TRINQUET • *Cisbio Bioassays, F-30200, Codolet, France*
XARALABOS VARELAS • *Department of Biochemistry, Boston University School of Medicine, Boston, MA, USA*
MARION VEREECKE • *IGF, Univ Montpellier, CNRS, INSERM, Montpellier, France*
CLAIRE VOL • *IGF, Univ Montpellier, CNRS, INSERM, Montpellier, France*

GENG WU • *State Key Laboratory of Microbial Metabolism, School of Life Sciences and Biotechnology, The Joint International Research Laboratory of Metabolic and Developmental Sciences, Shanghai Jiao Tong University, Shanghai, China*
JIAJIE XU • *Department of Molecular Genetics and Cell Biology, The University of Chicago, Chicago, IL, USA; Committee on Development, Regeneration and Stem Cell Biology, The University of Chicago, Chicago, IL, USA*
BIN ZHAO • *Life Sciences Institute and Innovation Center for Cell Signaling Network, Zhejiang University, Hangzhou, Zhejiang, China*
DAWANG ZHOU • *State Key Laboratory of Cellular Stress Biology, Innovation Center for Cell Signaling Network, School of Life Sciences, Xiamen University, Xiamen, Fujian, China*
DIANA ZINDEL • *IGF, Univ Montpellier, CNRS, INSERM, Montpellier, France*

Part I

The Hippo Pathway: Drosophila Genetics Lead the Way

Chapter 1

The Power of *Drosophila* Genetics: The Discovery of the Hippo Pathway

Rewatee Gokhale and Cathie M. Pfleger

Abstract

The Hippo Pathway comprises a vast network of components that integrate diverse signals including mechanical cues and cell surface or cell-surface-associated molecules to define cellular outputs of growth, proliferation, cell fate, and cell survival on both the cellular and tissue level. Because of the importance of the regulators, core components, and targets of this pathway in human health and disease, individual components were often identified by efforts in mammalian models or for a role in a specific process such as stress response or cell death. However, multiple components were originally discovered in the *Drosophila* system, and the breakthrough of conceiving that these components worked together in a signaling pathway came from a series of *Drosophila* genetic screens and fundamental genetic and phenotypic characterization efforts. In this chapter, we will review the original discoveries leading to the conceptual framework of these components as a tumor suppressor network. We will review chronologically the early efforts that established our initial understanding of the core machinery that then launched the growing and vibrant field to be discussed throughout later chapters of this book.

Key words Hippo, Warts, Sav, Mats, Yorkie, Drosophila, Signaling, Genetics

1 The *Drosophila* Model System

The animal model *Drosophila melanogaster* (fruit flies) enables cost-effective and large-scale experiments in an in vivo context. More than a century of concerted efforts have advanced the *Drosophila* research system and developed a series of powerful genetic tools. The quick developmental time period, entailing growth from a fertilized egg to a reproductive adult in 10 days at 25 °C, enables efficient implementation of those tools. There is high conservation of the signal transduction cascades between *Drosophila* and humans including many genes associated with disease. Even in the early days of genome analysis in 2000, it was clear that at least 60% of genes associated with human disease had clear homologs in *Drosophila* [1] and that number has grown substantially with increased genomic information and advances in understanding disease etiology. This

Alexander Hergovich (ed.), *The Hippo Pathway: Methods and Protocols*, Methods in Molecular Biology, vol. 1893, https://doi.org/10.1007/978-1-4939-8910-2_1,

evolutionary conservation allows *Drosophila* signaling insights to advance our fundamental knowledge of signaling networks across diverse organisms. Thus, the *Drosophila* model affords itself to rapid, phenotype-driven screens to identify mutations in core components of signaling processes highly relevant across species. This chapter will review the use of *Drosophila* in identifying a new signaling pathway, the Hippo Pathway, now considered a bona fide tumor suppressor pathway with roles in diverse processes important in development and in preventing disease. We apologize that we cannot review all the discoveries in the field or all mechanistic insights from each model system; we will focus on the seminal works that were crucial in building the framework of core machinery to cement this as a signaling pathway in *Drosophila*.

1.1 Drosophila *Genetic Tools and* Drosophila *Screens*

Drosophila research employs a number of powerful genetic tools innovated in this system or adapted from other genetic models. Undeniably, the ability to manipulate gene expression and genotype spatially and/or temporally affords creation of diverse contexts in which to elucidate the role of specific gene products; applied genome-wide, they afford us the ability to identify new components in fundamental processes such as growth, proliferation, patterning, and cell death. Two important tools that have advanced our ability to manipulate the genome and gene expression include the Gal4/UAS system [2] to modulate gene expression and the FLP/FRT system [3, 4] to induce mitotic recombination (Fig. 1).

Gal4 is a transcriptional regulator that recognizes upstream activating sequences (UAS) to promote gene expression. Coupling *gal4* expression to a promoter or enhancer element to drive its expression at a desired developmental period and/or in a desired tissue allows us to manipulate expression of a gene product that has been coupled to UAS. Thus, introducing a temporally and spatially expressed Gal4 will induce gene expression of a UAS transgene (Fig. 1a). With the advent of RNAi technology, coupling UAS to constructs to direct production of dsRNA or inverted repeats extends this tool from inducing expression of a gene product to the ability to reduce expression of a gene product.

Mosaic analysis is another useful tool that allows us to study adjacent populations of cells of different genotypes. Historically, one method to create mosaic tissues utilized the endogenous response to X-rays that would induce recombination at some frequency. When a cell carried two different alleles of a gene at a specific locus, random X-ray-induced recombination event proximal to the centromere would create two daughter cells—one homozygous for one allele, the other homozygous for the other allele, each distinct from the heterozygous mother cell. The FLP recombinase recognizes FLP recombination target (FRT) sequences to induce mitotic recombination [3, 4]. Expressing FLP spatially and temporally using promoter or enhancer elements (as described for Gal4) or randomly using heat shock (hs) promotes

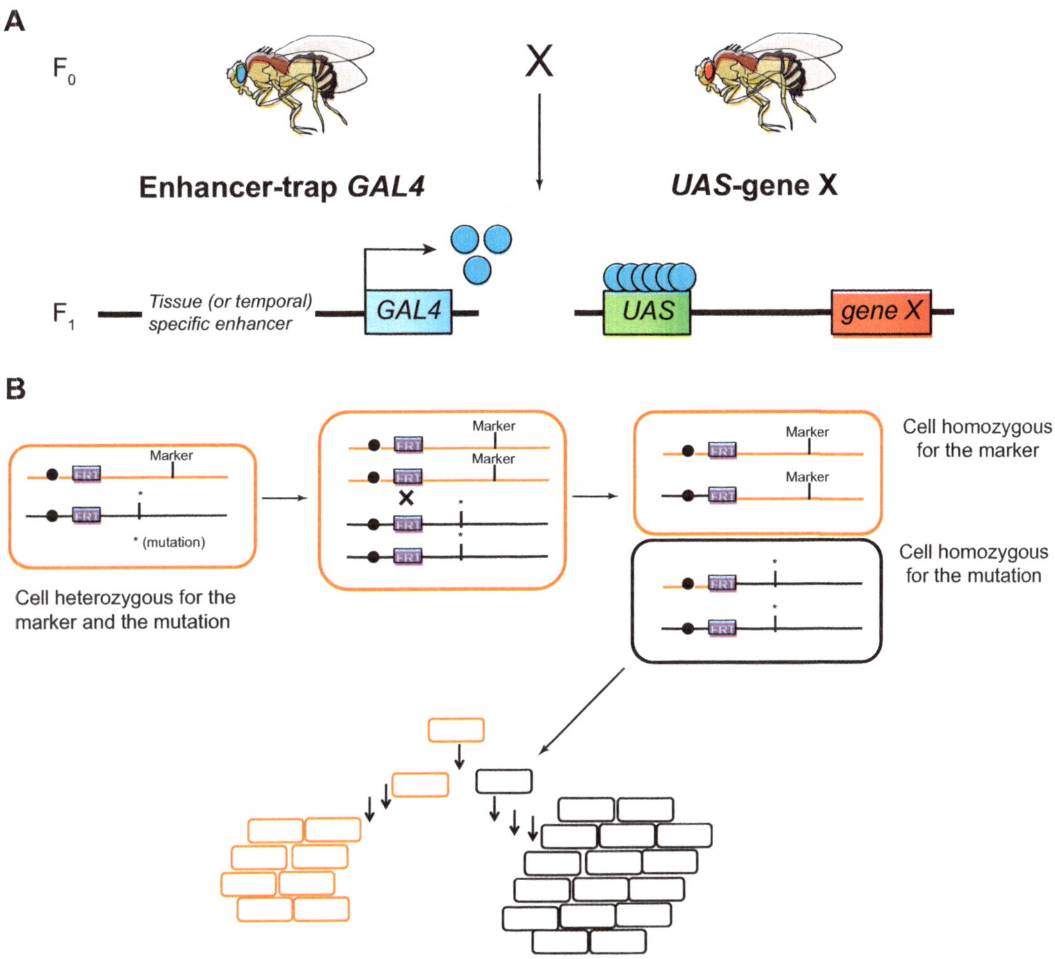

Fig. 1 Genetic tools and screening strategies. (**a**) The Gal4/UAS system controls gene expression temporally and spatially. Tissue-specific expression of gal4 is directed by specific promoters or enhancers. Gal4 recognizes upstream activating sequences (UAS). When a line containing a gene or inverted repeat under the control of UAS is crossed to a line carrying a tissue-specific gal4, progeny inheriting these elements will express the gene or inverted repeat according to the expression pattern defined by the enhancer. Using this tool, gene expression or knockdown can be controlled spatially and temporally. (**b**) The FLP/FRT system to generate mosaic tissue by mitotic recombination. The FLP recombinase recognizes FLP recombination target (FRT) sequences and induces recombination during mitosis so that a cell heterozygous for a mutation divides to give rise to two daughter cells, one homozygous wild-type and the other homozygous for the mutation. By placing a marker (such as a ubiquitously expressed GFP, lacZ, and/or pigment) on the wild-type chromosome, one can follow the presence or absence of the marker to identify wild-type or mutant tissue. If the mutant tissue has an advantage, for example an increased rate of proliferation or a decreased rate of cell death, mutant tissue will be overrepresented after a period of time

mitotic recombination at FRT sites. The resulting daughter cells are homozygous distal to the FRT (Fig. 1b). Thus, FLP/FRT tools allow us to induce recombination events to create mosaic tissues precisely. Importantly, the ability to generate mitotic recombination across an entire tissue with the FLP/FRT system was a profound advance at the heart of the discovery and characterization of many key components of the Hippo Pathway.

1.2 Genetic Screens

1.2.1 Classic Screens

Seminal genetic screening efforts in *Drosophila* include the classic screen by Nüsslein-Volhard and Wieschaus [5]. This screen isolated embryonic lethal mutations that were then examined for developmental abnormalities such as segmentation defects. This approach identified regulators of embryonic patterning including components of some of the most well-studied signal transduction pathways and processes that have individually become thriving independent fields today such as *hedgehog*. This screening strategy (simplified screening strategy shown in Fig. 2a) entails multiple generations of crosses and results in balanced, true breeding lines of mutants that can be scored for homozygous phenotypes including embryonic lethality, patterning defects, and more.

1.2.2 Modifier Screens

A popular genetic screening approach that builds on primary screen efforts or can follow directly from on obvious phenotype of interest is the genetic modifier screen. Genetic modifier screens utilize direct crosses to a genetic background with a phenotype of interest to isolate dominant modifiers (Fig. 2b) or multigenerational schemes to isolate recessive modifiers on a genome scale to identify individual mutations that enhance or suppress the phenotype in question. Modifier screens have traditionally built on primary screening efforts by (1) identifying genes that affect processes of interest or identifying other components of pathways discovered in primary screens and (2) helping to order the steps of signaling.

Decades of work thus characterized classical developmental signaling pathways including hedgehog signaling, Wingless/Wnt signaling, EGFR/Ras signaling, Notch signaling, PI3K/AKT signaling, and others in parallel to work in other model systems.

1.2.3 Mosaic Screens

Mutations in a number of important signaling components result in embryonic or larval lethality. Such mutant alleles could therefore be isolated by lethality screens as described above. Indeed, as noted, hedgehog was identified in the classic Nüsslein-Volhard and Wieschaus screen [5]. A limitation of these screens is that phenotypes associated with specific roles in proliferation, growth, or patterning in later stages might be missed without more labor-intensive efforts to characterize these specific phenotypes in those later developmental contexts. The advent of Gal4/UAS and FLP/FRT tools to modulate gene expression and genotype temporally and spatially extended the ability of large-scale screens to discover important signaling factors whose phenotypes might have eluded earlier screening efforts (genetic strategy for mosaic screens outlined in Fig. 2c).

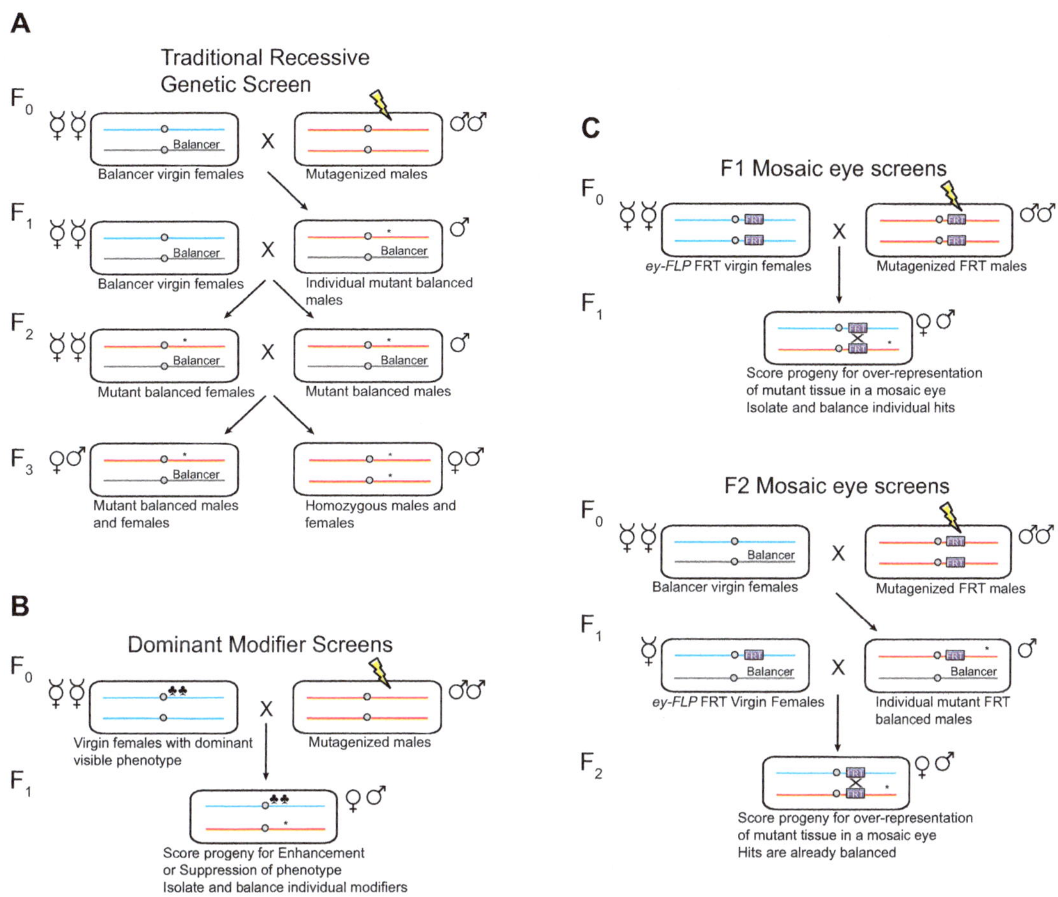

Fig. 2 Schematic of different screening strategies. (**a–c**) Comparison of traditional multigenerational genetic screens of recessive phenotypes (**a**) to dominant modifier screens (**b** panel) or mosaic screens (**c** panel). (**a**) In a traditional recessive screen, multiple generations are needed to generate homozygotes to screen for a phenotype. This strategy can be used to identify lethal mutants (where homozygotes would die, so would not be observed) or other phenotypes such as patterning abnormalities. An advantage of this strategy is that the scheme generates true breeding lines of any hits. (**b**) Modifier screens isolate mutations that modify a phenotype of interest. Dominant modifier screens (depicted) isolate mutants where a single mutant allele is sufficient to enhance or suppress the phenotype of interest. These mutant alleles must then be isolated and balanced to generate true breeding lines. Recessive modifier screens (not depicted) would utilize a multi-generation approach (as depicted in **a**) to identify recessive modifiers. (**b**) Mosaic screens (outlined for mosaic eye screens) can generate homozygous tissue in first-generation (F1) progeny. The ability to screen for recessive phenotypes in one generation accelerates the ability to identify mutants. F1 mosaic eye screen hits must then be balanced to create true breeding lines. In some cases, mosaic eye phenotypes are severe and can cause lethality, or the isolated mutants of interest can be difficult to breed. For this reason, many groups also utilized "F2" mosaic screens

Mutagenesis screens that utilized FLP/FRT tools to create homozygous mutant tissue adjacent to homozygous wild-type tissue in an otherwise heterozygous fly allowed for the comparison of the relative proportion of mutant to wild-type tissue. Thus, homozygous mutant tissue phenotypes could be observed in later developmental periods or adulthood in a heterozygous fly, thereby circumventing embryonic and larval lethality in a homozygous fly. A series of such screens utilized FLP expressed early in the developing eye under the control of *eyeless* (*eyFLP*). This allowed for swaths of homozygous tissue generated in one generation to replace multigenerational recessive modifier screens described above. This strategy also allowed sufficient rounds of division such that even a subtle advantage of the mutant cells could be easily detected by visualizing the adult eye.

These mosaic eye screens identified a variety of mutant alleles that conferred a growth or proliferation advantage. The alleles isolated based on an initial selection for overrepresentation of mutant tissue in a mosaic eye were primarily mutant alleles of genes that normally serve as negative regulators of growth and proliferation and/or positive regulators of cell death. For example, *archipelago*/Fbw7 was identified as a tumor suppressor whose mutation resulted in increased levels of cell cycle regulator cyclin E, transcriptional regulator myc, and signaling protein Notch to result in phenotypes of cell cycle reentry and tissue overgrowth [6, 7]. Mutations in the *Drosophila* Apaf1 homolog *ark* also gave overrepresentation of mutant tissue in a mosaic due to phenotypes of impaired cell death [8]. Tuberous sclerosis complex gene *Tsc1* and *Tsc2* mutations not only led to overrepresentation of mutant tissue compared to wild-type tissue but also dramatic organ overgrowth and increased cell size phenotypes [9]. Thus, these mutant phenotype characterizations revealed differences in the overgrowth phenotypes when examined at the tissue or cellular level including different downstream events such as dysregulation of cyclin E, specific components of other signaling cascades such as myc or Notch, and distinct effects on cell size, mass accumulation, and/or apoptosis.

2 Historical Overview: Identifying the Hippo Pathway Core Machinery

In 2002, mosaic eye screens by the Hariharan and Halder groups identified novel mutations in a gene called *Salvador* (*sav*, also referred to as *Shar-pei*) which caused phenotypes affecting proliferation, mass accumulation, cell death, and overall organ size. *Sav* mutant cells were overrepresented in a mosaic eye, showed cell cycle reentry and ectopic proliferation, and exhibited cell death resistance phenotypes [10, 11]. Characterization of *sav* mutant tissue to explain these phenotypes mechanistically showed increased levels of cyclin E (which could in part explain the cell cycle reentry and ectopic proliferation phenotypes) and also increased levels of apoptosis inhibitor DIAP1 (which would in part explain the cell death resistance phenotypes). Clearly, these alleles were distinct from those that only resulted in cell death resistance (such as *ark* alleles) or that resulted in increased cell size phenotypes (such as *Tsc1* and *Tsc2* alleles). Curiously, these same screening efforts identified another overgrowth complementation group that phenocopied *sav* very closely [10]. This complementation group contained alleles of the tumor suppressor *warts* (*wts*, also referred to as *lats*), a kinase originally identified in *Drosophila* in 1995 [12, 13]. Genes with like phenotypes that represent distinct complementation groups often represent genes involved in the same process or pathway. Given the similarity of phenotypes at the tissue level, Tapon et al. tested if these two genes interacted genetically and biochemically. Indeed, overexpression of Sav protein alone or Wts protein alone had minimal phenotypes, but co-overexpressing Sav and Wts together promoted obvious organ size reduction. Moreover, Sav and Wts co-immunoprecipitated when expressed in cells, reflecting a physical interaction. This early characterization of Sav and identification that Sav interacted with known tumor suppressor Wts was a key observation leading to the discovery that these proteins functioned as part of a core complex in a new pathway (Fig. 3 and Table 1).

The following year, in 2003, screening efforts by five labs independently identified a gene whose mutant phenotype bore striking resemblance to mutations in *sav* and *wts*—a gene called *hippo* (*hpo*, also referred to as *dMST*), a kinase homologous to mammalian Mst1 and MST2 kinases (Fig. 3 and Table 1) [14–18]. The characterization of mutant or RNAi alleles revealed the same ectopic proliferation, tissue overgrowth, cell death resistance, accumulation of cyclin E, and accumulation of DIAP1 phenotypes as *sav* and *wts* mutant tissues. Moreover, the genetic and biochemical analyses revealed both genetic and physical interactions between Hpo and Sav and Wts. In addition to the overlapping characterizations, the diversity of alleles and approaches revealed additional mechanistic insights including intriguing findings that DIAP1 was dysregulated transcriptionally in mutant tissue but also that Hippo kinase activity promoted phosphorylation of DIAP1 protein. This revealed that the Sav-Wts-Hpo complex regulated DIAP1 both transcriptionally and post-translationally.

Of these three proteins, Sav contained multiple candidate protein-protein interaction domains, and both Hpo and Wts possessed kinase activity. Therefore, at this early time in the field, many speculated that Sav served as a scaffold protein that could promote the interaction between Hpo and Wts and that this complex could function to propagate kinase activity akin to the kinase cascade in MAPK signaling. Biochemical experiments in these studies showed that Hpo autophosphorylated and phosphorylated Sav protein and promoted phosphorylation of Wts protein. Observations that *hpo* phenotypes were generally less severe than *wts* phenotypes and that Hpo overexpression on its own resulted in severe growth restriction

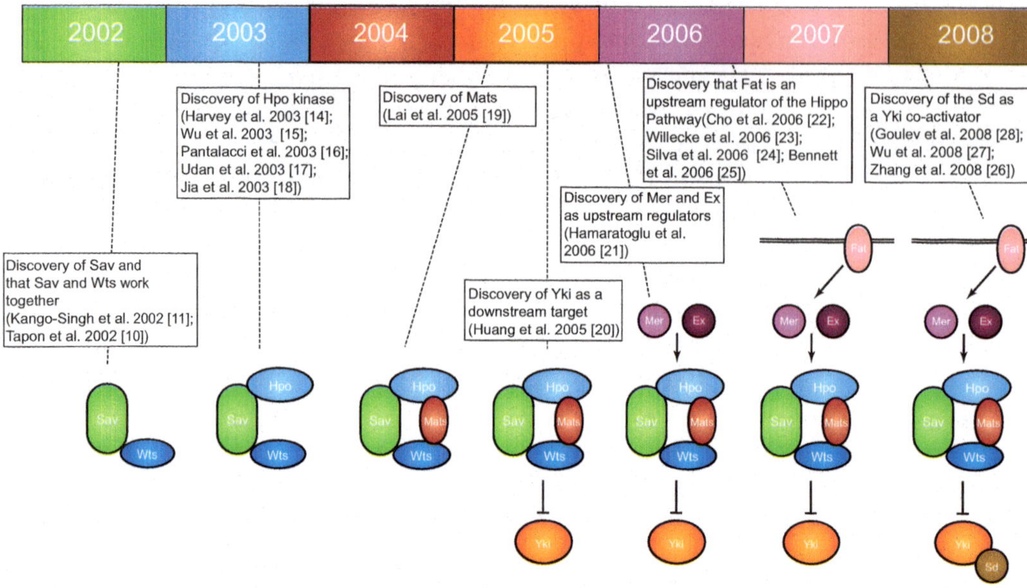

Fig. 3 Timeline denoting key events in piecing together upstream, core, and downstream nodes of the Hippo Pathway. Timeline of the first 6 years of the pathway starting with discovery the of Sav component. The timeline indicates key components as they came together to our current conception of the pathway. Below the timeline, a schematic indicates our growing model of how this network came together from discovery of core components to our understanding of how a signal would be transduced from upstream steps to a downstream transcriptional output. This simplified schematic is not exhaustive and does not reflect all discoveries in the pathway in this time. For additional reports during this time and after these years, please refer to Table 1

phenotypes compared to the weak phenotypes of Sav and Wts overexpression (unless co-expressed) together with the described biochemical analysis were considered consistent with a model that Hpo promoted Wts activity and that Wts acted downstream of Hpo (Fig. 3).

In MAPK signaling and a number of other traditional signaling pathways, an extracellular ligand, whether a soluble factor (such as EGF) or a transmembrane ligand (such as the Notch ligand Delta which resides in a neighboring cell), is received by a transmembrane receptor that is then transduced directly or through a series of steps to define downstream biological outputs. These outputs typically include a transcriptional response and may also include non-transcriptional effectors. A putative scaffold bringing together two kinases and resulting in transcriptional control of cyclin E and DIAP1 raised a number of important questions for the inchoate Sav-Wts-Hpo field: (1) what were the ultimate downstream effectors of the Sav-Wts-Hpo complex? (2) Was there an upstream signal or ligand that bound a receptor to activate the complex to transduce a signal? (3) Did this constitute a core complex in a new signaling pathway?

Table 1
Curated timeline of discoveries and insights in the *Drosophila* Hippo Pathway field

Year	Reports	Key discovery or insight
1995	Xu et al. 1995 [12]; Justice et al. 1995 [13]	• Discovery of Warts (Wts) • Characterization overgrowth/tumor suppressor phenotypes
2002	Tapon et al. 2002 [10]; Kango-Singh et al. 2002 [11]	• Discovery of Salvador (Sav) • Characterization of overgrowth/tumor suppressor phenotypes for *sav* mutants (ectopic proliferation/cell cycle reentry, cell death resistance, increased mass accumulation, increased cyclin E protein, increased cyclin E transcription, increased DIAP1 protein) • Sav protein interacts with Wts protein • Sav interacts genetically with Wts • Sav and Wts overexpression restricts tissue growth
2003	Harvey et al. 2003 [14]; Wu et al. 2003 [15]; Pantalacci et al. 2003 [16]; Udan et al. 2003 [17]; Jia et al. 2003 [18]	• Discovery of Hippo (Hpo) • Hippo phenocopies Sav and Wts • The core cassette phosphorylates DIAP1 and promotes its degradation • The core complex regulates DIAP1 transcription • Hpo interacts with Sav physically and genetically • Hpo promoted Sav and Wts phosphorylation
2005	Lai et al. 2005 [19]	• Discovery of Mob as tumor suppressor (Mats) • Mats phenocopies Sav, Wts, and Hpo • Mats acts as a Wts cofactor
2005	Huang et al. 2005 [20]	• Discovery of Yorkie (Yki) as an inhibitory target of the core cassette • Yki activation phenocopies many aspects of loss of the core components • Yki regulates transcription of cyclin EE and DIAP1
2005	Mikeladze-Dvali et al. 2005 [63]	• Warts and Melted act opposite each other in determining photoreceptor fate • This represented a bistable loop • This was a postmitotic role for the pathway
2006	Hamaratoglu et al. 2006 [21]	• Merlin (Mer) and Expanded (Ex) act upstream of the Sav/Wts/Hpo core cassette • *mer* and *ex* double mutant tissue phenocopies *hpo*, *sav*, *wts* individual mutant tissue

(continued)

Table 1 (continued)

Year	Reports	Key discovery or insight
		• Yki promotes transcription of *mer* and *ex*, establishing feedback
2006	Colombani et al. 2006 [74]	• Ionizing radiation activates Hpo via p53
2006	Emoto et al. 2006 [47]	• Hpo regulates dendritic tiling and maintenance
2006	Thompson and Cohen 2006 [75]; Nolo et al. 2006 [76]	• Growth regulator Bantam, a microRNA, is a Yki target regulated by the Hippo Pathway
2006	Cho et al. 2006 [22]; Willecke et al. 2006 [23]; Silva et al. 2006 [24]; Bennett et al. 2006 [25]	• The Fat cadherin promotes Wts activity • Fat inhibits Dachs to relieve Dachs inhibition of Wts (independent of other pathway components) • Fat acts via Ex to promote activity of Wts through the core cassette
2006	Polesello et al. 2006 [77]	• RASSF acts with Hippo to restrict growth parallel to Hippo acting with Wts to restrict growth
2007	Tyler et al. 2007 [78]	• Cells with decreased Hippo Pathway activity are resistant to elimination by cell competition
2007	Wei et al. 2007 [79]	• Mats associates with Hpo and is a target of Hpo kinase acting downstream of Hpo • Mats phosphorylation increases its association with Wts
2007	Pellock et al. 2007 [80]	• Ex and Mer individually regulate Hippo Pathway-controlled processes such as proliferation, apoptosis, and Wg differently
	2007–2008 Pellock et al. 2007 [80]; Shimizu et al. 2008 [81]	• Cyclin A and cyclin B accumulate upon loss of pathway components
2008	Zhang et al. 2008 [26]; Wu et al. 2008 [27]; Goulev et al. 2008 [28]	• Identification of Sd as a Yki-binding partner to regulate transcription
2008	Dutta and Baehrecke 2008 [43]	• Wts regulates autophagy independent of Yki
2007–2008	Dong et al. 2007 [31]; Oh and Irvine 2008 [32];	• Detailed molecular characterization of phosphorylation sites on Yki and regulation of Yki localization by phosphorylation • Confirmation that *Drosophila* 14-3-3 binds phosphorylated Yki

(continued)

Table 1 (continued)

Year	Reports	Key discovery or insight
2009	Badouel et al. 2009 [82]; Oh et al. 2009 [83]	• Ex, Hpo, and Wts proteins can each bind Yki directly and can sequester Yki in the cytoplasm independent of Yki phosphorylation
2009	Peng et al. 2009 [84]	• Yki can interact with DNA-binding transcription factors homothorax and teashirt to promote transcription
2009–2012	Genevet et al. 2009 [49]; [Hamaratoglu et al. 2009 [85]; Chen et al. 2010 [86]; Grzeschik et al. 2010 [87]; Ling et al. 2010 [88]; Parsons et al. 2010 [89]; Robinson et al. 2010 [90]; Hafezi et al. 2012 [91]; Verghese et al. 2012 [50]	• Junctional components and apical basal polarity regulators including Crumbs, aPKC, Lgl, and Scribble regulate the activity of the Hippo Pathway
2010	Baumgartner et al. 2010 [92]; Genevet et al. 2010 [93]; Yu et al. 2010 [94]	• Kibra physically associates with Mer, promotes Mer/Ex association, and acts with Mer and Ex to promote Hippo Pathway activity
2010	Das Thakur et al. 2010 [95]	• Ajuba LIM proteins negatively regulate the Hippo Pathway
2010	Ren et al. 2010 [33]	• Additional genetic and biochemical mechanistic analysis of 14-3-3-mediated regulation of Yki localization and activity
2010	Riebero et al. 2010 [96]	• The *Drosophila* PP2A phosphatase dSTRIPAK regulates Hpo phosphorylation to inhibit Hpo activity during development
2010	Neto-Silva et al. 2010 [66]; Ziosi et al. 2010 [67]	• Yki regulates the transcription of *myc* • Myc makes cells supercompetitive upon loss of Hippo signaling, explaining earlier reports regarding a role in cell competition • Myc regulates Yki in a feedback circuit
2010	Staley and Irvine 2010 [35]; Ren et al. 2010 [36]; Karpowicz et al. 2010 [37]; Shaw et al. 2010 [38]	• The Hippo Pathway regulates stem cell proliferation and regeneration in the intestine
2011	Oh and Irvine 2011 [97]	• Dpp signaling component Mad also acts with Yki to direct transcriptional outputs including growth regulator Bantam
2011	Sansores-Garcia et al. 2011 [51]	• Increased F-actin or loss of actin capping proteins A and B inhibits Hippo Pathway activity and leads to overgrowth

(continued)

Table 1
(continued)

Year	Reports	Key discovery or insight
2011	Rauskolb et al. 2011 [61]	• Identification of Zyxin downstream of Fat as a protein that interacts with Dachs to inhibit Wts • Zyxin in this role links the pathway to regulation by mechanical strain
2011	Fernández et al. 2011 [44]	• F-actin and capping proteins regulate Hippo Pathway activity • The Hippo Pathway regulates the actin cytoskeleton in a Yki-independent manner
2011	Boggiano et al. 2011 [98]; Poon et al. 2011 [99]	• The kinase TAO1 activates the Hippo Pathway core cassette
2011	Jukam and Desplan 2011 [64]	• Roles for specific components of the pathway in feedback and in neuronal specification and maintenance
2012	Verghese et al. 2012 [100]	• The Hippo Pathway regulates the caspase Dronc
2012	Herranz et al. 2012 [101]	• Capicua is involved in EGFR/Hippo cross talk
2012	Yue et al. 2012 [102]	• Cell adhesion molecule Echinoid regulates the Hippo Pathway
2012	Kagey et al. 2012 [103]	• Nonautonomous regulation of Yki by hedgehog receptor Patched
2012	Ye et al. 2012 [104]	• The Hippo Pathway negatively regulates AKT
2013	Wehr et al. 2013 [57]	• Salt-inducible kinases promote phosphorylation and inhibition of -Sav protein leading to activation of Yki
2013	Marcinkevicius and Zallen 2013 [45]	• The Hippo Pathway regulates planar polarity in a Yki-independent manner
2013	Sansores-Garcia et al. 2013 [105]; Sidor et al. 2013 [106]	• MASK functions as a cofactor to promote Yki activity
2013	Oh et al. 2013 [107]	• Yki associates widely with chromatin and chromatin remodeling complexes implicating a broad role in transcriptional regulation
2013	Reddy and Irvine 2013 [108]	• Hippo signaling is regulated by EGFR via Ajuba family proteins, adding additional depth to our knowledge of EGFR/Hippo Pathway cross talk

(continued)

Table 1 (continued)

Year	Reports	Key discovery or insight
2013	Koontz et al. 2013 [109]	• Sd acts as a transcriptional repressor; Yki interaction with Sd antagonizes this default repression to activate transcription
2013	Lucas et al. 2013 [42]	• The Hippo Pathway inhibits Enabled via Wts-mediated phosphorylation to activate F-actin capping protein activity to regulate actin during collective cell migration
2013	Huang et al. 2013 [110]	• Parl inhibits Hpo and Sav to promote Yki activity and tissue growth
2013	Sun and Irvine 2013 [111]	• JNK promotes Yki activity by promoting Ajuba family protein interaction with Warts
2013–2015	Yin et al. 2013 [61]; Sun et al. 2013 [62]	• Mer and Ex recruits Wts to the membrane to facilitate activation by Hpo
2013	Ilanges et al. 2013 [46]	• Alcohol interacts with the Hippo Pathway in a Yki-independent way
2013	Jukam et al. 2013 [65]	• Analysis of multiple feedback circuits for controlling growth or neural fate and the Wts-Melted bistable switch
2014	Rauskolb et al. 2014 [53]	• Cytoskeletal tension inhibits Hippo signaling
2014	Milton et al. 2014 [71]	• The Hippo Pathway regulates hematopoiesis
2014	Sing et al. 2014 [73]	• Fat and the Hippo Pathway regulate mitochondria and metabolism
2015	Gaspar et al. 2015 [54]	• Zyxin interacts with Enabled to regulate actin and inhibit Ex
2015	Deng et al. 2015 [55]; Fletcher et al. 2015 [56]; Wong et al. 2015 [57]	• Spectrin regulates the Hippo Pathway
2015	Aerne et al. 2015 [59]	• Hpo protects Sav from targeting by the Herc4 ubiquitin ligase
2015	Zhang et al. 2015 [112]	• Taiman forms a complex with Yki to direct specific transcriptional outputs
2015	Zheng et al. 2015 [113], Li et al. 2015 [114]	• The kinases happyhour/MAP4K3 and misshapen/MAP4K4 act in the Hippo Pathway

(continued)

Table 1 (continued)

Year	Reports	Key discovery or insight
2015	Keder et al. 2015 [40]	• The Hippo Pathway regulates asymmetric cell division by Wts phosphorylation of Canoe and Bazooka
2015	Dewey et al. 2015 [41]	• The Hippo Pathway regulates the orientation of cell division by Wts phosphorylation of Mud
2015	Hirabayashi et al. 2015 [60]	• Dietary sugar regulates Hippo Pathway activity via salt-inducible kinases
2016	Liu et al. 2016 [72]	• The Hippo Pathway regulates innate immunity
2016	Ding et al. 2016 [39]	• The Hippo Pathway regulates neural stem cell quiescence
2016	Jahanshahi et al. 2016 [48]	• Discovery of Rae1 as a downstream pathway target involved in pathway regulation of organ size and cyclin B

This table delineates the timeline of discovery of core components, upstream regulators, downstream targets, and several other mechanistic insights helping to build a comprehensive model of the Hippo Pathway and its roles in development and growth control in *Drosophila*. The year, published reports, and some of the highlights or advances from each study are listed. We apologize that we could not review in depth each of these studies or list every published report in the field. This table aims to review some of the key advances as well as some of the curious findings to highlight the dynamic field of research and diverse roles of the pathway.

3 Establishing Key Upstream and Downstream Events in the Hippo Pathway

The next few years saw a flurry of activity as multiple *Drosophila* research groups tried various approaches including mosaic screens, modifier screens, two hybrid screens, and other avenues to identify additional components that interacted with this core complex, to answer these fundamental questions, and to add molecular detail (some of these efforts are summarized in Table 1). Early in 2005, Lai et al. reported that *Mob as tumor suppressor* (*Mats*), the fly homolog of the Mob/Mob1 gene, was a new component of the pathway. A *Mats* allele was a spontaneous mutation identified in the Lai lab that phenocopied *hpo*, *sav*, and *wts* mutants leading to the report that Mats served as a Wts coactivator [19]. This assembly of Sav, Hpo, Mats, and Wts proteins established what we came to consider the core complex or the core kinase cassette of the pathway (Fig. 3 and Table 1).

Later that year, the field dramatically advanced with the identification of Yorkie (Yki, homologous to the mammalian YAP), a transcriptional coactivator, as a target inhibited by the core kinase cassette by direct phosphorylation by Wts (Fig. 3 and Table 1). Yki overexpression resulted in overgrowth phenotypes that resembled mutation in *hpo*, *sav*, *mats*, and *wts* [20], and Yki controlled transcription of cyclin E and DIAP1. This finding was a seminal point in the field; for this to be a signaling pathway, there would likely be a downstream transcriptional output affected by the core cassette. Like many transcriptional regulators, Yki lacked a DNA-binding domain. It was generally assumed that Yki must act by partnering with a DNA-binding protein to initiate transcription of its targets many of whose identification would come in the next couple of years.

Yki filled the void in the pathway schematic for the downstream transcriptional output one expects in a signaling pathway. However, the question of upstream receptors that bound ligand to initiate signaling to that transcriptional output remained. Shortly after identification of Yki, the Halder group identified the FERM domain membrane-associated proteins Merlin and Expanded as upstream regulators of the core cassette (Fig. 3 and Table 1) [21]. *Mer* and *ex* double mutant tissue phenocopied loss of *hpo*, *sav*, or *wts*, and complementing genetic and biochemical studies showed that they function upstream of the core cassette to promote Wts phosphorylation and activity. Although not transmembrane receptors that act to send a cue initiated by binding to an extracellular ligand, this was the first report of upstream membrane-associated factors that could affect the activity of the core cassette to define downstream outputs. Interestingly, expression of *mer* and *ex* was also controlled transcriptionally by Yki activity, indicating a negative feedback mechanism common in signaling pathways.

At the end of that same year, multiple groups identified the Fat cadherin as an additional upstream regulator activating the pathway via multiple mechanisms. As with the previously described studies, phenotypic characterization coupled with genetic interaction and biochemical studies revealed that Fat inhibits a protein called Dachs which itself binds and inhibits Wts. Thus Fat can promote Wts activity independently of the other components. These studies also indicated a role for Fat upstream of Ex to promote Wts activity through the core cassette (Fig. 3, Table 1) [22–25]. This discovery linked the pathway to planar cell polarity signaling (via known planar polarity components Fat and its ligand Dachsous) and also provided the first transmembrane component to the pathway.

After another year, multiple groups concluded the search for a Yki DNA-binding partner with the identification of scalloped (Sd, the *Drosophila* representative of the TEAD/TEF family of transcription factors) [26–28] (Fig. 3 and Table 1). Using multiple strategies including unbiased two-hybrid screening for Yki-binding partners as well as candidate approaches based on work on the mammalian homolog YAP, these groups confirmed that Sd co-immunoprecipitated with Yki from cells and could promote nuclear localization of Yki. Functional characterization revealed the Sd acted with Yki to promote transcription of Hippo Pathway targets, including work that identified Sd-binding sites in the DIAP1 promoter region. Importantly, Sd activity was important for the overgrowth seen upon mutation in the core components.

This work over the course of half a dozen years from initial efforts (1) to identify Sav and connect Sav to known tumor suppressor Wts through (2) to establishing that Yki/Sd acted as a key transcriptional output inhibited by the core cassette built the fundamental framework of the Hippo Pathway. Increased proliferation and growth may lead to cell autonomous overgrowth phenotypes within a tissue, but organ size control mechanisms still engage to limit the size of the overall tissue or organ (for review, *see* Reference [29]). Some refer to this control as an "organ size checkpoint," the molecular mechanisms of which remain unresolved. Over these years as the core machinery of the Hippo Pathway was identified and characterized, the tissue overgrowth phenotypes made clear that the Hippo Pathway played a role in organ size control or in the organ size checkpoint. These seminal developments in this time period launched a field and cemented the idea of these components functioning as an emerging signaling pathway with roles in development, organ size control, and tumor suppression.

4 Building a Picture of the Broader Signal Transduction Network

This basic framework we described that was built over those years is still in place today. Tremendous and growing efforts from a number of systems working throughout this time and in subsequent years yielded mechanistic detail into the interactions between the components of the pathway and revealed insights into the biological outputs and roles of the pathway in different contexts. A variety of the seminal discoveries and other insights into the role and components of the Hippo Pathway in the *Drosophila* system are surveyed in Table 1. From the early days of this work, the embrace of biochemical approaches by *Drosophila* geneticists and the bridges from the *Drosophila* model to groups focused on cancer biology and cancer genetics helped vault over hurdles to advance these efforts without delay. In fact, a number of advances followed rapidly due (1) to the efforts of *Drosophila* groups to corroborate their findings in mammalian models and also due (2) to remarkable efforts in mammalian systems on key components which predated the *Drosophila* work or was ongoing independently to drive our fundamental understanding of the pathway (for a brief review of components in the mammalian system, *see* Subheading 5). As an example, identification of Sd as a Yki-binding partner came in part from recognition of YAP interaction with the TEAD/TEF factors in mammalian systems. Moreover, YAP interaction with 14-3-3 [30] facilitated detailed *Drosophila* mechanistic studies of Yki to reveal specific Wts-mediated phosphorylation sites on Yki that created 14-3-3 binding sites. This revealed that one mode of inhibition was the ability of Wts activity to promote 14-3-3-mediated sequestration of Yki in the cytoplasm where it could not act to promote transcription [31–33]. The high conservation of this important signaling cascade allowed researchers utilizing these distinct approaches to integrate key findings across their systems to build a highly complex model as well as to highlight species-specific differences [34]. For additional discussion of the Hippo Pathway in other model systems, please see other chapters in this book.

4.1 Downstream Effectors

As noted, with the initial characterization of *sav*, *wts*, and *hpo* mutant alleles, it was clear that the pathway targeted cyclin E transcriptionally and DIAP1 through both transcription and post-translational mechanisms. Discovery of Yki as a key target of the pathway was a key advance in the field that helped not only elucidate the regulation of proliferation and cell death by the core machinery but that also linked the core machinery to these transcriptional events. Yki thus become the second post-translational substrate known to be regulated by the core kinases of the pathway. In the coming years, multiple efforts characterizing Yki established a multitude of Yki-mediated roles for the pathway not only in organ size, proliferation, growth, and cell survival but also in other processes such as a role in stem/progenitor cell proliferation and fate [35–39].

Diverse efforts in the field also continued to pursue more direct targets of the Wts kinase and to identify other downstream processes affected by the Hippo Pathway core cassette (Table 1) including direct targeting of Mud, Canoe, and Bazooka to regulate asymmetric cell division and orientation of cell division [40, 41] and Enabled to regulate actin [42] [Lucas et al. 2013]. Such efforts also uncovered Yki-independent regulation of autophagy [43], the actin cytoskeleton [44], planar polarity [45], response to alcohol [46], dendritic tiling and maintenance [47], and Rae1-mediated organ size control [48]. In some cases, these may be global processes regulated by the pathway, and in others, they may represent context-specific processes.

4.2 Upstream Regulation

Intriguingly, unlike the signaling systems in which a primary receptor binds a specific ligand, the Hippo Pathway utilizes a multitude of diverse upstream inputs (Table 1). The core kinase cassette has been characterized to be activated by numerous upstream regulators including membrane-associated and transmembrane proteins such as described for Mer, Ex, and Fat, as well as by distinct junctional complexes and polarity regulator complexes in the membrane [49, 50], by the actin and spectrin cytoskeletal networks and by tension and mechanical strain [44, 51–57]. To augment the regulation of targeting the core cassette, other activities influence specific components, for example, by targeting the stability of specific components in the core machinery independent of other inputs [58–60]. Focused studies taking into account specific components and their interactions with each other also furthered our understanding of the importance of Wts recruitment to the membrane by Mer or Ex for its activation by Hpo [61, 62].

As we gain a better understanding of both the upstream and downstream machinery guiding the activity of the core cassette in different contexts, it is compelling that a number of processes regulated by the core kinase cassette then in turn regulate the activity of the core cassette (Table 1). The core cassette regulates Yki which in turn regulates transcription of upstream components *mer* and *ex* [21]; the pathway regulates the actin cytoskeleton which in turn regulates the activity of the core cassette [44, 51]. The pathway regulates Rae1 which in turn feeds back to regulate levels of Mer, Hpo, and Wts [48]. Wts and Melted act in a bistable loop [63–65]. Yki promotes *myc* transcription, but then Myc can act to inhibit Yki activity [66, 67]. A paradigm is emerging that this pathway is characterized by extensive feedback circuitry, perhaps reflecting how important it is to maintain a precise range of pathway activity to ensure the appropriate downstream outputs.

5 The Hippo Pathway in Mammals

The Hippo Pathway is highly conserved from flies to humans, and the components of the core machinery reviewed in this chapter have counterparts in mammalian systems. Upstream regulator Mer corresponds to human Mer/NF2, a gene whose mutation is associated with neurofibromatosis type 2. Scaffold protein Sav corresponds to Sav1/hWW45. Other components presumably underwent duplication events: Hpo is represented by two kinases, Mst1 and Mst2; Wts by Lats1 and Lats2; Hippo Pathway target Yki by mammalian YAP and TAZ; and its binding partner Sd by the TEAD family of transcription factors TEAD1–4. Mammalian efforts have established roles for the Hippo Pathway in normal tissue homeostasis and have implicated the pathway in a variety of pathological conditions including cancer, neurodegeneration, and cardiovascular health/heart disease. For additional review of the pathway in mammalian systems, please *see* References [68–70] and other chapters in this book.

6 Concluding Thoughts and Open Avenues

As this field moves forward, multiple questions have arisen regarding potential roles for the Hippo Pathway in health and disease. In *Drosophila*, the pathway has been implicated in hematopoiesis [71], immunity [72], regulation of mitochondria and metabolism [73], and a number of other important processes. The high conservation of many components and how they interact established a global role for the core machinery. Other studies implicate stage-specific, tissue-specific, or context-specific inputs and outputs of the pathway. What other roles does the pathway play in development and adulthood? The pathway is clearly important in tumor suppression; why are pathway components rarely mutated in cancer? What is the precise role of the pathway in organ size control (e.g., an instructive role, a permissive role, etc.) and what are the underlying molecular mechanisms? How does the core machinery affect not only cell autonomous behavior but also convey information across the entire tissue? Multiple reports revealed avenues of cross talk with other important networks including EGFR/Ras, Wnt, AKT, Notch, and myc signaling systems. How does the core machinery interpret the variety of inputs from its upstream regulators and cross talk with other signaling pathways and then translate that information correctly to define the appropriate biological response?

In this chapter, we have described the discovery and origins of this field using *Drosophila* genetics. Ongoing avenues of investigation in a variety of systems, open questions, innovative strategies and tools, and greater mechanistic depth regarding the Hippo Pathway network will be expanded upon throughout the coming chapters in this book.

Acknowledgments

We would like to thank the *Drosophila* community and the broader Hippo Pathway field. We also apologize that we could not cover every advance in the field in this chapter. This review chapter was meant to review the early events in the conception of the field.

References

1. Rubin GM, Yandell MD, Wortman JR, Gabor Miklos GL, Nelson CR, Hariharan IK et al (2004) Comparative genomics of the eukaryotes. Science 287:2204–2215
2. Brand AH, Perrimon N (1993) Targeted gene expression as a means of altering cell fates and generating dominant phenotypes. Development 118:401–415
3. Dang DT, Perrimon N (1992) Use of a yeast site-specific recombinase to generate embryonic mosaics in Drosophila. Dev Genet 13:367–375
4. Harrison DA, Perrimon N (1993) Simple and efficient generation of marked clones in Drosophila. Curr Biol 3:424–433
5. Nüsslein-Volhard C, Wieschaus E (1980) Mutations affecting segment number and polarity in Drosophila. Nature 287:795–801
6. Moberg KH, Bell DW, Wahrer DC, Haber DA, Hariharan IK (2001) Archipelago regulates Cyclin E levels in Drosophila and is mutated in human cancer cell lines. Nature 413:311–316
7. Moberg KH, Mukherjee A, Veraksa A, Artavanis-Tsakonas S, Hariharan IK (2004) The Drosophila F box protein archipelago regulates dMyc protein levels in vivo. Curr Biol 14:965–974
8. Mills K, Daish T, Harvey KF, Pfleger CM, Hariharan IK, Kumar S (2006) The Drosophila melanogaster Apaf-1 homologue ARK is

required for most, but not all, programmed cell death. J Cell Biol 172:809–815
9. Tapon N, Ito N, Dickson BJ, Treisman JE, Hariharan IK (2001) The Drosophila tuberous sclerosis complex gene homologs restrict cell growth and cell proliferation. Cell 105:345–355
10. Tapon N, Harvey KF, Bell DW, Wahrer DC, Schiripo TA, Haber D et al (2002) Salvador promotes both cell cycle exit and apoptosis in Drosophila and is mutated in human cancer cell lines. Cell 110:467–478
11. Kango-Singh M, Nolo R, Tao C, Verstreken P, Hiesinger PR, Bellen HJ et al (2002) Shar-pei mediates cell proliferation arrest during imaginal disc growth in Drosophila. Development 129:5719–5730
12. Xu T, Wang W, Zhang S, Stewart RA, Yu W (1995) Identifying tumor suppressors in genetic mosaics: the Drosophila lats gene encodes a putative protein kinase. Development 121:1053–1063
13. Justice RW, Zilian O, Woods DF, Noll M, Bryant PJ (1995) The Drosophila tumor suppressor gene warts encodes a homolog of human myotonic dystrophy kinase and is required for the control of cell shape and proliferation. Genes Dev 9:534–546
14. Harvey KF, Pfleger CM, Hariharan IK (2003) The Drosophila Mst ortholog, Hippo, restricts growth and cell proliferation and promotes apoptosis. Cell 114:457–467
15. Wu S, Huang J, Dong J, Pan D (2003) Hippo encodes a Ste-20 family protein kinase that restricts cell proliferation and promotes apoptosis in conjunction with Salvador and Warts. Cell 114:445–456
16. Pantalacci S, Tapon N, Léopold P (2003) The Salvador partner Hippo promotes apoptosis and cell-cycle exit in Drosophila. Nat Cell Biol 5:921–927
17. Udan RS, Kango-Singh M, Nolo R, Tao C, Halder G (2003) Hippo promotes proliferation arrest and apoptosis in the Salvador/Warts pathway. Nat Cell Biol 5:914–920
18. Jia J, Zhang W, Wang B, Trinko R, Jiang J (2003) The Drosophila Ste20 family kinase dMST functions as a tumor suppressor by restricting cell proliferation and promoting apoptosis. Genes Dev 17:2514–2519
19. Lai ZC, Wei X, Shimizu T, Ramos E, Rohrbaugh M, Nikolaidis N et al (2005) Control of cell proliferation and apoptosis by mob as tumor suppressor, mats. Cell 120:675–685
20. Huang J, Wu S, Barrera J, Matthews K, Pan D (2005) The Hippo signaling pathway coordinately regulates cell proliferation and apoptosis by inactivating Yorkie, the Drosophila Homolog of YAP. Cell 122:421–434
21. Hamaratoglu F, Willecke M, Kango-Singh M, Nolo R, Hyun E, Tao C et al (2006) The tumour-suppressor genes NF2/Merlin and Expanded act through Hippo signalling to regulate cell proliferation and apoptosis. Nat Cell Biol 8:27–36
22. Cho E, Feng Y, Rauskolb C, Maitra S, Fehon R, Irvine KD (2006) Delineation of a fat tumor suppressor pathway. Nat Genet 38:1142–1150
23. Willecke M, Hamaratoglu F, Kango-Singh M, Udan R, Chen CL, Tao C et al (2006) The fat cadherin acts through the Hippo tumor-suppressor pathway to regulate tissue size. Curr Biol 16:2090–2100
24. Silva E, Tsatskis Y, Gardano L, Tapon N, McNeill H (2006) The tumor-suppressor gene fat controls tissue growth upstream of expanded in the Hippo signaling pathway. Curr Biol 16:2081–2089
25. Bennett FC, Harvey KF (2006) Fat cadherin modulates organ size in Drosophila via the Salvador/Warts/Hippo signaling pathway. Curr Biol 16:2101–2110
26. Zhang L, Ren F, Zhang Q, Chen Y, Wang B, Jiang J (2008) The TEAD/TEF family of transcription factor scalloped mediates Hippo signaling in organ size control. Dev Cell 14:377–387
27. Wu S, Liu Y, Zheng Y, Do J, Pan D (2008) The TEAD/TEF family protein scalloped mediates transcriptional output of the Hippo growth-regulatory pathway. Dev Cell 14:388–398
28. Goulev Y, Fauny JD, Gonzalez-Marti B, Flagiello D, Silber J, Zider A (2008) SCALLOPED interacts with YORKIE, the nuclear effector of the Hippo tumor-suppressor pathway in Drosophila. Curr Biol 18:435–441
29. Leevers SJ, McNeill H (2005) Controlling the size of organs and organisms. Curr Opin Cell Biol 17:604–609
30. Basu S, Totty NF, Irwin MS, Sudol M, Downward J (2003) Akt phosphorylates the yes-associated protein, YAP, to induce interaction with 14-3-3 and attenuation of p73-mediated apoptosis. Mol Cell 11:11–23
31. Dong J, Feldmann G, Huang J, Wu S, Zhang N, Comerford SA et al (2007) Elucidation of a universal size-control mechanism in Drosophila and mammals. Cell 130:1120–1133
32. Oh H, Irvine KD (2008) In vivo regulation of Yorkie phosphorylation and localization. Development 135:1081–1088

33. Ren F, Zhang L, Jiang J (2010) Hippo signaling regulates Yorkie nuclear localization and activity through 14-3-3 dependent and independent mechanisms. Dev Biol 337:303–312
34. Bossuyt W, Chen CL, Chen Q, Sudol M, McNeill H, Pan D et al (2014) An evolutionary shift in the regulation of the Hippo pathway between mice and flies. Oncogene 33:1218–1228
35. Staley BK, Irvine KD (2010) Warts and Yorkie mediate intestinal regeneration by influencing stem cell proliferation. Curr Biol 20:1580–1587
36. Ren F, Wang B, Yue T, Yun EY, Ip YT, Jiang J (2010) Hippo signaling regulates Drosophila intestine stem cell proliferation through multiple pathways. Proc Natl Acad Sci U S A 107:21064–21069
37. Karpowicz P, Perez J, Perrimon N (2010) The Hippo tumor suppressor pathway regulates intestinal stem cell regeneration. Development 137:4135–4145
38. Shaw RL, Kohlmaier A, Polesello C, Veelken C, Edgar BA, Tapon N (2010) The Hippo pathway regulates intestinal stem cell proliferation during Drosophila adult midgut regeneration. Development 137:4147–4158
39. Ding R, Weynans K, Bossing T, Barros CS, Berger C (2016) The Hippo signalling pathway maintains quiescence in Drosophila neural stem cells. Nat Commun 7:10510
40. Keder A, Rives-Quinto N, Aerne BL, Franco M, Tapon N, Carmena A (2015) The hippo pathway core cassette regulates asymmetric cell division. Curr Biol 25:2739–2750
41. Dewey EB, Sanchez D, Johnston CA (2015) Warts phosphorylates mud to promote pins-mediated mitotic spindle orientation in Drosophila, independent of Yorkie. Curr Biol 25:2751–2762
42. Lucas EP, Khanal I, Gaspar P, Fletcher GC, Polesello C, Tapon N et al (2013) The Hippo pathway polarizes the actin cytoskeleton during collective migration of Drosophila border cells. J Cell Biol 201:875–885
43. Dutta S, Baehrecke EH (2008) Warts is required for PI3K-regulated growth arrest, autophagy, and autophagic cell death in Drosophila. Curr Biol 18:1466–1475
44. Fernández BG, Gaspar P, Brás-Pereira C, Jezowska B, Rebelo SR, Janody F (2011) Actin-capping protein and the Hippo pathway regulate F-actin and tissue growth in Drosophila. Development 138:2237–2246
45. Marcinkevicius E, Zallen JA (2013) Regulation of cytoskeletal organization and junctional remodeling by the atypical cadherin fat. Development 140:433–443
46. Ilanges A, Jahanshahi M, Balobin DM, Pfleger CM (2013) Alcohol interacts with genetic alteration of the Hippo tumor suppressor pathway to modulate tissue growth in drosophila. PLoS One 8:e78880
47. Emoto K, Parrish JZ, Jan LY, Jan YN (2006) The tumour suppressor Hippo acts with the NDR kinases in dendritic tiling and maintenance. Nature 443:210–213
48. Jahanshahi M, Hsiao K, Jenny A, Pfleger CM (2016) The Hippo pathway targets Rae1 to regulate mitosis and organ size and to feed back to regulate upstream components Merlin, Hippo, and Warts. PLoS Genet 12: e1006198
49. Genevet A, Polesello C, Blight K, Robertson F, Collinson LM, Pichaud F et al (2009) The Hippo pathway regulates apical-domain size independently of its growth-control function. J Cell Sci 122:2360–2370
50. Verghese S, Waghmare I, Kwon H, Hanes K, Kango-Singh M (2012) Scribble acts in the Drosophila fat-hippo pathway to regulate warts activity. PLoS One 7:e47173
51. Sansores-Garcia L, Bossuyt W, Wada K, Yonemura S, Tao C, Sasaki et al (2011) Modulating F-actin organization induces organ growth by affecting the Hippo pathway. EMBO J 30:2325–2335
52. Rauskolb C, Pan G, Reddy BV, Oh H, Irvine KD (2011) Zyxin links fat signaling to the hippo pathway. PLoS Biol 9:e1000624
53. Rauskolb C, Sun S, Sun G, Pan Y, Irvine KD (2014) Cytoskeletal tension inhibits Hippo signaling through an Ajuba-Warts complex. Cell 158:143–156
54. Gaspar P, Holder MV, Aerne BL, Janody F, Tapon N (2015) Zyxin antagonizes the FERM protein expanded to couple F-actin and Yorkie-dependent organ growth. Curr Biol 25:679–689
55. Deng H, Wang W, Yu J, Zheng Y, Qing Y, Pan D (2015) Spectrin regulates Hippo signaling by modulating cortical actomyosin activity. elife 4:e06567
56. Fletcher GC, Elbediwy A, Khanal I, Ribeiro PS, Tapon N, Thompson BJ (2015) The Spectrin cytoskeleton regulates the Hippo signalling pathway. EMBO J 34:940–954
57. Wong KK, Li W, An Y, Duan Y, Li Z, Kang Y et al (2015) β-Spectrin regulates the hippo signaling pathway and modulates the basal actin network. J Biol Chem 290:6397–6407
58. Wehr MC, Holder MV, Gailite I, Saunders RE, Maile TM, Ciirdaeva E et al (2013) Salt-inducible kinases regulate growth through the Hippo signalling pathway in Drosophila. Nat Cell Biol 15:61–71

59. Aerne BL, Gailite I, Sims D, Tapon N (2015) Hippo stabilises its adaptor Salvador by antagonising the HECT ubiquitin ligase Herc4. PLoS One 10:e0131113
60. Hirabayashi S, Cagan RL (2015) Salt-inducible kinases mediate nutrient-sensing to link dietary sugar and tumorigenesis in Drosophila. elife 4:e08501
61. Yin F, Yu J, Zheng Y, Chen Q, Zhang N, Pan D (2013) Spatial organization of Hippo signaling at the plasma membrane mediated by the tumor suppressor Merlin/NF2. Cell 154:1342–1355
62. Sun S, Reddy BVVG, Irvine KD (2015) Localization of Hippo signalling complexes and Warts activation in vivo. Nat Commun 6:8402
63. Mikeladze-Dvali T, Wernet MF, Pistillo D, Mazzoni EO, Teleman AA, Chen YW et al (2005) The growth regulators warts/lats and melted interact in a bistable loop to specify opposite fates in Drosophila R8 photoreceptors. Cell 122:775–787
64. Jukam D, Desplan C (2011) Binary regulation of Hippo pathway by Merlin/NF2, Kibra, Lgl, and melted specifies and maintains postmitotic neuronal fate. Dev Cell 21:874–887
65. Jukam D, Xie B, Rister J, Terrell D, Charlton-Perkins M, Pistillo D et al (2013) Opposite feedbacks in the Hippo pathway for growth control and neural fate. Science 342:1238016
66. Neto-Silva RM, de Beco S, Johnston LA (2010) Evidence for a growth-stabilizing regulatory feedback mechanism between Myc and Yorkie, the Drosophila homolog of Yap. Dev Cell 19:507–520
67. Ziosi M, Baena-López LA, Grifoni D, Froldi F, Pession A, Garoia F et al (2010) dMyc functions downstream of Yorkie to promote the supercompetitive behavior of hippo pathway mutant cells. PLoS Genet 6: e1001140
68. Meng Z, Moroishi T, Guan KL (2006) Mechanisms of Hippo pathway regulation. Genes Dev 30:1–17
69. Yu FX, Zhao B, Guan KL (2015) Hippo pathway in organ size control, tissue homeostasis, and cancer. Cell 163:811–828
70. Watt KI, Harvey KF, Gregorevic P (2017) Regulation of tissue growth by the mammalian Hippo signaling pathway. Front Physiol 8:942
71. Milton CC, Grusche FA, Degoutin JL, Yu E, Dai Q, Lai EC et al (2014) The Hippo pathway regulates hematopoiesis in Drosophila melanogaster. Curr Biol 24:2673–2680
72. Liu B, Zheng Y, Yin F, Yu J, Silverman N, Pan D (2016) Toll receptor-mediated Hippo signaling controls innate immunity in Drosophila. Cell 164:406–419
73. Sing A, Tsatskis Y, Fabian L, Hester I, Rosenfeld R, Serricchio M et al (2014) The atypical cadherin fat directly regulates mitochondrial function and metabolic state. Cell 158:1293–1308
74. Colombani J, Polesello C, Josué F, Tapon N (2006) Dmp53 activates the Hippo pathway to promote cell death in response to DNA damage. Curr Biol 16:1453–1458
75. Thompson BJ, Cohen SM (2006) The Hippo pathway regulates the bantam microRNA to control cell proliferation and apoptosis in Drosophila. Cell 126:767–774
76. Nolo R, Morrison CM, Tao C, Zhang X, Halder G (2006) The bantam microRNA is a target of the hippo tumor-suppressor pathway. Curr Biol 16:1895–1904
77. Polesello C, Huelsmann S, Brown NH, Tapon N (2006) The Drosophila RASSF homolog antagonizes the hippo pathway. Curr Biol 16:2459–2465
78. Tyler DM, Li W, Zhuo N, Pellock B, Baker NE (2007) Genes affecting cell competition in Drosophila. Genetics 175:643–657
79. Wei X, Shimizu T, Lai ZC (2007) Mob as tumor suppressor is activated by Hippo kinase for growth inhibition in Drosophila. EMBO J 26:1772–1781
80. Pellock BJ, Buff E, White K, Hariharan IK (2007) The Drosophila tumor suppressors expanded and Merlin differentially regulate cell cycle exit, apoptosis, and wingless signaling. Dev Biol 304:102–115
81. Shimizu T, Ho LL, Lai ZC (2008) The mob as tumor suppressor gene is essential for early development and regulates tissue growth in Drosophila. Genetics 178:957–965
82. Badouel C, Gardano L, Amin N, Garg A, Rosenfeld R, Le Bihan T et al (2009) The FERM-domain protein expanded regulates Hippo pathway activity via direct interactions with the transcriptional activator Yorkie. Dev Cell 16:411–420
83. Oh H, Reddy BV, Irvine KD (2009) Phosphorylation-independent repression of Yorkie in fat-Hippo signaling. Dev Biol 335:188–197
84. Peng HW, Slattery M, Mann RS (2009) Transcription factor choice in the Hippo signaling pathway: homothorax and yorkie regulation of the microRNA bantam in the progenitor domain of the Drosophila eye imaginal disc. Genes Dev 23:2307–2319

85. Hamaratoglu F, Gajewski K, Sansores-Garcia-L, Morrison C, Tao C, Halder G (2009) The Hippo tumor-suppressor pathway regulates apical-domain size in parallel to tissue growth. J Cell Sci 122:2351–2359
86. Chen CL, Gajewski KM, Hamaratoglu F, Bossuyt W, Sansores-Garcia L, Tao C et al (2010) The apical-basal cell polarity determinant Crumbs regulates Hippo signaling in Drosophila. Proc Natl Acad Sci USA 107:15810–15815
87. Grzeschik NA, Parsons LM, Allott ML, Harvey KF, Richardson HE (2010) Lgl, aPKC, and Crumbs regulate the Salvador/Warts/Hippo pathway through two distinct mechanisms. Curr Biol 20:573–581
88. Ling C, Zheng Y, Yin F, Yu J, Huang J, Hong Y et al (2010) The apical transmembrane protein Crumbs functions as a tumor suppressor that regulates Hippo signaling by binding to expanded. Proc Natl Acad Sci USA 107:10532–10537
89. Parsons LM, Grzeschik NA, Allott ML, Richardson HE (2010) Lgl/aPKC and Crb regulate the Salvador/Warts/Hippo pathway. Fly 4:288–293
90. Robinson BS, Huang J, Hong Y, Moberg KH (2010) Crumbs regulates Salvador/Warts/Hippo signaling in Drosophila via the FERM-domain protein expanded. Curr Biol 20:582–590
91. Hafezi Y, Bosch JA, Hariharan IK (2012) Differences in levels of the transmembrane protein Crumbs can influence cell survival at clonal boundaries. Dev Biol 368:358–369
92. Baumgartner R, Poernbacher I, Buser N, Hafen E, Stocker H (2010) The WW domain protein Kibra acts upstream of Hippo in Drosophila. Dev Cell 18:309–316
93. Genevet A, Wehr MC, Brain R, Thompson BJ, Tapon N (2010) Kibra is a regulator of the Salvador/Warts/Hippo signaling network. Dev Cell 18:300–308
94. Yu J, Zheng Y, Dong J, Klusza S, Deng WM, Pan D (2010) Kibra functions as a tumor suppressor protein that regulates Hippo signaling in conjunction with Merlin and Expanded. Dev Cell 18:288–299
95. Das Thakur M, Feng Y, Jagannathan R, Seppa MJ, Skeath JB, Longmore GD (2010) Ajuba LIM proteins are negative regulators of the Hippo signaling pathway. Curr Biol 20:657–662
96. Ribeiro PS, Josué F, Wepf A, Wehr MC, Rinner O, Kelly G et al (2010) Combined functional genomic and proteomic approaches identify a PP2A complex as a negative regulator of Hippo signaling. Mol Cell 39:521–534
97. Oh H, Irvine KD (2011) Cooperative regulation of growth by Yorkie and Mad through bantam. Dev Cell 20:109–122
98. Boggiano JC, Vanderzalm PJ, Fehon RG (2011) Tao-1 phosphorylates Hippo/MST kinases to regulate the Hippo-Salvador-Warts tumor suppressor pathway. Dev Cell 21:888–895
99. Poon CL, Lin JI, Zhang X, Harvey KF (2011) The sterile 20-like kinase Tao-1 controls tissue growth by regulating the Salvador-Warts-Hippo pathway. Dev Cell 21:896–906
100. Verghese S, Bedi S, Kango-Singh M (2012) Hippo signalling controls Dronc activity to regulate organ size in Drosophila. Cell Death Differ 19:1664–1676
101. Herranz H, Hong X, Cohen SM (2012) Mutual repression by bantam miRNA and Capicua links the EGFR/MAPK and Hippo pathways in growth control. Curr Biol 22:651–657
102. Yue T, Tian A, Jiang J (2012) The cell adhesion molecule echinoid functions as a tumor suppressor and upstream regulator of the Hippo signaling pathway. Dev Cell 22:255–267
103. Kagey JD, Brown JA, Moberg KH (2012) Regulation of Yorkie activity in Drosophila imaginal discs by the Hedgehog receptor gene patched. Mech Dev 129:339–349
104. Ye X, Deng Y, Lai ZC (2012) Akt is negatively regulated by Hippo signaling for growth inhibition in Drosophila. Dev Biol 369:115–123
105. Sansores-Garcia L, Atkins M, Moya IM, Shahmoradgoli M, Tao C, Mills GB et al (2013) Mask is required for the activity of the Hippo pathway effector Yki/YAP. Curr Biol 23:229–235
106. Sidor CM, Brain R, Thompson BJ (2013) Mask proteins are cofactors of Yorkie/YAP in the Hippo pathway. Curr Biol 23:223–228
107. Oh H, Slattery M, Ma L, Crofts A, White KP, Mann RS et al (2013) Genome-wide association of Yorkie with chromatin and chromatin-remodeling complexes. Cell Rep 3:309–318
108. Reddy BV, Irvine KD (2013) Regulation of Hippo signaling by EGFR-MAPK signaling through Ajuba family proteins. Dev Cell 24:459–471
109. Koontz LM, Liu-Chittenden Y, Yin F, Zheng Y, Yu J, Huang B et al (2013) The Hippo effector Yorkie controls normal tissue growth by antagonizing scalloped-mediated default repression. Dev Cell 25:388–401

110. Huang HL, Wang S, Yin MX, Dong L, Wang C, Wu W et al (2013) Par-1 regulates tissue growth by influencing hippo phosphorylation status and hippo-Salvador association. PLoS Biol 11:e1001620
111. Sun G, Irvine KD (2013) Ajuba family proteins link JNK to Hippo signaling. Sci Signal 6:ra81
112. Zhang C, Robinson BS, Xu W, Yang L, Yao B, Zhao H et al (2015) The ecdysone receptor coactivator Taiman links Yorkie to transcriptional control of germline stem cell factors in somatic tissue. Dev Cell 34:168–180
113. Zheng Y, Wang W, Liu B, Deng H, Uster E, Pan D (2015) Identification of Happyhour/MAP4K as alternative Hpo/Mst-like kinases in the Hippo kinase Cascade. Dev Cell 34:642–655
114. Li S, Cho YS, Yue T, Ip YT, Jiang J (2015) Overlapping functions of the MAP4K family kinases Hppy and Msn in Hippo signaling. Cell Discov 1:15038

Chapter 2

Drosophila Genetics: The Power of Genetic Mosaic Approaches

Mardelle Atkins

Abstract

Drosophila melanogaster has been a central player in the discovery of the Hippo pathway and in understanding its in vivo functions. From a technique standpoint, the Flp-FRT system for the generation of genetic mosaics has been a principle tool. It has broadly been used in the discovery of Hippo pathway members in mutagenesis screens, in the analysis of target gene expression, and in genetic epistasis. Here we briefly introduce this tool, summarize its use in the Hippo pathway field, and provide a protocol for the generation of Flp-FRT clones in imaginal discs with dissection and staining for reporter gene expression to characterize candidate Hippo pathway genes.

Key words *Drosophila*, Hippo signaling, Mitotic clones, Protocol, Reporter analysis, Larval dissection

1 Introduction

1.1 Mosaic Analysis and the Flp-FRT System

The first isolated *Drosophila* mutants were viable and showed phenotypes in adults that affected traits such as eye color, body color, bristle morphology, or wing shape. Later, lethal mutations were also recovered and analyzed. However, early homozygous lethality of a mutant precludes the analysis of a gene's function later in development or in adults. Overcoming this limitation required the generation of conditional systems that allowed the deletion (knockout) or suppression of a gene in only a fraction of cells or in a temporally restricted manner or both. In *Drosophila*, one of the most powerful tools for this purpose is the generation of genetic mosaics by Flp-FRT-mediated mitotic recombination.

Mitotic recombination can occur in cells when double-stranded breaks occur after the duplication of the chromosomes in preparation for mitosis. In *Drosophila*, homologous chromosomes align during mitosis, and the repair of these breaks is preferentially done using the homologous chromosome as a template [1]. Thus, in a heterozygous animal, for example, repair of these breaks can result in an exchange of DNA between the mutation containing and

Alexander Hergovich (ed.), *The Hippo Pathway: Methods and Protocols*, Methods in Molecular Biology, vol. 1893,
https://doi.org/10.1007/978-1-4939-8910-2_2,

wild-type chromosome arms. Depending upon the segregation outcome, the heterozygous condition can be maintained in the descendant cells. However, it is also possible to recover a homozygous mutant cell and a wild-type sister. As these two cells subsequently proliferate, they will yield two genetically distinct populations, respectively, referred to as the "clone" and the "twin-spot." As a result, the homozygous mutant phenotype can be analyzed in a population of cells at a later stage of development. Importantly for the analysis of mutations that affect growth rates or cell fitness, normal twin-spot clones are generated in the same event, permitting comparison of the normal to mutant tissue growth.

In *Drosophila*, mitotic clones were initially generated using ionizing radiation to induce double-stranded breaks in heterozygous cells. This method, however, is somewhat inefficient, yields breaks in random locations, and also triggers tissue damage. These problems were overcome by the introduction of a yeast-derived site-specific recombinase (Flippase or FLP) and led to the establishment of a genetic toolkit in *Drosophila* for the generation of genetic mosaics [2, 3]. The general outline of FLP-FRT-mediated mitotic recombination is presented in Fig. 1a. Briefly, the flippase enzyme (Flp) is expressed under the direction of a tissue-specific or tissue-inducible promoter in proliferating tissues. Flp catalyzes recombination at target sites termed FRT. Importantly, FRT sites were introduced at positions close to the centromeres for each of the major chromosome arms; thus recombination with these sites "flips" nearly the entire chromosome. Therefore, for each chromosome arm, the vast majority of mutations can be made homozygous. Depending on the promoter used to drive the expression of Flp, clones can be induced at specific times or locations in the developing fly. Table 1 summarizes commonly used Flp constructs. Stocks for conducting Flp-FRT experiments are publically available for order from the Bloomington Drosophila Stock Center.

For a useful analysis, the genotype of each cell must be unambiguously known. In some instances this may be possible by direct staining against the protein of interest or by its phenotype. However, with the Flp-FRT system, a great number of homozygous viable, visible markers have been developed or incorporated for this purpose. For visualization in adults, this is most commonly a visible marker for eye color, body color, or bristle morphology. In larvae, constructs that constitutively express beta-galactosidase or fluorescent molecules (GFP or RFP) are most commonly used. In the traditional Flp-FRT system, the wild-type chromosome is positively marked, and mutant cells are identified by loss of the marker. With the fluorescent marker chromosome, the double dosage of the twin-spot is also distinct from the heterozygous tissue,

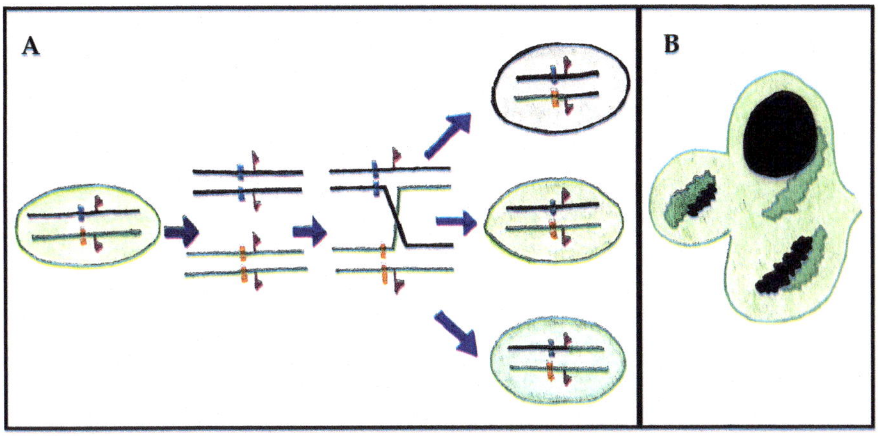

Fig. 1 Generating loss of function mitotic clones. (**a**) A schematic of the generation of mitotic clones. From left to right: Mutant chromosome (black lines with blue centromere) and GFP marked wild-type chromosome (green with orange centromere). FRT indicated as magenta flag. Cells entering mitosis duplicate and align their chromosomes. Flippase expression catalyzes recombination between FRT sites on homologous chromosomes. The following double-stranded break repair chromosomes are segregated to daughter cells. There are three potential outcomes: a homozygous mutant cell lacking GFP expression (top, gray shade), a conservative outcome that restores the heterozygous condition (light green, one copy of GFP), or the "twin-spot" cell containing two copies of the GFP wild-type chromosome (dark green). In cells that continue to express flippase, subsequent divisions can result in the formation of additional clones and twin-spots. (**b**) Illustration of potential clone and twin-spot outcomes in the analysis of clone growth. Clockwise from bottom: Clone and twin-spot are approximately the same size indicating the mutant grows normally. Clone is smaller than twin-spot indicating that clone cells grow poorly. Clone is larger than twin-spot and has a growth advantage. Additionally, unlike the other clones, this clone has smooth borders, which can indicate changes in cell adhesion, a common phenotype of *hpo* or *wts* clones

Table 1
An overview of commonly used flippase constructs

Flp construct	Location of activity	Timing of activity
hs-flp	Ubiquitous	User induces by heat shock
ey-flp	Proliferating cells of eye-antennal imaginal disc, genital disc, brain	First to early third instar larvae
ey(3.5)-flp	Proliferating cells of eye-antennal imaginal disc, genital disc	First to early third instar larvae
Ubx-flp	Wing and haltere imaginal disc	First to early third instar larvae
Ovo-flp	Adult ovary	Adult female
UAS-flp	Dependent upon Gal4 driver expression	Dependent upon Gal4 driver

facilitating growth analysis, as mentioned above (Fig. 1b). Since its initial integration as a research tool in *Drosophila*, the Flp-FRT system has been harnessed for many different strategies to induce clones in *Drosophila*, both through trans-recombination events, as described here, and also in tools for cis-recombination-mediated strategies. For a recent, thorough review of its many forms, we recommend the FlyBook chapter "Mosaic Analysis in *Drosophila*" [4]. In addition to describing the major techniques, this chapter also includes links to publically available stocks and descriptions of common markers.

1.2 Flp-FRT: Discovery and Analysis of the Hippo Pathway

The development of the FLP-FRT system added an extremely powerful and today ubiquitously used genetic tool to the *Drosophila* researcher's toolbox. Firstly, it permitted researchers to perform high-throughput mosaic screens in adults. The initial development of the *hs-flp* system led to the discovery of *warts*, the *Drosophila LATS1/2* homolog in 1995 [5, 6]. Alleles of *salvador*, *hippo*, *expanded*, *crumbs*, and *kibra* were later recovered in *ey-flp*-based Flp-FRT screens and subsequently linked into the Hippo pathway [7–17].

In addition to their utility in mutagenesis screens, mitotic clones have proven a valuable tool for analyzing the Hippo pathway's composition and feedbacks. For example, the use of the *ey-flp* technique permitted analysis of *fat*, *merlin*, and *mats* mitotic clones, which was in turn critical to their identification as Hippo pathway members [14, 18–22].

Finally, the ability to clearly mark clones in developing larvae has several uses beyond analysis of growth rate. Perhaps most obviously, changes in target gene expression can be visualized in vivo, in intact tissues, during the growth phase. As the Hippo pathway core was discovered, target genes including *cyclin E*, *expanded*, *diap-1*, *myc*, and *bantam* were also identified [14, 23–26]. Interestingly, because many/most of the Hippo pathway target genes are ubiquitously expressed at relatively uniform levels, having the internal control of the wild-type tissue expression level makes immunohistochemistry semiquantitative in this system. This has been particularly useful in Hippo pathway analysis because mutation of the pathway members frequently increases target gene expression, and this change would be difficult to discern in tissues that are wholly mutant. Furthermore, changes in clone morphology, which can correlate with changes in cell adhesion, can be easily detected. Finally, because the boundaries between normal and mutant populations are clearly defined, it is easy to assess if changes in gene expression occur in a cell autonomous or nonautonomous fashion. This has been particularly useful in analysis of some of the upstream regulators of the pathway including alpha-catenin, Fat, and Dachsous and of exploring the function of the Hippo pathway during cell competition [26–35].

2 Materials

2.1 Before You Start

Plan your cross strategy to ensure the correct genotype can be unambiguously identified, and obtain necessary fly stocks (*see* **Note 1**).

2.2 Drosophila Culture

1. Red Star Active dry baker's yeast (e.g., Flystuff 62-103 or grocery store).
2. Suitable vials or bottles and closures for culture. We use wide S vials (e.g., Genesee 32-121-BF), but there are several comparable suppliers.
3. Culture Media: There are also many good recipes for the preparation of fly media. We use a very rich media variant of the German food recipe. This recipe yields 1300 wide vials (can be scaled as needed): 19 L distilled water, 1 kg corn meal (Flystuff 62-100), 400 g dry yeast (Flystuff 62-103), 780 mL molasses (Flystuff 62-117), 112 g agar (Flystuff 66-103), 270 g dextrose (Flystuff 62-113), 150 mL propionic acid (Fisher A258-500), 22.5 g hydroxybutyrate (Sigma H3635), and 225 mL ethanol.
4. Cheesecloth or six to ten cotton kitchen towels.
5. A cooking pot and heat source sufficient for boiling your media. For aliquoting food, for small batches clean, empty ketchup bottles are very effective. For large batches such as described here, a peristaltic pump (e.g., Major Science, MU-D01).

2.3 Dissecting Larvae

1. Two sets of Dumont Forceps #5.
2. Sylgard plate—This is a 140 mm petri dish that has been ½ filled with a Sylgard resin and charcoal powder mixture and allowed to cure.
3. 70% (v/v) EtOH diluted in dH_2O.
4. 1× PBS (137 mM NaCl, 2.68 mM KCl, 10.14 mM Na_2HPO_4, 1.76 mM KH_2PO_4 at pH 7.4).
5. 2.0 mL round bottom Eppendorf tubes (1 per sample) (*see* **Note 2**).
6. Ice and ice bucket.

2.4 Staining for LacZ Reporter Expression

1. 1× phosphate-buffered saline (PBS).
2. PBT: 1× PBS +0.3% (v/v) Triton X-100.
3. PBNT: PBT + 5% (v/v) heat inactivated, filtered normal donkey serum.
4. 37% formaldehyde (hazardous, handle and dispose according to institutional regulations).

5. 3.7% formaldehyde in PBS (dilute 1/10 from 37% stock just before use).
6. Fluorescence-stabilizing mounting medium such as VECTASHIELD (H1200) (*see* **Note 3**).
7. For LacZ reporter detection, we use mouse anti-beta-galactosidase (Promega, Z3781) and Alexa Fluor 555-conjugated secondary antibodies (Thermo Fisher A-31570).

2.5 Mounting Imaginal Discs for Analysis

There are several strategies for mounting discs. We use minutien stainless steel insect pins mounted in microdissecting needle holders. The most critical elements for good mounting dissections are (1) oblique lighting, which is achieved by using gooseneck illuminators, (2) a dark background, and (3) sharp tools. Mounting dissections are performed in VECTASHIELD containing DAPI (Vector Labs, H1200, H1000 if you prefer no DAPI) on Superfrost slides, and dissected tissue is mounted on Superfrost slides beneath 18 mm square coverslips sealed with clear nail polish.

3 Methods

Generation of Flp-FRT Mitotic Clones for *LacZ* Reporter Analysis in Imaginal Disc Epithelia in *Drosophila melanogaster*

All fly husbandry steps are performed at 25 °C unless otherwise noted. The goal of this protocol is to generate mitotic clones of (candidate) Hippo pathway mutations and stain larval clones to examine the expression of target gene *LacZ* reporters such as *ex-LacZ*, *diap1-LacZ* (aka *th-LacZ*), *fj-lacZ*, *cycE-LacZ*, and *myc-LacZ*.

3.1 Preparation Steps and Setting Up of Crosses

1. Make fly media. Bring 12 L of water to a boil. In a separate container, make a paste of 4 L distilled water, corn meal, dry yeast, and molasses. Stir thoroughly and set aside. To the 12 L of boiling water, add agar and dextrose. Stir vigorously to prevent the agar from clumping. Carefully add your reserved molasses paste mixture. Boil for 20 min. Add 3 L of water. Boil an additional 10 min. Remove the mixture from the heat and cool for 30 min. To the cooled mixture, add propionic acid, hydroxybenzoate, and ethanol. This mixture should be aliquoted (approx. 30 mL/vial) while still fluid. Cover vials with cheesecloth or a light kitchen towel, and allow drying for 12–24 h before plugging. Store vials in sealed plastic bags up to 4 weeks at 4 °C. Chilled food should be warmed to at least 16 °C before placing flies into the vial (*see* **Note 4**).
2. Collect and separate virgin females and young males of the correct genotypes under anesthesia (*see* **Note 5**). Store collected females at 18–22 °C with a very small amount of

granulated dry yeast for 2–5 days. Store collected males on fresh media, but do not add yeast.

3. To set the cross, add a small amount of yeast (five to ten grains of dry yeast) to a fresh vial along with five to eight virgins and three to five males (*see* **Notes 6** and **7**). We typically set three to ten crosses depending on the frequency of desired genotype (*see* **Notes 8** and **9**). Place crosses at appropriate temperature (*see* **Note 10**).

3.2 Flipping the Cross

After approximately 24 h, check the crosses. If sufficient eggs have been laid (>50), create a second "flip" of the cross by transferring the adults to fresh, yeasted vials (*see* **Note 11**). Optimally this is done without anesthetizing the flies again. Return the first and second flips to the appropriate temperature. Repeat this process every 24 h to generate additional flips (*see* **Note 12**). If using heat shock flippase to generate clones, *see* **Note 13**.

3.3 Collecting Larvae for Dissection

On day 5 or 6 after setting the cross, larvae should be visible crawling up onto the wall of the vial. These are referred to as third instar wandering larvae and are the most typical stage dissected for analysis of clones in imaginal discs. Carefully collect larvae from the wall of the vial. Do not squeeze or poke the larvae enough to injure them. Place larvae into chilled 1× PBS as they are collected (*see* **Note 14**).

3.4 Selecting the Correct Larvae

Remove and replace PBS to clean larvae and make it easier to visualize markers (*see* **Note 15**). To limit the dissection of incorrect genotypes, use the microscope to sort larvae. For best results, remove the larvae of the undesirable genotype(s) to limit opportunities to damage tissue in the desired larvae.

3.5 Setup for Dissection

Clean dust/debris from the Sylgard plate using a Kimwipe (or similar lint-free wipe) and 70% ethanol. Arrange three to five drops of PBS (approx. 100 μL each) on the plate, at least 1 inch apart. Label a 2.0 mL round bottom Eppendorf tube for each sample, and add 800 μL of 1× PBS to each tube. Place tube (s) with PBS in ice (*see* **Note 16**). Place aliquots of 1× PBT and 1× PBNT in the ice bucket to chill.

3.6 Collecting Imaginal Discs: Rough Dissection

1. Carefully transfer three to five of the desired larvae carefully to each drop of PBS.
2. Select a larva to dissect and orient its head toward your dominant hand (Fig. 2a). Using the other hand, pin the larvae with forceps to the plate by placing the forceps 30–50% of the way down the body from the head (Fig. 2b).

3. To recover wing, haltere, leg, and eye imaginal discs from the same dissection, pinch the cuticle just anterior to the pinning forceps, and smoothly pull apart the larva (Fig. 2b).
4. Remove the posterior half of the larva from the drop, leaving the anterior "head complex" (Fig. 2c).
5. Holding onto the mouth hooks, grab the edge of the torn cuticle, and using a push/pull movement, turn the dissected head complex inside out by pushing the mouth hooks posteriorly while simultaneously tugging the cuticle edge anteriorly (Fig. 2d; and *see* **Note 17**).
6. Carefully remove excess tissue including the gut, fat, and salivary glands (Fig. 2e).
7. Transfer the everted head complex to the prepared 2.0 mL round bottom Eppendorf tube. Dissections can be added to the tube for up to 30 min before proceeding to fixation.

3.7 Staining Imaginal Discs for LacZ Reporter Analysis

1. Add 200 μL of freshly prepared 3.7% formaldehyde solution and gently mix by tilting the tube and/or tapping. Do not invert. Check to ensure all head complexes are submerged. Incubate 20 min on ice.
2. Carefully remove the fix and discard according to institutional regulations. Replace with 1 mL 1× PBS. Tap the tube gently until you can see that the individual complexes do not stick together (*see* **Note 18**). Incubate on ice for 10 min.
3. Carefully remove the 1× PBS and replace with 1× PBT (*see* **Note 19**). Incubate on ice for 10 min.
4. Carefully remove the 1× PBT and replace with 200 μL of PBNT. Block in PBNT for at least 10 min (*see* **Note 20**).
5. Dilute mouse anti-beta-galactosidase antibody 1:1000 in PBNT.
6. Remove PBNT from dissections and replace with antibody cocktail. Incubate in primary antibody overnight at 4 °C.
7. Remove the primary antibody. Wash sample three times for 10 min each in 200 μL PBNT at room temperature. Add secondary antibody (diluted 1:600 in PBNT), and incubate in a dark location at room temperature for 2 h.
8. Remove the secondary antibody. Wash as follows: 1× PBNT, 1× PBT, and 1× PBS for 10 min each at room temperature. Place samples in a dark location (such as a drawer) for incubation.
9. Postfix in freshly diluted 4% formaldehyde for 10 min at room temperature. Return samples to dark.
10. Replace with PBS. Wash three times in 1× PBS for 10 min each at room temperature in the dark. Tap the tube to ensure that

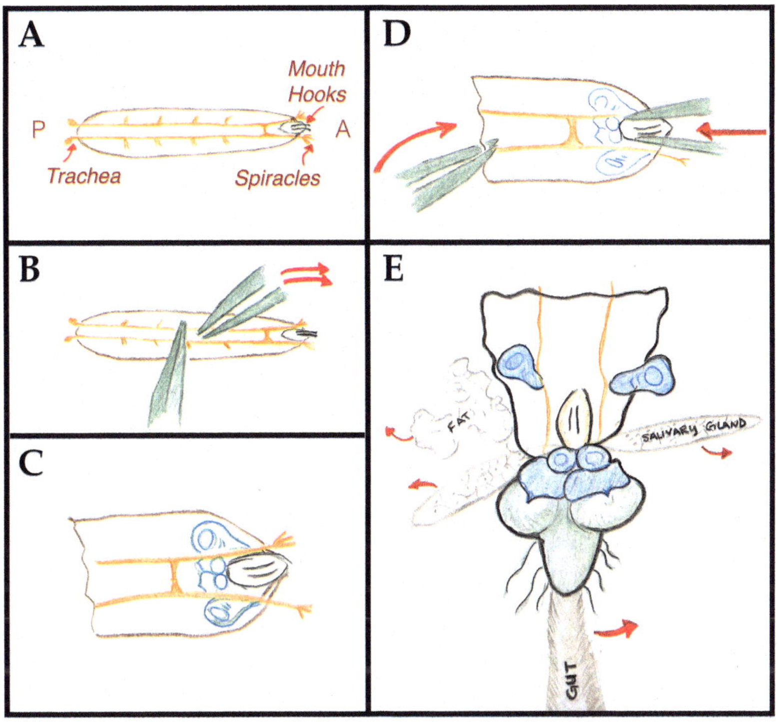

Fig. 2 Illustrations of larval dissection steps. (**a**) Larva. Anterior (A) and posterior (P) are indicated; position is for a right-handed dissector. The trachea (yellow in all panels) runs the length of the dorsal side and terminates in the anterior in spiracles (red arrow) to the right and left of the mouth hooks (black). Spiracle morphology can help stage larvae. Roughly, in and closed is early third instar, in and open is mid-third instar, and open and outside is late third instar. (**b**) Forceps position for dissection is shown. Red arrows indicate direction of pull. (**c**) Isolated head complex. For simplification only position of wing and eye-antennal discs is drawn (blue). (**d**) Preparing to evert the head complex, arrows show direction of forceps movements for the push-pull of eversion. (**e**) Everted head complex showing wing and eye-antennal discs (blue), mouth hooks, and commonly associated tissues. The fat body (opaque white in actual dissection) and salivary glands should be removed. Gut material will extend below the brain (gray structure) and should also be removed. If present, the brain should not be removed at this time, to prevent accidental damage to the eye-antennal discs. However, sometimes the brain does not stay attached in the dissection, which is not a problem if the eye-antennal discs can be seen attached to the base of the mouth hooks

head complexes are separated from each other after the first wash.

11. Completely remove PBS; leaving PBS at this point will thin the mounting media and appears to adversely affect tissue integrity. Replace with one to two drops of VECTASHIELD mounting media (H1200). Store overnight at 4 °C to allow the DAPI and mounting media to fully penetrate the tissue.

1. Cut the tip of a 200 μL pipette tip. Use this tip to transfer the stained head complexes to a clean Superfrost slide.

3.8 Collecting Imaginal Discs: Fine Dissection and Mounting

2. Adjust gooseneck lighting such that disc tissue appears white and solid.
3. Locate the wing and haltere discs along the tracheae. Using the minutien pins, carefully cut the discs away from the tracheae. Move the isolated discs to a clean area of the VECTASHIELD.
4. Using your pins or forceps, separate the cuticle from the complex of the mouth hooks, eye-antennal discs, and brain.
5. Cut the optic lobes free from the CNS. Separate the eye-antennal disc from the optic lobes by cutting the optic stalk. Cut at the base of the mouth hooks to release the eye-antennal discs.
6. Collect imaginal discs to one area of the VECTASHIELD drop using the side of your needle, carefully avoiding poking holes in the discs. Push debris away.
7. Using a P20 pipette, transfer the discs in 9 μL of VECTASHIELD onto a clean Superfrost slide (*see* **Note 21**).
8. Use the dissecting needles to gently arrange the discs on the slide. We preferentially place the apical surface upwards.
9. Carefully place an 18 mm square coverslip onto the sample to avoid bubbles. Once the VECTASHIELD has spread to the edges of the slip, seal with clear nail polish.
10. Store slides flat at least 2 h at 4 °C before imaging with confocal microscopy to avoid having samples drift during imaging. Samples prepped by this method can be stably stored at 4 °C for imaging for at least 1 year (*see* **Note 22**).

4 Notes

1. If you are wholly unfamiliar with using *Drosophila* or with mosaic analysis, I recommend you consult two excellent sources. First, if you are new to *Drosophila* work, an excellent introductory manual has been generated by Roote and Prokop [36], and updated versions can be downloaded at dx.doi.org/10.6084/m9.figshare.106631.
2. We have found that the round bottom tubes provide more consistent staining results than the conical 1.5 mL tubes, though some users have reported that the 1.5 mL tubes work well if the samples are rocked. In our experience, larger volumes of wash solution are required when rocking dissections to prevent the samples from being damaged.
3. We preferentially use VECTASHIELD containing DAPI as visualizing the DAPI channel can confirm tissue integrity as well as provide information on relative cell size. We find that Ubi-GFP or Ubi-RFP marked clones typically do not require

antibody amplification of the signal for imaging with confocal. If using a fluorescently labeled protein expressed at endogenous levels, we find you do typically need to amplify the signal with a secondary for good imaging on standard detectors.

4. Common media problems: Many of the ingredients can be purchased (for cheaper) through your on-campus food vendor, but we have seen problems from some manufacturers or lots for both cornmeal and for molasses. It is best to do a small batch test before buying in bulk. During media prep there are three common pitfalls: (1) If the medium is not boiled sufficiently, the agar will not fully melt, and food will separate on cooling. (2) If not enough water is added to account for water lost through boiling, then the food will be too dry and will pull away from the sides of the vials; conversely too much water can result in food that liquefies in the top layer and drown larvae. (3) The food must cool sufficiently to add the last ingredients (<70 °C), but not so much that it sets before the vials are filled. Finally, excess moisture (droplets or mist) in the vials can trap and kill your flies. Allowing vials to cure before plugging will limit this problem, as will allowing vials to fully warm to room temperature before use. If the medium becomes too loose/watery during culture, adding a bit of Kimwipe or additional yeast granules to the vial can help stiffen the food, reducing the number of adults that become trapped.
5. Adult flies must be anesthetized for sorting and setting crosses. Anesthetization is typically done with carbon dioxide or ether, though agents such as FlyNap (Carolina Biologicals) are also effective, particularly if you are working with relatively small numbers of flies. For carbon dioxide anesthetization, there are commercially available fly pads that allow a diffuse regulated flow of CO2 to maintain anesthetization while flies are examined under the microscope. There are also protocols for anesthetizing flies by chilling, but our experience is that it is challenging to keep the flies knocked out well enough to sort markers with this method without the risk of killing or sterilizing the flies.
6. Before setting crosses, check that the food in the vials housing females has no signs of larvae. Virgin females can and typically do lay eggs on the food, which will be visible. Young larvae can be difficult to see with the naked eye so a check under the microscope is useful. If larvae are seen, the vial should not be used for setting the cross. Females from multiple collections can be pooled for setting crosses.
7. The number of animals in a vial must be managed as overcrowding can cause tissues to be smaller than normal and will cause some larvae in the vial to become delayed. Too few

animals can lead to weak or sick cultures. Individual crosses must be optimized, but a good starting point for generating clones is five to eight virgins with three to five males if you are using large vials. Typically virgin females should be fed yeast and be aged 2–7 days without males. Young males (2–5 days) are preferable for good crosses. Adding a small amount of dry yeast to the cross vials will increase egg production and larval survival.

8. Sufficient larvae of the same stage are needed for a good analysis. The number of crosses necessary to ensure that 10–50 larvae of the same genotype are available at the same time for dissection will depend on the frequency of the genotype in the vial. For crosses where few animals are needed and the genotype is 50%, as few as three crosses may be set in parallel. For rare animals or large numbers, this must be scaled.
9. Depending on the flippase used, the size, frequency, and location of clones vary widely. Clones that are rare or which cause position-dependent effects will greatly increase the number of animals that must be analyzed.
10. Temperature can affect outcomes. This has two main aspects. (1) Temperature affects growth rate. (2) The Gal4 system is temperature dependent; thus, altering the temperature can modulate the strength of Gal4 phenotypes. Standard temperature is 25 °C, but lower temperatures can reduce Gal4 activity, decreasing strength of UAS expression. Similarly, stronger expression is observed at temperatures above 25 °C.
11. Adults should be transferred to new media every 24–48 h to limit the number of larvae that will develop in each vial and to increase the uniformity of the developmental stage in the vial. If provided with yeast in each vial, adults can typically be transferred 5–10 times before the females will stop laying. If you are new to transferring flies and nervous about losing them all with the bang transfer method, it is best to remember that their natural response is to run upward (negative gravitaxis). Thus, you can hold an open (uncrowded) vial parallel to the bench with few escapers. To transfer the adults, it is best to tap them to the side of the vial and then quickly transfer the disoriented flies to the new prepared vial by joining the two ends together and banging the flies down into the new vial. Avoid banging the flies down in the cross vial, or they will get stuck in the food. Additional young animals can be added to later flips to keep the cross going. It is important to note that if too much yeast is added to the vials, it will become sticky and trap adults, decreasing the efficiency of the cross.
12. Additional flips can be used for collection of adult progeny, additional rounds of dissection and staining, or increasing

scored animal numbers for quantification assays. It is of critical importance to remember that the progeny are the target population for the experiment, not the adults that you are turning. You must transfer the adults to prevent crowding of the progeny, and you must return the vial of eggs/larvae to the correct temperature for the experiment. This may sound obvious, but we have observed many novices dutifully transfer the adults, and discard the previous flip, only to wonder some days later why they never have larvae.

13. For experiments using heat shock flippase, it is typical to transfer larvae at 24 h and to heat shock 24–48 h after the transfer. This is to ensure that all the larvae will have imaginal discs undergoing mitosis (necessary for recombination). Heat shocks are typically performed at 37–38 °C for 15–60 min. The duration and timing of the shock is empirically determined to yield a desirable number of clones per disc on average. Later heat shocks will yield smaller clones as the cells have fewer remaining divisions for the clones to grow. Longer heat shocks will induce more frequent clones. If the desired genotype has a mutation that slows the development of that genotype, such as Minute, then the timing of the heat shock should be adjusted accordingly.

14. Larvae will survive well for up to an hour or more but can eventually drown if left in the PBS too long. Cold PBS will slow larval movements and reduce their ability to escape the holding dish. If you are using glass 9 well plates, this can be particularly helpful as the wells are not so deep. If there are either too many larvae or not enough PBS, the larvae can and do escape the wells. If larvae are frequently escaping, increasing the PBS volume or decreasing the number of larvae in the well or droplet can stop this problem. However, on the Sylgard plate, drops that are too large (generally >250 μL) cannot hold their surface tension during the dissection, and once the surface tension is broken, the larvae will rapidly crawl away. This should certainly be considered before collecting multiple genotypes to the same plate.

15. For dissecting and analyzing larval tissues, it is important to note the developmental stage but also to be able to definitively know the genotype. We utilize a variety of visible markers for the purpose of genotyping our larvae including gender, *yellow*, *Tubby*, Ubi-GFP or Ubi-RFP, and fluorescently labeled balancer chromosomes.

16. We have found that using a small foam floating rack is quite effective for organizing and stabilizing multiple tubes in the ice.

17. Everting the dissection permits the discs to be freely exposed to the solutions while staying attached to the cuticle. Failure to evert the dissection can have few consequences, but in the case of large dissections, we find that a significant number of discs will become crushed or fixed to the cuticle preventing their analysis. The cuticle overlaying the discs also seems to sometimes impair fixation and staining. To handle the dissections, use forceps to grasp the cuticle or mouth hooks to avoid damaging the discs inadvertently.
18. Fixed head complexes should rapidly settle to the bottom after the tube is tapped. If they tend to float, they are not sufficiently fixed, and staining will be poor. For changing washes without losing head complexes, lift tube to eye level and tilt it about 45°. See the disc complexes settle. Place the pipette tip touching the bottom of the tube, above the head complexes, and withdraw the solution until 50–100 μL remains. Tilt the tube slowly down to almost 90° while removing the remaining solution.
19. We have found that properly fixed imaginal discs can be stored in PBT for some months at 4 °C before proceeding with the staining. In cases of some difficult crosses or instances when the user is not certain how many animals will need to be analyzed, it can be expedient to store extra dissections at this stage.
20. We have observed that blocking can be done for 10–20 min on ice or up to overnight at 4 °C without compromising the staining if samples are well fixed.
21. This volume will give a very flat mount of the tissue, which is ideal in certain imaging contexts. For more depth in the sample or for severely overgrown discs, more volume may be necessary. Typically 12-16 μL will be appropriate. In our experience, volumes over 20 μL with this size coverslip result in samples that are highly folded or even floating, making imaging difficult or impossible. For very thick samples, you may want to use a larger volume and a 22 mm coverslip, which is a bit heavier and can help to flatten hyperplastic discs. Alternatively, if more depth is desired, mouth hooks or brains can be placed around the samples, as they are rigid enough to hold the coverslip up a bit.
22. It is important to analyze both larval and adult phenotypes where possible. To optimize time, we typically analyze adults from the first flip and larvae from subsequent flips of the same cross. We also typically examine pupae to determine if there is significant lethality in this stage and if so then at what developmental stage.

References

1. Rong YS, Golic KG (2003) The homologous chromosome is an effective template for the repair of mitotic DNA double-strand breaks in Drosophila. Genetics 165:1831–1842
2. Golic KG, Lindquist S (1989) The FLP recombinase of yeast catalyzes site-specific recombination in the Drosophila genome. Cell 59:499–509
3. Xu T, Rubin GM (1993) Analysis of genetic mosaics in developing and adult Drosophila tissues. Development 117:1223–1237
4. Germani F, Bergantinos C, Johnston L (2018) Mosaic analysis in Drosophila. Genetics 208:473–490. https://doi.org/10.1534/genetics.117.300256
5. Xu T, Wang W, Zhang S et al (1995) Identifying tumor suppressors in genetic mosaics: the Drosophila lats gene encodes a putative protein kinase. Development 121:1053–1063
6. Justice RW, Zilian O, Woods DF et al (1995) The Drosophila tumor suppressor gene warts encodes a homolog of human myotonic dystrophy kinase and is required for the control of cell shape and proliferation. Genes Dev 9:534–546
7. Kango-Singh M, Nolo R, Tao C et al (2002) Shar-pei mediates cell proliferation arrest during imaginal disc growth in Drosophila. Development 129:5719–5730
8. Tapon N, Harvey KF, Bell DW et al (2002) salvador Promotes both cell cycle exit and apoptosis in Drosophila and is mutated in human cancer cell lines. Cell 110:467–478. https://doi.org/10.1016/S0092-8674(02)00824-3
9. Harvey KF, Pfleger CM, Hariharan IK (2003) The Drosophila Mst ortholog, hippo, restricts growth and cell proliferation and promotes apoptosis. Cell 114:457–467
10. Jia J, Zhang W, Wang B et al (2003) The Drosophila Ste20 family kinase dMST functions as a tumor suppressor by restricting cell proliferation and promoting apoptosis. Genes Dev 17:2514–2519. https://doi.org/10.1101/gad.1134003
11. Udan RS, Kango-Singh M, Nolo R et al (2003) Hippo promotes proliferation arrest and apoptosis in the Salvador/Warts pathway. Nat Cell Biol 5:914–920. https://doi.org/10.1038/ncb1050
12. Wu S, Huang J, Dong J, Pan D (2003) Hippo encodes a Ste-20 family protein kinase that restricts cell proliferation and promotes apoptosis in conjunction with salvador and warts. Cell 114:445–456. https://doi.org/10.1016/S0092-8674(03)00549-X
13. Pantalacci S, Tapon N, Léopold P (2003) The Salvador partner Hippo promotes apoptosis and cell-cycle exit in Drosophila. Nat Cell Biol 5:921–927. https://doi.org/10.1038/ncb1051
14. Hamaratoglu F, Willecke M, Kango-Singh M et al (2006) The tumour-suppressor genes NF2/Merlin and Expanded act through Hippo signalling to regulate cell proliferation and apoptosis. Nat Cell Biol 8:27–36. https://doi.org/10.1038/ncb1339
15. Genevet A, Wehr M, Brain R et al (2010) Kibra is a regulator of the Salvador/Warts/Hippo signaling network. Dev Cell 18:300–308. https://doi.org/10.1016/j.devcel.2009.12.011
16. Ling C, Zheng Y, Yin F et al (2010) The apical transmembrane protein Crumbs functions as a tumor suppressor that regulates Hippo signaling by binding to Expanded. Proc Natl Acad Sci U S A 107:10532–10537. https://doi.org/10.1073/pnas.1004279107
17. Yu J, Zheng Y, Dong J et al (2010) Kibra functions as a tumor suppressor protein that regulates Hippo signaling in conjunction with Merlin and Expanded. Dev Cell 18:288–299. https://doi.org/10.1016/j.devcel.2009.12.012
18. Willecke M, Hamaratoglu F, Kango-Singh M et al (2006) The fat cadherin acts through the hippo tumor-suppressor pathway to regulate tissue size. Curr Biol 16:2090–2100. https://doi.org/10.1016/j.cub.2006.09.005
19. Silva E, Tsatskis Y, Gardano L et al (2006) The tumor-suppressor gene fat controls tissue growth upstream of expanded in the hippo signaling pathway. Curr Biol 16:2081–2089. https://doi.org/10.1016/j.cub.2006.09.004
20. He Y, Emoto K, Fang X et al (2005) Drosophila Mob family proteins interact with the related tricornered (Trc) and warts (Wts) kinases. Mol Biol Cell 16:4139–4152. https://doi.org/10.1091/mbc.E05-01-0018
21. Lai Z-C, Wei X, Shimizu T et al (2005) Control of cell proliferation and apoptosis by mob as tumor suppressor, mats. Cell 120:675–685. https://doi.org/10.1016/j.cell.2004.12.036
22. Bennett FC, Harvey KF (2006) Fat cadherin modulates organ size in Drosophila via the Salvador/Warts/Hippo signaling pathway. Curr Biol 16:2101–2110. https://doi.org/10.1016/j.cub.2006.09.045
23. Huang J, Wu S, Barrera J et al (2005) The Hippo signaling pathway coordinately regulates cell proliferation and apoptosis by

inactivating Yorkie, the Drosophila Homolog of YAP. Cell 122:421–434. https://doi.org/10.1016/j.cell.2005.06.007

24. Nolo R, Morrison CM, Tao C et al (2006) The bantam microRNA is a target of the hippo tumor-suppressor pathway. Curr Biol 16:1895–1904. https://doi.org/10.1016/j.cub.2006.08.057
25. Thompson B, Cohen S (2006) The Hippo Pathway regulates the bantam microRNA to control cell proliferation and apoptosis in Drosophila. Cell 126:767–774. https://doi.org/10.1016/j.cell.2006.07.013
26. Neto-Silva RM, de Beco S, Johnston LA (2010) Evidence for a growth-stabilizing regulatory feedback mechanism between Myc and Yorkie, the Drosophila homolog of Yap. Dev Cell 19:507–520. https://doi.org/10.1016/j.devcel.2010.09.009
27. Katsukawa M, Ohsawa S, Zhang L et al (2018) Serpin facilitates tumor-suppressive cell competition by blocking toll-mediated Yki activation in Drosophila. Curr Biol 28:1756–1767.e6. https://doi.org/10.1016/j.cub.2018.04.022
28. Suijkerbuijk SJ, Kolahgar G, Kucinski I, Piddini E (2016) Cell competition drives the growth of intestinal adenomas in Drosophila. Curr Biol 26:428–438. https://doi.org/10.1016/j.cub.2015.12.043
29. Menéndez J, Pérez-Garijo A, Calleja M, Morata G (2010) A tumor-suppressing mechanism in Drosophila involving cell competition and the Hippo pathway. Proc Natl Acad Sci U S A 107:14651–14656. https://doi.org/10.1073/pnas.1009376107
30. Yang C-CC, Graves HK, Moya IM et al (2015) Differential regulation of the Hippo pathway by adherens junctions and apical-basal cell polarity modules. Proc Natl Acad Sci U S A 112:1785–1790. https://doi.org/10.1073/pnas.1420850112
31. Chen C-LL, Schroeder MC, Kango-Singh M et al (2012) Tumor suppression by cell competition through regulation of the Hippo pathway. Proc Natl Acad Sci U S A 109:484–489. https://doi.org/10.1073/pnas.1113882109
32. Hafezi Y, Bosch JA, Hariharan IK (2012) Differences in levels of the transmembrane protein Crumbs can influence cell survival at clonal boundaries. Dev Biol 368:358–369. https://doi.org/10.1016/j.ydbio.2012.06.001
33. Tyler DM, Li W, Zhuo N et al (2007) Genes affecting cell competition in Drosophila. Genetics 175:643–657. https://doi.org/10.1534/genetics.106.061929
34. Robinson BS, Huang J, Hong Y, Moberg KH (2010) Crumbs regulates Salvador/Warts/Hippo signaling in Drosophila via the FERM-domain protein Expanded. Curr Biol 20:582–590. https://doi.org/10.1016/j.cub.2010.03.019
35. Simon MA, Xu A, Ishikawa HO, Irvine KD (2010) Modulation of fat:dachsous binding by the cadherin domain kinase four-jointed. Curr Biol 20:811–817. https://doi.org/10.1016/j.cub.2010.04.016
36. Roote J, Prokop A (2013) How to design a genetic mating scheme: a basic training package for Drosophila genetics. G3 (Bethesda) 3:353–358. https://doi.org/10.1534/g3.112.004820

Chapter 3

Drosophila Genetics: Analysis of Tissue Growth in Adult Tissues

Alexander D. Fulford and Paulo S. Ribeiro

Abstract

Drosophila melanogaster has been widely used in the study of developmental growth control and has been instrumental in the discovery and delineation of many signalling pathways that contribute to this growth, in particular the Hippo pathway. Quantitative analysis of adult tissue size has remained a vital tool in the study of tissue growth. This chapter will describe how to dissect, image, and quantify two tissues commonly used to measure growth, the *Drosophila* wing and eye.

Key words *Drosophila*, Adult eye, Adult wing, SEM, Tissue growth

1 Introduction

The use of *Drosophila melanogaster* as a model organism to study tissue growth has significantly enhanced our knowledge into the genetics and molecular control of developmental growth [1]. In particular, studies using the fruit fly have identified and delineated many of the now well-known developmental signalling pathways, which are typically highly conserved [2]. The Hippo pathway, the focus of this book, is a prime example of the power of *Drosophila* genetics, since the identification of its main components and of their genetic interactions would not have been possible without the seminal work of many laboratories, which performed mosaic genetic screens to identify novel tumor suppressor genes [3–5].

Two adult tissues in particular are widely used to study tissue growth, the eye and the wing. These tissues are specified during embryogenesis and form monolayered epithelial structures during the larval stage called imaginal discs [6]. During larval development, imaginal discs undergo rapid proliferation until pupariation where much of the larval tissue undergoes dramatic reorganization, allowing the formation of the adult organ structures [7]. The adult structure of the wing, including veins and hairs, fully develops

Alexander Hergovich (ed.), *The Hippo Pathway: Methods and Protocols*, Methods in Molecular Biology, vol. 1893, https://doi.org/10.1007/978-1-4939-8910-2_3,

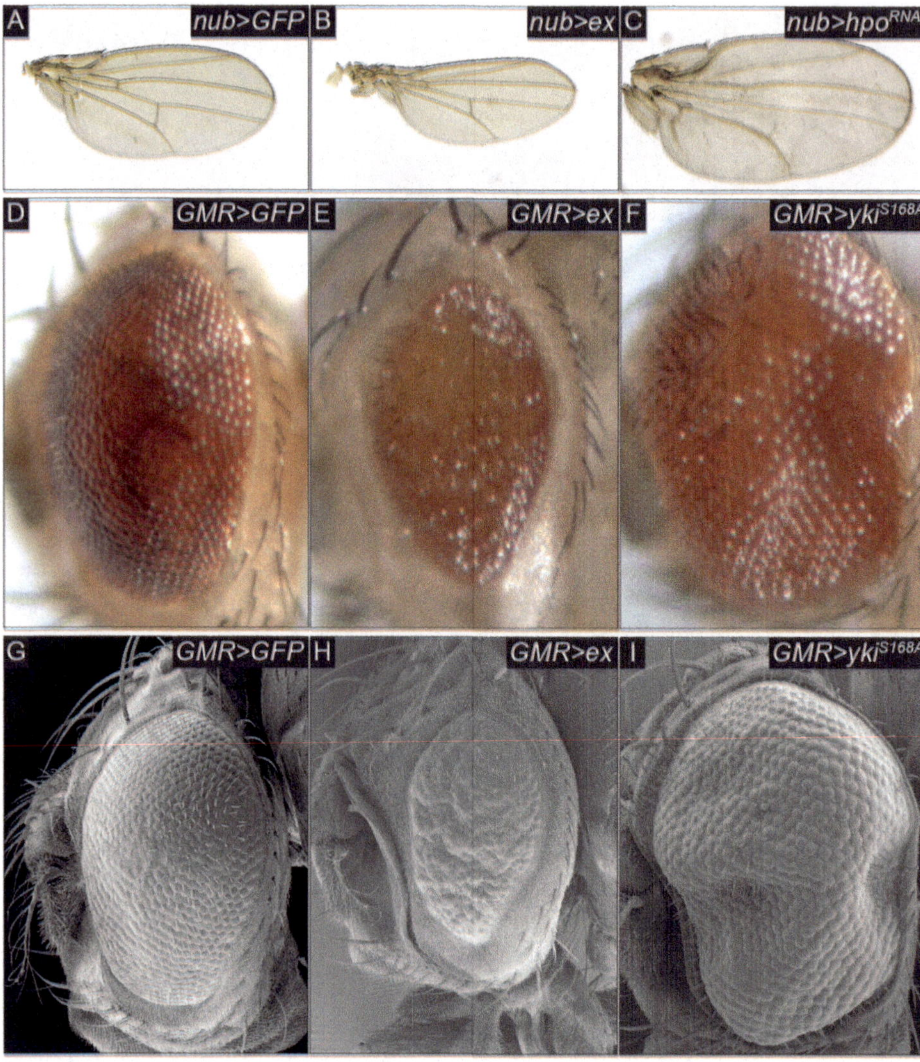

Fig. 1 *Drosophila* adult tissues processed for tissue growth analysis. (**a–c**) Adult wings expressing *UAS-GFP* (**a**), *UAS-ex* (**b**), or *UAS-hpo*^{RNAi} (**c**) under the control of *nubbin-Gal4*. (**d–f**) Adult eyes expressing *UAS-GFP* (**d**), *UAS-ex* (**e**), or *UAS-yki*^{S168A} (**f**) under the control of *GMR-Gal4*. (**g–i**) Scanning electron micrographs of adult eyes expressing *UAS-GFP* (**g**), *UAS-ex* (**h**), or *UAS-yki*^{S168A} (**i**) under the control of *GMR-Gal4*

during pupariation when the wing expands and secretes cuticle, which after eclosion, comprises most of the adult wing [8] (Fig. 1a). The compound eye of an adult *Drosophila* consists of several hundred individual structures called ommatidia, each forming a hexagonal lattice of eight photoreceptor cells, four cone cells, and two primary pigment cells, in addition to sensory bristle cells and secondary and tertiary pigment cells [9] (Fig. 1d, g).

The rapid development of these organs is highly useful in the context of studying growth, as minor perturbations in growth signalling can result in a profound overall effect on the adult eye and wing sizes. For example, compared to control tissue (Fig. 1a, d, g), manipulation of the Hippo pathway by increasing signalling activity through overexpression of the *expanded (ex)* gene results in tissue undergrowth (Fig. 1b, e, h). In contrast, decreasing pathway activity through expression of RNA interference (RNAi) targeting *hippo (hpo)* or through overexpression of the hyperactive *yorkie (yki)* S168A mutant results in tissue overgrowth (Fig. 1c, f, i). Imaginal disc development is also remarkably autonomous, which makes these tissues, and the adult structures they originate, ideally suited for study, as the systemic effect of potentially disruptive genetic manipulations are often limited to the structures in question [10]. Therefore, an additional advantage of analyzing adult eyes and wings is the fact that, in the majority of cases, tissue-specific disruption of gene function does not lead to organism lethality, even in instances where the tissue affected is severely affected.

The manipulation of genes to study adult tissue growth commonly uses the tractable Gal4/UAS bipartite expression system [11] combining tissue- or compartment-specific control of Gal4 expression with the enormous array of available UAS-transgenes, to either induce gain or loss of function phenotypes. Once adults have emerged, the desired tissue is dissected, mounted, and imaged, and subsequent analysis provides a quantitative measure of adult tissue growth. The nature of eye and wing development, allied with the ease of manipulation, dissection, and analysis, makes the adult eye and wing excellent tissues for the study of tissue growth, either by light microscopy or scanning electron microscopy (SEM).

2 Materials

All general equipment and SEM-specific equipment are stored at room temperature.

2.1 General Equipment

1. 1.5 mL microcentrifuge tubes.
2. 10, 30, 50, 70, and 100% (v/v) ethanol prepared in deionized water.
3. Clear bottom dissecting dish or Depression glass spot plate (available from Corning).
4. CO_2 source designed for anesthetizing *Drosophila*.
5. Computer with ImageJ software (NIH).
6. Cover slips, 24 × 32 mm.
7. Double-sided sticky tape.
8. Euparal mounting media.
9. Incubator capable of reaching 60 °C—for example, HIS25 (Boekel Scientific).

10. Isopropanol.
11. Paperclips.
12. Pasteur pipettes.
13. Slide scanner—for example, Pannoramic 250 Flash (3DHISTECH).
14. Standard microscope slides.
15. Stereo dissecting microscope—for example, M80 (Leica).
16. Stereomicroscope fitted with a camera—for example, SteREO Lumar.V12 (Zeiss).
17. Super thin dissecting forceps—for example, No.5 (Dumont Biology).

2.2 SEM-Specific Equipment

1. Aluminum specimen stubs, 0.5 in. (available from Agar Scientific).
2. Carbon conductive tabs, 12 mm diameter (available from Agar Scientific).
3. Conductive silver paint (available from Agar Scientific).
4. Critical point dryer—for example, CPD300 (Leica).
5. Filter paper.
6. Microporous specimen capsules, 30 μm (available from Fisher Scientific).
7. Nutating mixer.
8. Parafilm.
9. Scanning electron microscope—for example, JSM-6700F (JEOL).
10. Sputter coater—for example, Q150R Rotary-Pumped Sputter Coater (Quorum Technologies).
11. Storage box for microporous specimen capsules.

3 Methods

Unless otherwise stated, all methods are performed at room temperature.

3.1 Preparation and Analysis of Adult Wings

1. Anesthetize *Drosophila* with CO_2, and, under a dissecting microscope, collect adult flies of the appropriate genotypes in labelled microcentrifuge tubes containing 70% ethanol.
2. Fill a dissecting dish with isopropanol using a Pasteur pipette.
3. Using the dissecting forceps, transfer adult flies from the microcentrifuge tube to the dissecting dish.
4. Select either male or female flies depending on the gender selected for analysis (*see* **Note 1**).

5. Under the dissecting microscope, and using one set of forceps, gently pin down the adult fly on one side by the thorax.
6. Using the other set of forceps, remove one wing from as close to the thorax as possible (*see* **Note 2**).
7. Repeat wing dissection (**steps 2–6**) for approximately 20 adults.
8. Once all flies have been dissected, label a microscope slide in pencil to prevent isopropanol dissolving the label.
9. Apply a thin coat of isopropanol to the microscope slide using the Pasteur pipette.
10. Transfer the wings from the dissecting dish to the microscope slide using the forceps.
11. Using the forceps, carefully arrange the wings into rows and columns contacting only the most proximal point of the wing if possible (*see* **Note 3**).
12. Allow the isopropanol to evaporate for approximately 1–2 min.
13. Apply Euparal mounting media in between the rows and columns of wings (*see* **Note 4**).
14. Place a coverslip over the wings (*see* **Note 5**).
15. Using a bent paperclip, fix the coverslip to the microscope slide, facilitating constant pressure between the coverslip and the microscope slide to remove air bubbles.
16. Bake the microscope slides at 60 °C for approximately 3 h to allow the Euparal to set and to harden (*see* **Note 6**).
17. Image the wings using an automated slide scanner—for example a Panoramic 250 Flash (3DHISTECH) machine. Alternatively, wings can be individually imaged using a stereomicroscope fitted with a camera. Regardless of the size of individual adult wings, ensure that magnification is the same for all images as this will allow easy comparison of different experiments.
18. Open images in the ImageJ software, and using the "polygon selections" tool, draw around the wing (*see* **Note 7**).
19. Run the "measure" function within the "analyze" menu to determine the area size.

3.2 Preparation and Analysis of Adult Eyes by Light Microscopy

1. Anesthetize *Drosophila* with CO_2, and, under a dissecting microscope, collect adult flies of the appropriate genotypes in labelled microcentrifuge tubes and store at −80 °C or on dry ice for at least 30 min.
2. Apply double-sided sticky tape to labelled microscope slides, taking care to avoid air bubbles.

3. Using the dissecting forceps, transfer frozen adult *Drosophila* to an empty microscope slide.
4. Under the dissecting microscope, select either males or females depending on the gender selected for analysis.
5. Carefully remove the whole head of the *Drosophila* using the dissecting forceps and discard the body.
6. Transfer the whole head onto the labelled microscope slide covered with double-sided sticky tape, and position the fly head along the sagittal plane, as if the fly is on its side (*see* Fig. 1d–f) or along the coronal plane so the whole head is visible (e.g., *see* Fig. 5 in ref. [12]). The positioning and orientation of the head will influence what can be measured and quantified. For instance, if the fly head is positioned on its side, only the size of the eye can be quantified. If the whole fly head is visible (either face on or top view), the size of both eyes (and potentially of the head capsule itself) can be measured and quantified.
7. Repeat head dissection (**steps 3–6**) for approximately 20 adults arranging the eyes in rows and columns, and image the tissue immediately after mounting.
8. Image eyes using a stereomicroscope fitted with a camera ensuring the magnification is the same for all images.
9. Open images in the ImageJ software, and using the "polygon selections" tool, draw around the eye.
10. Run the "measure" function within the "analyze" menu to determine the area size.

3.3 Preparation and Analysis of Adult Eyes by SEM

1. For analysis of adult eyes using scanning electron microscopy, anesthetize *Drosophila* with CO_2, and, under a dissecting microscope, collect males of the appropriate genotypes in labelled microcentrifuge tubes containing 10% ethanol (*see* **Note 8**).
2. Incubate flies in 10% ethanol for 24 h at room temperature on a nutating mixer.
3. Remove 10% ethanol, and replace with 30% ethanol solution. Incubate for at least 8 h at room temperature on a nutating mixer.
4. Repeat the procedure described (**steps 2** and **3**) above with solutions of 50, 70, and 100% ethanol (*see* **Note 9**).
5. Following the incubation with 100% ethanol solution, replace with fresh 100% ethanol, and store at 4 °C until sample processing and preparation for scanning electron microscopy.

6. Load samples onto an automated critical point drier—for example, a CPD300 machine (Leica)—and follow the instructions of the equipment protocol.
7. Following critical point drying, store samples at room temperature in a microporous specimen capsule (30 μm) prior to mounting and sputter coating (*see* **Note 10**).
8. Attach the conductive carbon tab to the aluminum specimen stub, and, using forceps, remove the upper plastic layer to expose the adhesive surface.
9. Using forceps, retrieve dried fly from storage container. Please note that samples are very fragile and, therefore, flies should be manipulated using forceps and captured by one of their wings to avoid damage to the tissue of interest.
10. Transfer the fly to an aluminum specimen stub used by the SEM, and attach it to the adhesive surface of a conductive carbon tab. Position the fly head along the sagittal plane, with the fly on its side, with one eye facing up (*see* Fig. 1g–i). The fly wing that is not bound by the forceps will attach first to the adhesive surface, and, after this, gently position the fly body to one side and attach it to the adhesive surface. The bound fly wing can either be detached from the fly body or also attached to the adhesive surface, taking care not to damage the fly eye or altering the positioning of the fly head.
11. Depending on the sample holder used, several flies of the same genotype can be attached to the same holder.
12. To ensure that flies are connected to the conductive material and the holder after coating, use a small amount of silver paint to paint over the back end of the fly connecting it with the metal stub. Ensure that the silver paint does not reach the fly eye and the structures that will be subsequently imaged.
13. Perform sputter coating using appropriate equipment—for example, a Q150R Rotary-Pumped Sputter Coater (Quorum Technologies). In the case of platinum coating, a 9 nm coating protocol is sufficient for adequate imaging of samples.
14. Following sputter coating, samples are ready to image in the SEM—for example, a JSM-6700F Scanning Electron Microscope (JEOL)—as per equipment protocol. Images of adult eyes should be acquired at the same magnification and, as much as possible, using the same eye positioning to allow comparison between genotypes.
15. Open images in the ImageJ software, and using the "polygon selections" tool, draw around the eye.
16. Run the "measure" function within the "analyze" menu to obtain the area (*see* **Note 11** for an alternative analysis approach).

4 Notes

1. Female *Drosophila* have slightly larger wings than males, so for this reason they are preferred for analysis.
2. Removal of wings from the flies should be done with care to not damage the wing structure. Wings should be detached from the thorax by the hinge, as close as possible to the notum. Note that the alula can frequently be damaged when detaching wings from the thorax. For this reason, both the alula and the costal cell of the wing are not included in the analysis of wing size.
3. While mounting adult wings in the microscope slide, if the procedure is not completed rapidly, the wings may start to dry up, and, consequently, they can start folding at the edges or crumpling. To ensure that wings remain flat, add more isopropanol using a Pasteur pipette. This will also aid the rearrangement of wings into rows and columns.
4. Other alternatives to Euparal mounting media exist, such as Canada Balsam [13] or Hoyer's medium [14].
5. Use Euparal sparingly when mounting. This will avoid the wing row and column arrangement from being disrupted once the coverslip is added, aiding image analysis. In addition, this prevents excess mounting media seeping from under the coverslip which can damage microscope objectives if using an automated slide scanner.
6. Once baked, slides can be stored almost indefinitely.
7. If a Gal4 driver is used that expresses the desired transgene in a specific wing compartment, measure both the compartment area and the total wing area to generate a compartment area/total area ratio.
8. Male flies are preferred for electron microscopy analysis. Complete dehydration of female flies can take longer than that of males, particularly if females have mated. Incomplete dehydration will result in sample explosion during critical point drying.
9. Serial incubation of adult tissue in increasing concentrations of ethanol (10, 30, 50, 70, and 100%) is vital for complete dehydration of samples, ensuring the maintenance of tissue integrity during critical point drying.
10. Following critical point drying, to avoid rehydration of samples by contact with atmospheric moisture, samples stored in microporous specimen capsules should also be stored in in boxes containing tissue or filter paper sealed with Parafilm.
11. In addition to measuring the area of the eye, number of ommatidia can be calculated as a measure of tissue growth.

Acknowledgments

We thank Maxine Holder and Nic Tapon for comments on the manuscript. This work was supported by funding from Cancer Research UK (C16420/A18066) and from The Academy of Medical Sciences/Wellcome Trust Springboard Award (SBF001/1018).

References

1. St Johnston D (2002) The art and design of genetic screens: Drosophila melanogaster. Nat Rev Genet 3(3):176–188. https://doi.org/10.1038/nrg751
2. Hariharan IK, Bilder D (2006) Regulation of imaginal disc growth by tumor-suppressor genes in Drosophila. Annu Rev Genet 40:335–361. https://doi.org/10.1146/annurev.genet.39.073003.100738
3. Justice R, Zilian O, Woods D, Noll M, Bryant P (1995) The Drosophila tumor suppressor gene warts encodes a homolog o-f human myotonic dystrophy kinase and is required for the control of cell shape and proliferation. Genes Dev 9:534–546
4. Harvey K, Tapon N (2007) The Salvador-Warts-Hippo pathway - an emerging tumour-suppressor network. Nat Rev Cancer 7 (3):182–191. https://doi.org/10.1038/nrc2070
5. Pan D (2010) The hippo signaling pathway in development and cancer. Dev Cell 19 (4):491–505. https://doi.org/10.1016/j.devcel.2010.09.011
6. Bate M, Arias AM (1991) The embryonic origin of imaginal discs in Drosophila. Development 112(3):755–761
7. Fristrom DK, Fristrom JW (1993) The metamorphic development of the adult epidermis. The Development of *Drosophila melanogaster*. Cold Spring Harbor Laboratory Press, New York
8. de la Loza MC D, Thompson BJ (2017) Forces shaping the Drosophila wing. Mech Dev 144 (Pt A):23–32. https://doi.org/10.1016/j.mod.2016.10.003
9. Cagan R (2009) Principles of Drosophila eye differentiation. Curr Top Dev Biol 89:115–135. https://doi.org/10.1016/s0070-2153(09)89005-4
10. Hariharan IK (2015) Organ size control: lessons from Drosophila. Dev Cell 34 (3):255–265. https://doi.org/10.1016/j.devcel.2015.07.012
11. Brand AH, Perrimon N (1993) Targeted gene expression as a means of altering cell fates and generating dominant phenotypes. Development:118
12. Kulaberoglu Y, Lin K, Holder M, Gai Z, Gomez M, Assefa Shifa B, Mavis M, Hoa L, Sharif AAD, Lujan C, Smith ESJ, Bjedov I, Tapon N, Wu G, Hergovich A (2017) Stable MOB1 interaction with Hippo/MST is not essential for development and tissue growth control. Nat Commun 8:695. https://doi.org/10.1038/s41467-017-00795-y
13. Stern DL, Sucena E (2012) Preparation and mounting of adult Drosophila structures in Canada balsam. Cold Spring Harb Protoc 2012(3):373–375. https://doi.org/10.1101/pdb.prot067389
14. Ashburner M, Golic KG, Scott Hawley R (1989) *Drosophila*: a laboratory handbook. Cold Spring Harbor Laboratory Press, New York

Chapter 4

Live Imaging of Hippo Pathway Components in Drosophila Imaginal Discs

Jiajie Xu, Ting Su, Sherzod A. Tokamov, and Richard G. Fehon

Abstract

Examining the subcellular localization of Hippo pathway components has helped elucidate the molecular mechanisms that regulate the pathway. Here we describe methods for performing live imaging of fluorescently tagged Hippo pathway components in *Drosophila* wing imaginal discs.

Key words Hippo pathway components, Live imaging, *Drosophila* wing imaginal discs

1 Introduction

Examining subcellular localization of Hippo pathway components has been an important aspect of studies of this conserved tissue growth control pathway. Current models propose that the Hippo pathway regulates Yorkie (Yki, a *Drosophila* orthologue of mammalian YAP/TAZ) activity by controlling its subcellular localization: when the pathway is active, Yki is retained in the cytoplasm, while in the absence of pathway activity, Yki accumulates in the nucleus. Therefore, the subcellular localization of Yki has been widely used as a readout for Hippo pathway activity [1–3]. Upstream regulators of the Hippo pathway as well as the core kinases have been shown to localize to the apical junctional region (AJR), a cortical region that extends from the lateral edge of the apical cortex through the adherens junctions, where they appear to organize into signaling complexes [4–6]. In addition, we recently identified the apical medial cortex as an additional subcellular domain for Hippo pathway regulators Merlin (Mer) and Kibra (Kib) [2].

Further analysis of subcellular localization of Hippo pathway components under different genetic and mechanical manipulations, especially in live tissue, could provide a better mechanistic understanding of the pathway. To achieve that goal, we have generated endogenously expressed, YFP-tagged transgenes of Mer,

Alexander Hergovich (ed.), *The Hippo Pathway: Methods and Protocols*, Methods in Molecular Biology, vol. 1893, https://doi.org/10.1007/978-1-4939-8910-2_4, © Springer Science+Business Media, LLC, part of Springer Nature 2019

Expanded, Hippo, Warts, and Yki, as well as GFP-tagged Kib using a combination of recombineering genomic transgenes and CRISPR/Cas-9 genome editing [2]. Here we describe methods for performing live imaging of *Drosophila* wing imaginal discs carrying these fluorescently tagged Hippo pathway components.

2 Materials

1. Dissection media: Schneider's *Drosophila* medium (e.g., Sigma #S9895) supplemented with 10% fetal bovine serum (FBS) (*see* **Note 1**).
2. Incubation chamber for imaging: Glass bottom microwell dish (e.g., MatTeK Corporation, Part No.: P35G-1.5-14-C) (*see* **Note 2**).
3. Phosphate-buffered saline (PBS): 155 mM NaCl, 15 mM Na_2HPO_4, 6 mM $NaH_2PO_4 \cdot H_2O$ (*see* **Note 3**).
4. 70% (v/v) EtOH in deionized water.
5. Supporting beads: 45–53 μm solid soda lime glass microspheres (e.g., Cospheric SLGMS-2.5 45–53 μm) (*see* **Note 4**).
6. Dissecting tools: Forceps (e.g., FST DUMONT No.5 1125-20), a tungsten needle, a diamond-tipped pen, and micropipettes.
7. Siliconized slides for dissection (*see* **Note 5**).
8. Laser scanning confocal microscope equipped with GaAsP detectors (e.g., Zeiss LSM 880 or Zeiss LSM 800).

3 Methods

Carry out all procedures at room temperature.

3.1 Preparing an Incubation Chamber for Imaging (Fig. 1a)

1. Prepare top coverslip by cutting a No.1.5 thickness coverslip into ~8 × 8 mm fragments using a diamond-tipped pen.
2. Add 2 mL dissection media into an incubation chamber.
3. Add 5–10 μL of resuspended supporting microsphere beads near the center of the cover glass.
4. Use forceps to clear the beads from the center of the cover glass to make space for the imaginal discs. Replace cover and set aside (*see* **Note 6**).

3.2 Dissecting Wing Imaginal Discs from Wandering Third Instar Larvae

1. Remove wandering third instar larvae from uncrowded vials using a damp paintbrush. Place in a 50 mL beaker with ~10 mL of water.
2. Rinse several times with distilled water to remove debris.

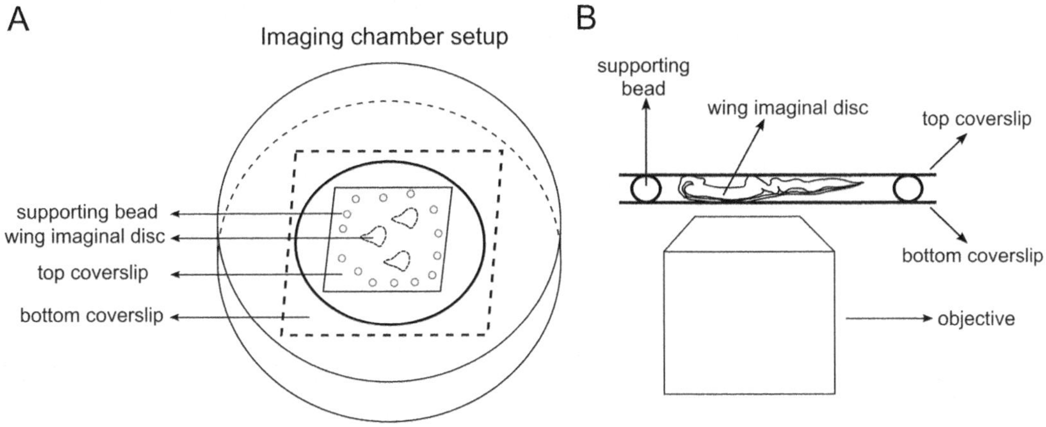

Fig. 1 Setup for live imaging of Hippo pathway components on an inverted microscope. (**a**) Imaging chamber setup. Wing imaginal discs are placed between two coverslips with the apical side facing downwards. The top coverslip is supported by glass microspheres. (**b**) Illustration showing cross-sectional views of the imaginal chamber

3. Wash 1 min in 70% EtOH to clean larvae.
4. Rinse several times in dH_2O.
5. To dissect larvae, place in a drop of dissection media on a siliconized slide.
6. Rotate the larva so that the dorsal side (tracheal tubes) is up. Identify the anterior (mouth hooks) and posterior ends. Using the less coordinated hand, grasp the larva from the posterior end at about the middle of the body.
7. Using the more coordinated hand, grasp the dorsal cuticle approximately one quarter of the way posterior from the head. Tear the dorsal cuticle toward the anterior end of the larva.
8. The wing imaginal discs should "flop out" laterally when the cuticle is torn open. They have a teardrop shape and a translucent appearance and lie close to the longitudinal tracheal tube.
9. Using the more coordinated hand, gently grasp the entire wing disc with the forceps (place the body of the disc between the prongs of the forceps). Pull the disc laterally to separate it from its connection to the trachea. Transfer the dissected discs to a fresh drop of dissection media between the tips of your forceps.

3.3 Setting Up an Incubation Chamber for Imaging on an Inverted Microscope

1. Transfer dissected wing imaginal discs into the center of the bottom cover glass. This can be done by carrying the imaginal discs between the tips of the forceps. Alternatively, use a micropipette with a yellow tip that has been cut short to widen the opening to transfer the discs. Pre-wet the tip with dissecting medium to prevent the discs from sticking to the tip.

2. Orient the wing imaginal discs with a tungsten needle so that the apical sides are facing the bottom cover glass. The pouch portion of the wing imaginal discs is bowl-shaped. The convex side of the bowl is the apical side of the wing epithelium (Fig. 1b) (*see* **Note** 7).
3. Use forceps to grasp the coverslip piece. Place on top of the dissection media at the edge of the microwell dish to wet the bottom side. Then, flip it over to wet the other side. When both sides are wet, the coverslip piece will sink to the bottom of the chamber.
4. Use forceps to grasp the coverslip piece, gently lift it from the bottom while still immersed in the dissection media and move it over the wing imaginal discs in the center of the microwell dish.
5. Gently release the top coverslip piece from the forceps so that it gently comes to rest on the basal sides of the wing imaginal discs, supported by the microsphere beads.
6. Place cover over the microwell dish and immediately transfer it to the inverted microscope.
7. For imaging imaginal discs mounted in microwell dishes, we use a Zeiss LSM 880 laser scanning confocal microscope equipped with GaAsP detectors mounted on a Zeiss Axio Observer Z1 inverted stand. We use 40 and 63× oil immersion objectives with numerical apertures of 1.3–1.4. The 514 nm line from an argon ion laser is optimal for excitation of YFP, though the 488 nm line will work as well. Examples of YFP-tagged Hippo pathway components are shown in Fig. 2 (*see* **Notes 8–10**).

3.4 Imaging on an Upright Microscope

1. For imaging on an upright microscope stand, place ~50 μL of dissecting medium near the center of a clean standard microscope slide. Add 5–10 μL of resuspended supporting microspheres near the center of the medium.
2. Use forceps to clear the beads from the center of the drop to make space for the imaginal discs.
3. Transfer dissected wing imaginal discs into the center of the drop (*see* Subheading 3.3, **step 1**).
4. Orient the wing imaginal discs with a tungsten needle so that the apical sides are facing up. The pouch portion of the wing imaginal discs is bowl-shaped. The bottom side of the bowl is the apical side of the wing epithelium.
5. Gently place a 22 × 22 mm cover glass directly onto the imaginal discs in the drop of dissecting medium (putting one edge of the coverslip down first may cause the imaginal discs to

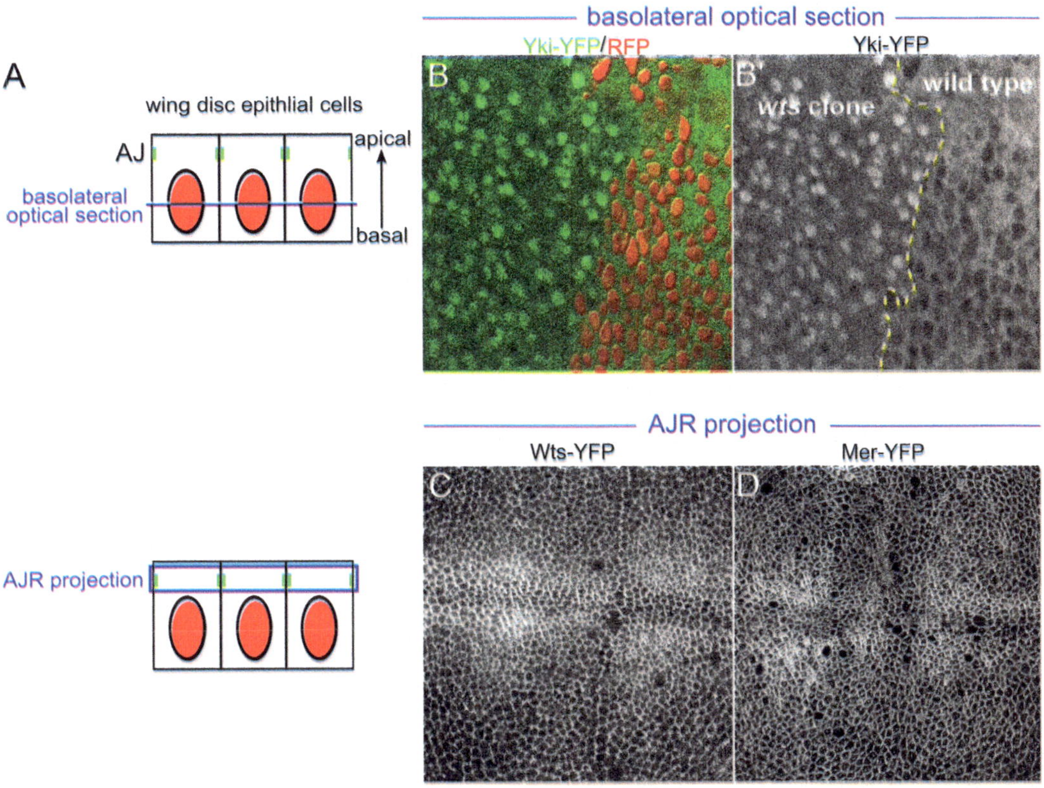

Fig. 2 Examples of Hippo pathway components imaged in live tissues. (**a**) Illustrations showing cross-sectional views of the imaginal epithelium and the approximate positions of basolateral and AJR images shown in B-D. As indicated in blue, basolateral images are single sections while AJR views are maximal projections of a small number of apical sections to compensate for curvature of the epithelium. (**b**) The effect of Hippo pathway inactivation on Yki subcellular localization. In basal sections of live tissues containing *wts* null mitotic clones (clone marked by the absence of RFP and a yellow dashed line), Yki-YFP is primarily cytoplasmic in normal imaginal tissue but is strongly nuclear in *wts* mutant cells (B-B′). (**c**, **d**) Apical projections showing the localization of endogenously expressed Wts-YFP (**c**) and Mer-YFP (**d**) in wing imaginal discs. Both Wts-YFP and Mer-YFP are enriched at the AJR. A portion of Mer-YFP also localizes to the apical medial region

invert). Avoid moving the cover glass once it has been released onto the slide.

6. Place a small drop of melted petroleum jelly on each corner of the cover glass to secure it onto the slide.
7. These preparations can be imaged on either an inverted or an upright microscope stand. For the latter, we use a Zeiss LSM 800 laser scanning confocal microscope mounted on a Zeiss Axio Imager M2 stand, with 40 and 63× oil immersion objectives. On this system, we use the 488 nm line to image both YFP- and GFP-labeled proteins (*see* **Notes 8–10**).

4 Notes

1. Dissection media is prepared according to manufacturer's instructions using Milli-Q deionized water and filter sterilized. Store at 4 °C and warm to room temperature before using.
2. The incubation chamber is a 35 mm petri dish with a No. 1.5 cover glass bottom for imaging on an inverted microscope.
3. PBS: Recipe for 1 L:

NaCl	9.0 g
Na_2HPO_4	2.0 g
$NaH_2PO_4{\cdot}H_2O$	0.83 g
dH_2O to 1 L	

4. The supporting beads prevent tissue damage by supporting the coverslip. Suspend these beads in PBS at 100 mg/mL.
5. To siliconize slides, place them (within cardboard box) in an empty vacuum desiccator with 1 mL dichloromethylsilane in the bottom of a 25–50 mL glass beaker. Pull a vacuum with a pump or aspirator and leave sealed overnight. Clean slides with soap and water before using. Perform this procedure in a fume hood.
6. The chamber is typically prepared just before dissections begin and is kept at room temperature.
7. It is important to orient the wing imaginal discs so that the apical side is facing the objective to provide optimal conditions for imaging the apical region of the disc epithelium, where pathway components primarily localize and function. If the discs are oriented the opposite way, both excitation light and emission light would need to travel through the entire tissue resulting in much weaker signal.
8. To maximize signal to noise for weakly expressed proteins, we often increase the pinhole to 1.5–2.0 airy units rather than increasing laser excitation light, to avoid photobleaching. Re-scanning samples often causes noticeable photobleaching and should be avoided.
9. Typically, we limit imaging time to no more than 15 min after mounting the tissues. However, we have imaged tissues for up to one hour.
10. Mechanical tension has been shown to regulate the Hippo pathway [3, 7, 8]. Our method of mounting the wing imaginal disc between coverslips inevitably applies mechanical force to the tissue. This should be kept in mind when interpreting results.

Acknowledgments

T.S. and J.X. were recipients of the Children's Tumor Foundation Young Investigator Award (2012-01-033, 2013-01-020, and 2014-01-020, respectively), and S.T. was supported by NIH training grant T32 GM007183. This work was supported by a grant from the NIH to R.G.F. (NS034783).

References

1. Oh H, Irvine KD (2008) In vivo regulation of Yorkie phosphorylation and localization. Development 135:1081–1088. https://doi.org/10.1242/dev.015255
2. Su T, Ludwig MZ, Xu J, Fehon RG (2017) Kibra and Merlin activate the Hippo pathway spatially distinct from and independent of expanded. Dev Cell 40:478–490.e3. https://doi.org/10.1016/j.devcel.2017.02.004
3. Rauskolb C, Sun S, Sun G et al (2014) Cytoskeletal tension inhibits Hippo signaling through an Ajuba-Warts complex. Cell 158:143–156. https://doi.org/10.1016/j.cell.2014.05.035
4. Boggiano JC, Fehon RG (2012) Growth control by committee: intercellular junctions, cell polarity, and the cytoskeleton regulate Hippo signaling. Dev Cell 22:695–702. https://doi.org/10.1016/j.devcel.2012.03.013
5. Sun S, Reddy BVVG, Irvine KD (2015) Localization of Hippo signalling complexes and Warts activation in vivo. Nat Commun 6:8402. https://doi.org/10.1038/ncomms9402
6. Chung H-LL, Augustine GJJ, Choi K-WW (2016) Drosophila Schip1 links expanded and Tao-1 to regulate Hippo signaling. Dev Cell 36:511–524. https://doi.org/10.1016/j.devcel.2016.02.004
7. Dupont S, Morsut L, Aragona M et al (2011) Role of YAP/TAZ in mechanotransduction. Nature 474:179–183. https://doi.org/10.1038/nature10137
8. Aragona M, Panciera T, Manfrin A et al (2013) A mechanical checkpoint controls multicellular growth through YAP/TAZ regulation by actin-processing factors. Cell 154(5):1047–1059. https://doi.org/10.1016/j.cell.2013.07.042

Chapter 5

Localization of Hippo Signaling Components in *Drosophila* by Fluorescence and Immunofluorescence

Cordelia Rauskolb and Kenneth D. Irvine

Abstract

Visualization of in vivo protein levels and localization is essential to analysis and elucidation of Hippo signaling mechanisms and its roles in diverse tissues. This is best done by imaging proteins using fluorescent labels. Fluorescent labeling of a protein can be achieved by direct conjugation to an intrinsically fluorescent protein, like GFP, or by use of antibodies conjugated to fluorescent dyes. Immunofluorescence imaging in *Drosophila* typically begins with dissection and fixation of a sample tissue, followed by a series of washes and incubations with primary antibodies, directed against proteins of interest, and dye-labeled secondary antibodies, directed against the primary antibodies. This may be followed by fluorescent dyes that label cellular components, such as DNA-labeling dyes to mark nuclei. After staining and washing is completed, samples are placed in a mounting media, transferred to a microscope slide, and imaged on a confocal microscope.

Key words Hippo, Yorkie, Antibody, Immunofluorescence, Confocal, GFP

1 Introduction

Detection of protein levels and localization in vivo has had a major impact on our understanding of diverse cellular processes, including Hippo signaling [1, 2]. Fluorescent labeling offers several advantages for detecting proteins, including high sensitivity, large linear range of detection, and the ability to image multiple proteins separately or in combination by using distinct fluorophores. At low resolution, fluorescence imaging enables investigation of tissue expression patterns and relative expression levels, whereas at high resolution, fluorescence imaging enables determination of subcellular localization, which is a crucial parameter of protein function and often dynamically modulated in signal transduction pathways.

The two main approaches to fluorescent labeling are genetic conjugation of a protein of interest to an intrinsically fluorescent protein [3] and indirect attachment of fluorescent labels using antibodies [4]. In practice, many experiments examining protein

Alexander Hergovich (ed.), *The Hippo Pathway: Methods and Protocols*, Methods in Molecular Biology, vol. 1893, https://doi.org/10.1007/978-1-4939-8910-2_5,

localization in *Drosophila* Hippo signaling use both of these approaches in combination within a single experiment. Through spectral separation of different fluorophores, the location of a protein of interest can then be compared to fluorescent markers of cell fate, genotype, gene expression, or subcellular localization. Detection by immunofluorescence bypasses the need to create and introduce unique genetic constructs but requires the creation and validation of antisera or antibodies with high avidity and specificity. Direct conjugation to intrinsically fluorescent proteins, like GFP, bypasses the need to create antibodies and often results in lower background noise but requires the creation and incorporation of appropriate genetic constructs, which must be functionally validated. Tagging with intrinsically fluorescent proteins can also enable live imaging approaches.

A typical *Drosophila* immunofluorescence staining experiment begins with the creation of flies of the appropriate genotype. The tissues to be analyzed are then dissected out of the animal and subject to histological fixation. Most analysis of Hippo signaling has been performed on imaginal discs [5], which are dissected out of developing larvae, and we have used the basic method described below to study the localization of many Hippo pathway components in discs [6–14]. The fixation and staining methods described below can also be applied to other tissues as well [15–18]. After fixation, samples are washed, incubated in primary antibodies directed against proteins of interest, washed, incubated in secondary antibodies that are directed against the primary antibodies and coupled to fluorescent dyes, washed again, incubated in fluorescent dyes for labeling cellular structures, and mounted on slides for microscopy. A wide-field fluorescence microscope can be used, but most studies employ confocal microscopes due to the improved spatial and spectral resolution they offer.

2 Materials

2.1 Tissue Dissection and Fixation

1. Flies (*see* **Note 1**).
2. Clear dissection dish (*see* **Note 2**).
3. Fine forceps (*see* **Note 3**).
4. 25% sucrose (*see* **Note 4**).
5. Ringers: 111.23 mM NaCl, 1.88 mM KCl, 2.38 mM $NaHCO_3$, 0.09 mM NaH_2PO_2–$2H_2O$, 0.82 mM $CaCl_2$–H_2O (*see* **Note 5**).
6. 4% paraformaldehyde (*see* **Note 6**).
7. Dissecting microscope (*see* **Note 7**).

2.2 Antibody Incubation

1. Nutating mixer (*see* **Note 8**).
2. 10× PBS (phosphate-buffered saline): 80.6 mM sodium phosphate, 19.4 mM potassium phosphate, 27 mM KCl and 1.37 M NaCl, pH 7.4 (*see* **Note 9**).
3. PBT: PBS + 0.1% (v/v) Triton X-100 + 1% (w/v) BSA + 0.01% (v/v) azide (*see* **Note 10**).
4. Primary antibodies (*see* **Note 11**).
5. Secondary antibodies (*see* **Note 12**).
6. Normal serum from the animal that is the host for the secondary antibodies (typically donkey, sheep, or goat) (*see* **Note 13**).
7. Hoechst 33342 (1 μg/mL in double distilled or deionized water (ddH_2O)) (*see* **Note 14**).
8. Phalloidin conjugated to a fluorescent molecule (i.e., Molecular Probes Alexa Fluor dyes) (*see* **Note 15**).

2.3 Mounting Tissues

1. Microscope slides (25 × 75 × 1.0 mm precleaned slides).
2. No. 1.5 cover glass (22 mm × 22 mm microscope cover glass).
3. VECTASHIELD Antifade Mounting Media (*see* **Note 16**).
4. Nail polish.

3 Methods

3.1 Tissue Dissection and Fixation

1. Scoop larvae from food using a metal spatula and gently mix in 25% sucrose (larvae float on sugar water, the food will sink) in a Petri dish. Alternatively, pour 25% sucrose into the fly vial, gently agitate the food while gently stirring with a spatula, and then pour the contents into a Petri dish.
2. Using dull forceps, pick larvae of the desired age and genotype. This can be done by either picking the desired larvae directly from the Petri dish (i.e., GFP-positive larvae can be directly picked from the sucrose/food) or all of the larvae can be transferred to a spot well dish filled with chilled Ringers and then selected for the correct genotype (this may be an easier method for visualization of some weaker fluorochromes, such as RFP). The larvae to be dissected should all ultimately be placed in a spot well dish filled with chilled Ringers.
3. Dissect the larvae in Ringers using forceps, while visualizing larvae through the dissecting microscope (Fig. 1). To tear a larva in half, use two forceps and pinch the larva at about one third back from the anterior end. The tips of the forceps should be close together; then move the forceps apart while still pinching the tissue. The larva will now be cleanly torn in half. Discard the posterior half. Remove the protruding guts and

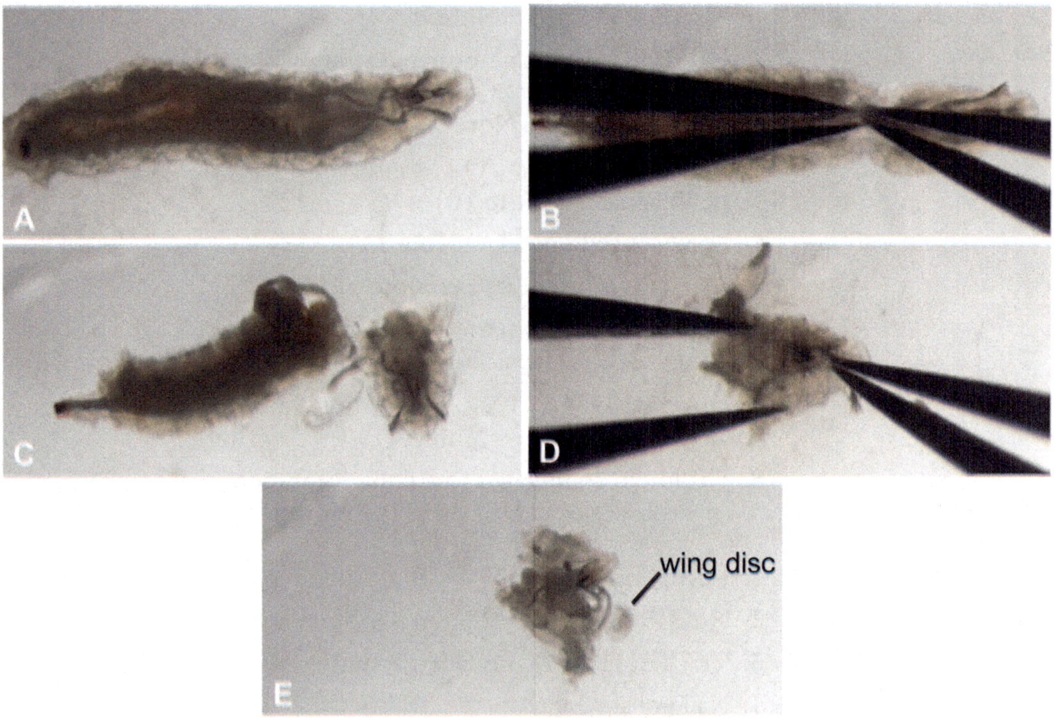

Fig. 1 Dissecting and inverting larvae for immunostaining. Images show steps in larval dissection, as described in Subheading 3.1, **step 3**. (**a**) *Drosophila* larva. (**b**) Positioning of forceps before larva is ripped in half. (**c**) Larva after the anterior end is separated from the posterior end. (**d**) Positioning of forceps before the larval head is inverted. (**e**) Larval head after inversion, with imaginal discs now outside the cuticle. A wing disc is visible

fat from the anterior end, although if one has torn the larva in the "perfect" spot, this won't be necessary. Invert the anterior end of the larva (*see* **Note 17**), being careful not to disturb any discs or trachea. Transfer inverted larva (with discs attached) immediately after inversion to approximately 1 mL Ringers on ice, using forceps to gently grab the larval cuticle. Transfer as many larvae as you can invert in 10–15 min, with a maximum of 10–15 larvae per tube.

4. To fix the larval tissues, remove Ringers from tube containing inverted larvae, and then add 4% paraformaldehyde. Place tube on a Nutator at room temperature for gentle mixing. Immediately start a timer. Typically, fixation time is for 15 min, with some exceptions (*see* **Note 18**).
5. Remove fixative with a Pasteur pipet and discard in hazardous liquid waste container.
6. Rinse larvae twice with Ringers by adding Ringers, closing and inverting the tube, allowing larval cuticles to settle, and removing the liquid.

Note: After this stage, the fixed imaginal discs can be stored in PBT for up to a few days or immediately used for antibody tissue staining (*see* **Note 19**).

3.2 Staining Tissues

1. Rinse fixed discs (larval cuticles) twice with PBT, then wash 2 × 20 min each in PBT on Nutator (*see* **Note 20**).
2. Incubate for 30 min in 180 μL PBT + 10 μL Donkey serum (total volume will be ~200 μL, because the larval tissue will be in ~ 10 μL) on Nutator.
3. Incubate with primary antibodies:

   180 μL PBT + 10 μL Donkey serum + primary antibody at appropriate dilution, to a total volume of 200 μL (*see* **Note 21**).

   Incubate overnight at 4 °C on Nutator (*see* **Note 22**).
4. Remove liquid. Rinse in PBT, and then wash four times 10–15 min each in PBT on Nutator.
5. Incubate for 30 min in 180 μL PBT + 10 μL Donkey serum on Nutator.
6. Incubate with secondary antibodies:

   180 μL PBT + 10 μL Donkey serum + secondary antibody at appropriate dilution (*see* **Note 23**). Wrap tubes in aluminum foil to keep solutions dark (*see* **Note 24**) and place tube on Nutator. Incubate 2 h RT (or overnight 4 °C).
7. Rinse in PBT, wash 3× 10–15 min each in PBT on Nutator. Keep tubes wrapped in aluminum foil. The immunostained tissue can now be stored in the dark (typically the Eppendorf tube is wrapped in aluminum foil) at 4 °C until ready for removal and mounting of the imaginal discs (*see* **Note 25**).
8. (Optional) Incubate with phalloidin to label F-actin (*see* **Note 26**).
9. (Optional) Incubate with Hoechst to label DNA (*see* **Note 27**).

3.3 Dissecting and Mounting Tissues

1. Place one larval cuticle onto a depression slide in some PBT. Dissection of imaginal discs requires a dissecting microscope. Using transmitted light, locate the larval tissue under the microscope. Using fine forceps, dissect the desired imaginal discs away from the larval cuticle. Imaginal discs are easily damaged and should be separated from the cuticle and other discs without actually grabbing onto the discs. Instead, place the tips of the forceps in between tissues to be separated and gently spread them apart. It may help to pin the larval cuticle with one forceps, and then use the other to gently remove the imaginal disc. Once the discs have been removed, discard the larval cuticle. Add a new larval cuticle into the same well as the previous, and repeat the dissection process. Once all desired discs have been removed from all of the larval tissues, one is ready to mount the imaginal discs.

2. To mount the imaginal discs onto a microscope slide, place the slide on a black surface. With a Pasteur pipet, suck up the discs in PBT, trying to minimize the amount of liquid. Gently squirt the discs and PBT into the center of the slide. Transfer the slide to the microscope, and while observing, remove some liquid from the slide with the Pasteur pipet. Using one forceps, in the "open" position, push the discs near each other, so that they are within an area that can be covered by the cover glass. For visualization of apical proteins (e.g., Jub, Wts), it is important that the imaginal discs are oriented with the apical side up. The wing disc is slightly convex, and the more curved side will be up. To orient the disc appropriately, use one forceps to try to flip the disc. Do not grasp the tissue, as this will destroy it, but rather capillary action is sufficient to "suck" up the disc, and then flip the disc. Once the discs are all appropriately oriented, remove the last of the liquid, and tip the slide so that the PBT gently runs to the edge of the slide. The discs should stay in place! With a Kimwipe remove any excess PBT. Place a drop (about 8 uL) of mounting media (VECTASHIELD) next to, but not on, the discs. Grab a cover glass with a pair of forceps, and holding it at a 45 angle, place near the VECTASHIELD drop, and then gently lower it, as the VECTASHIELD spreads across the discs. The amount of VECTASHIELD used needs to be such that it spreads under the entire cover glass, but not so much that it, and potentially the dissected imaginal discs, leaks out. Seal the edges of the coverslip with clear nail polish, to prevent mixing of the mounting media and immersion lens oil when imaging. The slides with the mounted tissue can be stored for a few days in the dark at 4 °C prior to imaging.

3.4 Imaging

Image the immunofluorescently stained imaginal discs using a confocal microscope with a 40× or 63× oil-immersion lens (Fig. 2) (*see* **Note 28**).

4 Notes

1. *Drosophila* stocks can be obtained from other laboratories or from public stock centers. Large numbers of fly stocks expressing tagged fluorescent proteins are available, and others are being actively generated both by large-scale projects and individual researchers. The main source of information about *Drosophila* stocks is FlyBase (http://flybase.org).
2. We use Pyrex 3 depression glass spot plates, which have depressions that are 22 mm wide and 7 mm deep, and can be quickly rinsed and then reused, but other transparent dishes could also be used. Considerations are transparency (larvae are dissected

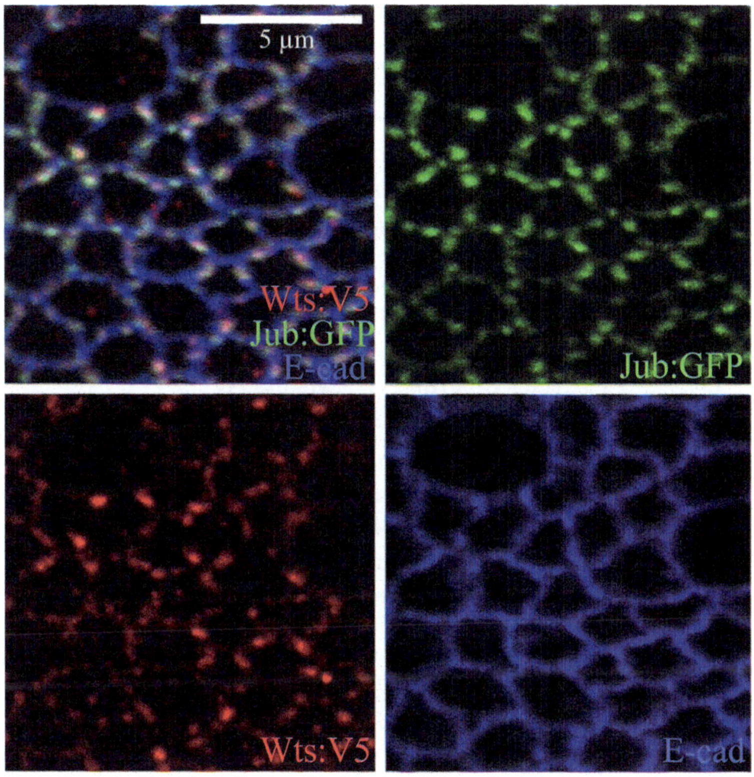

Fig. 2 Example of confocal microscopy of Hippo pathway components. An example of a confocal micrograph showing multicolor fluorescent labeling of Hippo pathway components in a *Drosophila* wing imaginal disc. Top left panel shows all three channels together, clockwise from top left show Jub localization, based on a GFP:Jub fusion protein; E-cadherin localization, based on antibody staining; and Wts localization, based on V5 antibody staining of a V5-tagged Wts genomic construct [1]

under transmitted light), volume (spot plates allow dissection in less than 1 mL of liquid), and accessibility of larvae to forceps for dissection).

3. High quality, fine tip forceps, like Dumont #5, are essential for the final stages of dissection, but they are expensive and easily become blunt or bent. Banged-up forceps should be used for earlier stages of dissection, like inverting larvae, and fine, undamaged forceps for later stages, like separating imaginal discs. Forceps can also be resharpened using a sharpening stone, fine sandpaper, or sent to a company specializing in sharpening used forceps.
4. Dissolve 125 g sucrose in 400 mL water, add water up to a final volume of 500 mL. Filter the solution using a 500 mL filter unit (0.2 μm CN membrane). Store at 4 °C. This 25% sucrose

solution does not need to be made from chemical grade reagents; a bag of sugar from the grocery store works fine.

5. Ringers is an isotonic salt solution used for maintaining tissue integrity during dissections. There are different formulations, the version we use is also known as Becker's Ringers. Dissolve the following in 800 mL ddH_2O. 6.5 g NaCl, 0.14 g KCl, 0.2 g $NaHCO_3$, 0.01 g NaH_2PO_2–$2H_2O$, and 0.12 g $CaCl_2$–$2H_2O$. Adjust volume to 1 L with ddH_2O. Stir until dissolved and then filter the solution using a 1 L Nalgene Rapid-Flow filter unit, 0.2 μm CN membrane. Store stock solution at 4 °C. 50 mL aliquots can be left at room temperature for several days.
6. For 500 mL, place 450 mL of ddH20 in a glass beaker, insert a thermometer, and heat to 60 °C using a hot plate in the hood, with a stir bar. While stirring, add 20 g of paraformaldehyde powder (weighed in the hood) to the heated water. Cover with aluminum foil, and maintain at 60 °C. Add five drops of 2 N NaOH. The solution should clear within a couple of minutes, albeit there may be a few fine particles that remain. Do not heat above 70 °C as the paraformaldehyde will break down. If necessary, add one 2 N NaOH drop at a time to get most of the powder in solution. Remove the thermometer, and add 50 mL 10× PBS. Stir. Remove from heat, allow solution to cool slightly, and then adjust pH to 7.2 (using a pH meter) by adding HCl, or if too low, NaOH. Adjust the final volume to 500 mL with ddH20. Filter the solution using a 500 mL Nalgene Rapid-Flow filter unit, 0.2 μm CN membrane. Make 20–40 mL aliquots, wrap each tube with aluminum foil, and freeze them at −20 °C. They will last for 4–6 months. Thaw each aliquot as needed, typically overnight at 4 °C. Each thawed tube of paraformaldehyde will be good for 1–2 weeks; any leftover needs to be discarded in liquid hazardous waste. When preparing paraformaldehyde fixative, wear protective clothing including lab coat, eyewear, and gloves, and work in a fume hood. If the pH is not properly adjusted, the wing imaginal discs will become swollen, with the peripodial membrane puffing up from the epidermal cells. Fixative that is no longer "good" will result in tissue disintegration.
7. Dissecting microscopes have a large working distance and typically have a magnification range from ~8× to 60×. For best visualization of imaginal discs, a dissecting microscope should be equipped with transmitted light. In order to genotype larvae with fluorescent markers (such as balancer chromosomes labeled by GFP), it should also be equipped for epi-fluorescence.

8. A Clay Adams Nutator (single-speed, 12 rpm) provides a gentle orbital motion that can be used for continuous gentle mixing.
9. PBS is a blend of phosphate buffers and saline solutions, typically prepared as a 10× concentrated stock. Filter with a Nalgene Rapid-Flow filter unit, 0.2 μm CN membrane, to sterilize. Alternatively, 10× PBS is commercially available.
10. For 500 mL, mix 50 mL 10× PBS, 5 mL 10% Triton X-100, 5 g BSA (bovine serum albumin Fraction V), and ddH_20 to 498 mL. Filter with a Nalgene Rapid-Flow filter unit, 0.2 μm CN membrane, to sterilize. Then add 2.5 mL of 2% sodium azide to the filtered solution. Sodium azide is not essential for immunofluorescent staining but is added as a preservative to inhibit microbiological growth in PBT. Gloves should be used and care should be taken in handling solutions with sodium azide, which is toxic.
11. Most primary antibodies against *Drosophila* proteins are made by individual labs. A list of sources for commercially available antibodies that work on *Drosophila* tissues is maintained at FlyBase (http://flybase.org/wiki/FlyBase:Antibodies). Another useful resource for *Drosophila* antibodies is the Developmental Studies Hybridoma Bank (http://dshb.biology.uiowa.edu/Antibody-Collections/Drosophila-antigens), a nonprofit NIH-supported resource that distributes over 200 monoclonal antibodies that recognize *Drosophila* proteins.
12. Secondary antibodies are antibodies from one animal that recognize antibodies of another animal. Fluorescently labeled secondary antibodies are available from many commercial sources. Our lab mostly uses secondary antisera from Jackson ImmunoResearch. Important considerations in choice of secondary antibodies include host species, fluorescent label, and cross-absorption. Secondary antibodies are made in larger animals (with more sera) like goat, sheep, or donkey. Secondary antisera from the same host should be used in a single experiment (i.e., if staining against mouse and rabbit primary antibodies, use donkey anti-mouse and donkey anti-rabbit, not donkey anti-mouse and goat anti-rabbit, because the secondary antibodies might cross-react with each other). The best fluorescent labels will depend upon your application and available equipment; we would typically use Alexa 405, Alexa 488, Cy3, and Alexa 647 for four-color immunofluorescent labeling. Secondary antibodies will normally be cross-absorbed so that they don't recognize IgGs of certain species (i.e., an anti-mouse IgG antisera will have been depleted of antibodies recognizing anti-rabbit IgGs). However, there can be substantial differences in the extent of cross-absorption performed for different reagents, so this should be checked carefully. Anti-

mouse IgG antisera may be available with or without cross-absorption against rat IgGs, which is essential if an experiment involves both mouse and rat primary antibodies, but will typically result in a weaker signal due to loss of antibodies.

13. Lyophilized powder can be purchased from Jackson ImmunoResearch. To reconstitute add 10 mL sterile ddH_20 to the powder. Aliquot 1 mL into Eppendorf tubes. Store one tube at 4 °C as the working solution; the others should be frozen at −20 °C until ready for use.
14. Hoechst 33342 is a cell-permeant dye that fluoresces when intercalated into nucleic acids; it is used to label nuclei. DAPI (4′,6-diamidino-2-phenylindole, dihydrochloride) and propidium iodide are alternative dyes for labeling nuclei.
15. Phalloidin is used to label F-actin and is available conjugated to a variety of fluorochromes, but we typically use the conjugate to Alexa 647.
16. A variety of mounting media are available, but we use nonhardening VECTASHIELD Antifade Mounting Medium (Vector Labs). Mounting media for immunofluorescence include chemicals that reduce photobleaching of fluorescent protein and fluorescent dyes and can be hardened or nonhardened.
17. Comments on inverting larvae: if you are inverting larvae for the first time, we recommend that you practice first using wild type. It may take a week to become proficient at inverting larvae, but do not become discouraged. Begin with the wandering third instar larvae that are crawling on the sides of the vial as these are larger and hence easier to invert. For tearing the larva in half: if the larva is ripped too far anterior, the imaginal discs will be lost, while tearing them too far posterior makes the inversion nearly impossible. In older larvae, the imaginal discs are large, such that there is a "clearing" at the anterior end of the larva. You want to tear the larva in half just slightly posterior to this region. A technique for successful inversion is as follows: imagine your lab glove is the anterior end of a larva. With your right-hand pointer and middle finger, pinch the finger end of the glove. With your left hand pointer and middle fingers, insert slightly into the open end of the glove and splay your fingers wide apart. Next step is to invert the glove which is done by moving the right pinched finger end toward the left hand and through the splayed left fingers (the left hand moves slightly right during this process). The glove should now be turned inside out. The procedure is similar for a larva but is done while looking through a dissecting microscope: after the larva is torn in half, orient the larva such that the anterior end is facing right. Use the right-hand forceps to pinch the anterior tip of the larva. Using the left forceps,

slightly insert them into the larval opening, and allow them to slightly splay open. With the same motion as with the glove scenario, move the right-hand forceps through the left splayed ones. The larval cuticle should be inside out now.

18. The typical duration for fixation that we use is 15 min at room temperature, but the optimal fixation time can vary depending upon the protein being examined. Prolonged fixation reduces GFP fluorescence, so for weak GFP signals as short a fixation time as possible should be used. For example, for genomic Wts: GFP we use only 8 min of fixation.
19. While in most cases fixed tissues can be stored for up to a few days, for sensitive antigens or weak GFP signals it is better to complete the staining and imaging as soon as possible.
20. To "rinse" means add liquid to the tube containing the fixed larval tissue, allow tissue to settle to bottom of the tube, remove liquid with a Pasteur pipet without disturbing the larval tissue (i.e., don't create a suction at the bottom of the tube, as then the imaginal discs may be sucked away!), and replace with fresh appropriate liquid. To "wash" means incubate the larval tissue with the appropriate solution for an extended period of time. The period of time for rinsing and washing can be extended, if convenient.
21. Add in order: PBT first, then serum, then antibodies. The appropriate dilution of the primary antibody will have to be determined by trial-and-error experiments for each antisera. Too high an antisera concentration is wasteful and often gives more non-specific (i.e., background) staining. If background staining persists despite trying a variety of dilutions, non-specific background antibodies (e.g., from other antibodies in the antisera) can be reduced by preabsorption of the antisera against fixed tissues. The procedure for preabsorption is to fix the tissue of interest as per protocol. Rinse the tissue once with PBT, and then wash twice with PBT for 20 min each with the tube placed on the Nutator. Remove the PBT from the tissue, make a 1:10 dilution of your antibody in PBT, and then add it to the fixed tissue. Incubate overnight at 4 °C on a Nutator. After this preabsorption step, the supernatant is ready for tissue staining. Remember that the starting dilution of the antibody is now 1:10.
22. Typically we incubate overnight at 4 °C, but for most antibodies 2 h at room temperature (25 °C) also works.
23. The appropriate dilution of 2° antibodies will need to be experimentally determined. For 2° antibodies from Jackson ImmunoResearch, we typically use 1.5 μL of each secondary antibody in a 200 μL staining volume. This sometimes needs to be adjusted for different batches of 2° antibodies, as there can

be some lot-to-lot variability. Consideration should also be given to the particular fluorochromes used for each 2° antibody. In our experience, brightest signals are obtained with the Alexa 488, so we would use that for the antibody stain that is expected to be the weakest. Conversely Alexa 405 generally gives a weak signal and so should only be used with primary antibodies that are expected to give a strong signal.

24. Brief exposure to light will not significantly affect fluorescence (e.g., while dissecting the imaginal discs), but prolonged exposure should be avoided by keeping tubes and slides in the dark as much as is practical.
25. Confocal imaging of the imaginal discs is best if conducted within 2–3 days of completion of the immunostaining and for weak signals may need to be conducted even sooner.
26. Phalloidin is used to detect filamentous actin in tissues. Methanol treatment of tissues disrupts filamentous actin, and despite the fact that phalloidin is dissolved in methanol, the following procedure will give excellent visualization of actin in the *Drosophila* imaginal disc. To stain with phalloidin, premix in a separate Eppendorf tube: 160 μL PBT + 20 μL phalloidin conjugated to a fluorescent dye (i.e., phalloidin-488, phalloidin-555, phalloidin-633, or phalloidin-647). Remove liquid from stained tissue, and replace with premix. Incubate at RT on Nutator for 40 min, tube wrapped in aluminum foil. Rinse in PBT, wash 3× 10–15 min in PBT on Nutator, tube wrapped in aluminum foil.
27. Stain 30 min with a 1:2000 dilution as follows:

    1 mL PBT + 0.5 μL Hoechst (kept in small brown bottle at 4 °C).

    Incubate for 30 min at RT on Nutator. Tubes wrapped in aluminum foil.

    Rinse in PBT, wash 3× 10–15 min in PBT on Nutator. Keep tubes wrapped in aluminum foil.
28. In theory, a wide-field fluorescence microscope, in combination with deconvolution software, could be used, but we have always used confocal microscopes. For some GFP-tagged proteins, expression levels may be very low, and thus detectors (such as the Leica HyD) with excellent signal-to-noise ratios may be necessary for appropriate imaging.

Acknowledgments

Research in the Irvine laboratory is supported by NIH grants GM78620 and GM121537.

References

1. Sun S, Irvine KD (2016) Cellular organization and cytoskeletal regulation of the hippo signaling network. Trends Cell Biol 26:694–704
2. Yu F-X, Zhao B, Guan K-L (2015) Hippo pathway in organ size control, tissue homeostasis, and cancer. Cell 163:811–828
3. Chalfie M (1995) Green fluorescent protein. Photochem Photobiol 62:651–656
4. Fritschy J-M, Härtig W (2001) Immunofluorescence. In: eLS. John Wiley & Sons, Ltd, Hoboken
5. Held LI (2002) Imaginal discs: the genetic and cellular logic of pattern formation. In: Developmental and cell biology series. Cambridge Press, Cambridge, p 460
6. Misra JR, Irvine KD (2016) Vamana couples fat signaling to the Hippo pathway. Dev Cell 39:254–266
7. Sun S, Reddy BVVG, Irvine KD (2015) Localization of Hippo Signaling complexes and Warts activation in vivo. Nat Commun 6:8402
8. Rauskolb C, Sun S, Sun G et al (2014) Cytoskeletal tension inhibits Hippo signaling through an Ajuba-Warts complex. Cell 158:143–156
9. Ambegaonkar AA, Pan G, Mani M et al (2012) Propagation of dachsous-fat planar cell polarity. Curr Biol 22:1302–1308
10. Rauskolb C, Pan G, Reddy BV et al (2011) Zyxin links fat signaling to the hippo pathway. PLoS Biol 9:e1000624
11. Mao Y, Kucuk B, Irvine KD (2009) Drosophila lowfat, a novel modulator of fat signaling. Development 136:3223–3233
12. Oh H, Irvine KD (2008) In vivo regulation of Yorkie phosphorylation and localization. Development 135:1081–1088
13. Feng Y, Irvine KD (2007) Fat and expanded act in parallel to regulate growth through warts. Proc Natl Acad Sci U S A 104:20362–20367
14. Mao Y, Rauskolb C, Cho E et al (2006) Dachs: an unconventional myosin that functions downstream of Fat to regulate growth, affinity and gene expression in Drosophila. Development 133:2539–2551
15. Ambegaonkar AA, Irvine KD (2015) Coordination of planar cell polarity pathways through Spiny legs. eLife 4
16. Reddy BV, Irvine KD (2011) Regulation of Drosophila glial cell proliferation by Merlin-Hippo signaling. Development 138:5201–5212
17. Staley BK, Irvine KD (2010) Warts and Yorkie mediate intestinal regeneration by influencing stem cell proliferation. Curr Biol 20:1580–1587
18. Reddy BVVG, Rauskolb C, Irvine KD (2010) Influence of Fat-Hippo and Notch signaling on the proliferation and differentiation of Drosophila optic neuroepithelia. Development 137:2397–2408

Chapter 6

Bimolecular Fluorescence Complementation (BiFC) in Tissue Culture and in Developing Tissues of *Drosophila* to Study Protein-Protein Interactions

Yurika Matsui and Zhi-Chun Lai

Abstract

Protein-protein interactions provide a common mechanism for regulating protein functions and also serve as the fundamental step of many biochemical reactions. To accurately determine the involvement and function of protein-protein interactions, it is crucial to detect the interactions with the minimum number of artifacts. In this chapter, we report the method of bimolecular fluorescence complementation (BiFC) in tissue culture and developing tissues of *Drosophila*, which allows the visualization of subcellular localization of protein-protein interactions in living cells.

Key words Bimolecular fluorescence complementation (BiFC), Fluorescence microscopy, Protein-protein interaction, Tissue culture, *Drosophila* wing imaginal discs

1 Introduction

Components of the Hippo pathway take various biochemical forms and protein conformations in order to transduce growth-regulatory signals [1]. In obtaining these features, these components often undergo protein-protein interactions. Therefore, assays that detect protein-protein interactions play a crucial part in gaining the molecular insight of the Hippo pathway. One of such assays is **Bi**molecular **F**luorescence **C**omplementation (BiFC), which allows us to detect the subcellular localization of protein-protein interactions in living cells via fluorescence microscopy [2].

In BiFC, we express two proteins of interest, one tagged with N-terminal half of Venus (VN), a derivative of yellow fluorescent protein (YFP), and the other tagged with C-terminal half of Venus (VC). If the proteins of interest form a complex, this will bring the N- and C-terminal halves of Venus into close proximity, emitting the fluorescent signal of YFP [3]. With the requirement of only three simple steps, i.e., tagging of proteins, expression of proteins,

Alexander Hergovich (ed.), *The Hippo Pathway: Methods and Protocols*, Methods in Molecular Biology, vol. 1893,
https://doi.org/10.1007/978-1-4939-8910-2_6,

and imaging of fluorescent signals, BiFC can serve as an informative tool for readily examining the large number of protein-protein interactions without disturbing the cellular environment through processes such as cell lysis. Despite this advantage, BiFC has some limitations as well. For instance, the formation of the fluorescent complex is irreversible and therefore, we cannot examine the dynamics of protein-protein interactions [3]. Also, overexpression of the tagged proteins may induce a non-specific interaction due to the high abundance of proteins at their localized site within a cell. Therefore, choosing a proper negative control is crucial for the interpretation of BiFC data. Alternatively, reducing the expression levels of the proteins to their endogenous levels by, for instance, optimizing the duration of transfection and/or the amount of transfected plasmids would help eliminate such false positives [3]. In this chapter, we will describe the method of BiFC both in tissue culture and in the wing imaginal discs of *Drosophila*, a model organism in which the Hippo pathway was first discovered.

2 Materials

2.1 Transfection of pBiFC Constructs into HEK293T Cells

1. pBiFC constructs (*see* **Note 1**).
2. Coverslips with 12 mm in diameter (*see* **Note 2**).
3. 0.4% (w/v) gelatin solution: Mix 0.4 g gelatin in100 mL ddH_2O and autoclave the mixture.
4. Two pairs of sterile forceps.
5. 24-well tissue culture plate.
6. P10/20/200/1000 micropipettes and sterile tips.
7. Human embryonic kidney 293 T (HEK293T) (e.g., ATCC® CRL-3216™).
8. Culture medium for HEK293T cells: DMEM containing 4.5 g/L glucose and 0.584 g/L L-glutamine (e.g., Corning™ 10013CV) and 10% fetal bovine serum (FBS).
9. Transfection reagents (*see* **Note 3**).
10. Tissue culture hood.
11. Incubator set at 5% CO_2 at 37 °C.

2.2 Staining of Transfected HEK293T Cells

1. Coverslips with transfected HEK293T cells.
2. 24-well tissue culture plate.
3. Two pairs of forceps.
4. Aluminum foil.
5. P10/20/200/1000 micropipettes and sterile tips.
6. 1× PBS (*see* **Note 4**).

7. 4% paraformaldehyde (1.33 M) (*see* **Note 5**).
8. 0.1% PBST: 1× PBS supplemented with 0.1% (v/v) Triton™X-100.
9. Kimwipes®.
10. A mini tray.
11. 0.1% PBST +5% FBS: 1× PBS supplemented with 0.1% Triton™X-100 and 5% (v/v) fetal bovine serum (FBS).
12. DAPI solution (a nucleic acid stain, the final concentration is 1.8 nM) (*see* **Note 6**).
13. Phalloidin solution (an F-actin marker, the final concentration is 33 nM) (*see* **Note 7**).
14. Stain solution (*see* **Note 8**).
15. Glass slides.
16. Glycerol.
17. Clear nail polish.
18. Inverted fluorescence microscope equipped with 405 nm laser for DAPI, 488 nm laser for BiFC, and the third laser for phalloidin.

2.3 Staining of *Drosophila* Wing Imaginal Discs

1. Third instar larvae of *Drosophila*: Larvae expressing both VN-tagged protein and VC-tagged protein in the wing (*see* **Note 9**).
2. Dissection microscope.
3. Two pairs of forceps.
4. 10-cm petri dish as a dissection stage.
5. 1× PBS.
6. 12-well glass spot plate.
7. P10/20/200/1000 micropipettes and sterile tips.
8. FBS.
9. Aluminum foil.
10. Parafilm®.
11. 4% paraformaldehyde.
12. 0.1% PBST.
13. 0.1% PBST +5% FBS.
14. Primary antibodies (*see* **Note 10**).
15. Secondary antibodies (*see* **Note 11**).
16. DAPI solution (a nucleic acid stain, the final concentration is 1.8 nM).
17. Phalloidin solution (an F-actin marker, the final concentration is 33 nM).

18. Stain solution.
19. Glycerol.
20. Coverslips.
21. Kimwipes®.
22. Clear nail polish.
23. Permanent marker pen.
24. Inverted fluorescence microscope equipped with 405 nm laser for DAPI, 488 nm laser for BiFC, and other lasers for phalloidin or other stainings, such as for genes of interest and E-cadherin.

3 Methods

Carry out all procedures at room temperature unless otherwise noted.

3.1 Transfection of pBiFC Constructs into HEK293T Cells

The following procedure needs to be done in an aseptic tissue culture hood.

1. One day before transfection, place a coverslip into a well of a 24-well plate with a pair of sterile forceps. Add 500 μL of 0.4% gelatin solution onto the coverslip and leave it for 15 min at room temperature. Remove the 0.4% gelatin solution from the well, and dry the coverslip for at least 1 h at room temperature. Place the gelatin-coated coverslip in a new well of the 24-well plate.
2. Seed 1.5×10^5 cells of HEK293T cells in 500 μL of DMEM containing glucose and L-glutamine supplemented with 10% FBS onto the coverslip placed in the 24-well plate. Culture the cells for about 24 h.
3. On the day of transfection, transfect 200–300 ng of each of pBiFC vectors (*see* **Note 12**). Proper positive and negative controls are essential for BiFC in addition to testing the protein interactions of interest. The pair of pBiFC-bJunVN173 and pBiFC-bFosVC155 can be used as a positive control with their fluorescent signal in the nucleus. The pair of pBiFC-VN173 and pBiFC-VC155 cannot be used for a negative control as this combination gives a fluorescent signal. We recommend using a pair of the protein of interest with a protein with the following characteristics as a negative control: (1) its subcellular localization is similar to the protein of interest, and (2) its absence of interaction with the protein of interest has previously been reported.
4. Incubate the cells in an incubator with 5% CO_2 at 37 °C for 40–48 h (*see* **Note 13**).

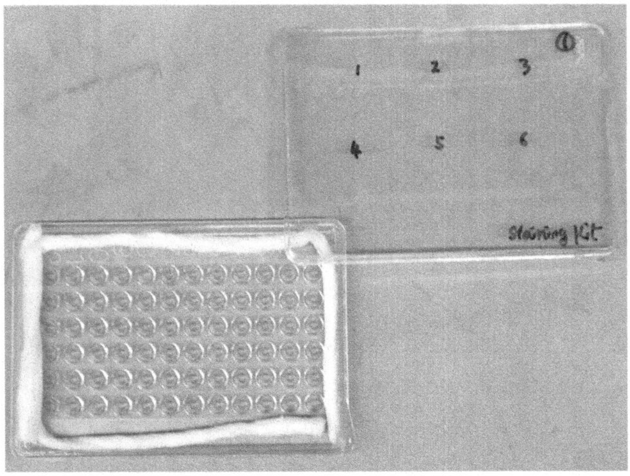

Fig. 1 A homemade humidified minitray for staining. A damp Kimwipe® is placed inside of the minitray at its four edges. The minitray needs to be closed with a lid and covered with aluminum foil during the incubation with fluorescent dyes

3.2 Staining of Transfected HEK293T Cells

1. Remove the culture medium and carefully wash the cells with 300–500 μL of 1× PBS once. Transfer the coverslip into a clean well of a 24-well plate (*see* **Note 14**). During the preparation of samples for imaging, cover the plate or mini tray with aluminum foil.
2. Add 30–50 μL of 4% paraformaldehyde onto the coverslip, and incubate for 10 min at room temperature.
3. Wash the cells for 3 min twice with 300–500 μL of 0.1% PBST (*see* **Note 15**).
4. Place damp Kimwipes® at the edges inside of a mini tray, and transfer the coverslip onto the mini tray (Fig. 1).
5. Add 20 μL of a stain solution containing a nuclear marker (e.g., DAPI) and a membrane marker (e.g., phalloidin) in 0.1% PBST supplemented with 5% FBS onto the center of the coverslip, and incubate at room temperature for 1 h.
6. Wash the cells for 3 min three times with 300–500 μL of 0.1% PBST.
7. Apply a drop of 3–5 μL of glycerol onto a glass slide, and place the coverslip, facing down, onto the glycerol.
8. Carefully wipe off the glycerol that comes out between the glass slide and the coverslip with Kimwipes®. Seal the coverslip with a clear nail polish and dry the nail polish for at least 30 min in dark. Slides covered with aluminum foil can be stored at 4 °C for a few days.
9. Capture the fluorescence images with an inverted fluorescence microscope. Minimally, three lasers are needed: 488 nm laser

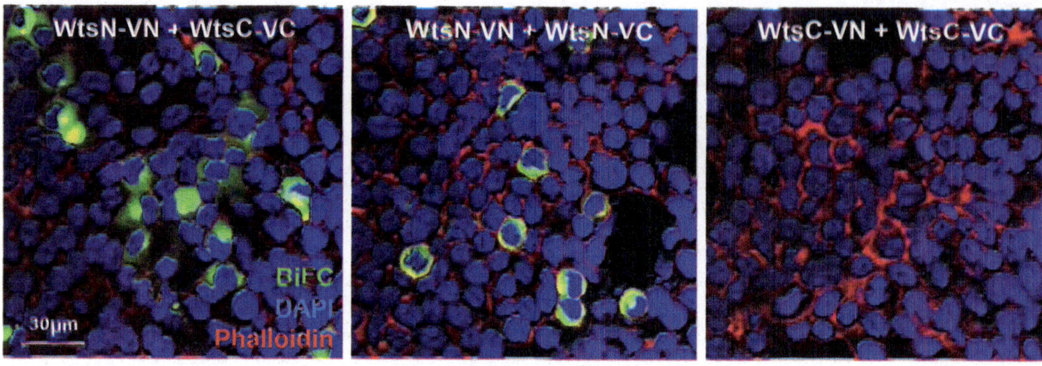

Fig. 2 Protein-protein interactions between fragments of *Drosophila* warts (Wts) protein in HEK293T cells. The N-terminal half of Wts (WtsN: 1–599 amino acids) and the C-terminal half of Wts (WtsC: 600–1105 amino acids) were tagged with VN or VC at their C-terminus. Interaction between WtsN and WtsC and homodimerization among WtsN or WtsC were tested with BiFC (green). DAPI (blue) and phalloidin (red) were used to stain the nucleus and F-actin, respectively

for BiFC (excitation/emission = 500 ± 10 nm/535 ± 15 nm), 405 nm laser for DAPI (excitation/emission = 358 nm/461 nm), and the third laser for phalloidin [3]. Optimize the parameters of the fluorescent microscope using the positive and negative controls (Fig. 2; *see* **Note 16**).

3.3 Staining of Drosophila *Wing Imaginal Discs*

During the staining procedure, observe collected wing discs under a dissection microscope in order to prevent any loss of wing discs by pipetting.

1. Add 50–100 μL of 1× PBS into a well of a 12-well glass spot plate. Coat the P10/20 micropipette tips with FBS in order to prevent wing discs from sticking to the side of the tips.
2. Dissect wing imaginal discs from third instar larvae of *Drosophila* under a dissection microscope and collect the discs in a well with 1× PBS at room temperature. During dissection and staining, cover the spot plate with aluminum foil.
3. Remove 1× PBS from the well containing the discs, and add 50 μL of 4% paraformaldehyde. Fix the discs at room temperature for 30 min.
4. Wash the discs for 3 min twice with 50 μL of 0.1% PBST. If antibody staining is unnecessary, move onto **step** 7.
5. Incubate the discs in 30–50 μL of primary antibodies diluted in 0.1% PBST +5% FBS at room temperature for 2 h or 4 °C overnight. In order to prevent evaporation of primary antibodies, tightly cover the spot plate with Parafilm® (*see* **Note 17**).
6. Remove the primary antibody solution and wash the discs for 3 min three times with 50 μL of 0.1% PBST.

7. Incubate the discs in 30–50 μL of secondary antibodies diluted in 0.1% PBST +5% FBS at room temperature for 1 h. With or without the secondary antibodies, stain the discs with a nuclear marker (e.g., DAPI) and a membrane marker (e.g., phalloidin) diluted in 0.1% PBST +5% FBS at room temperature for 1 h.
8. Remove the stain solution containing DAPI and phalloidin with or without the secondary antibodies, and wash the discs for 3 min three times with 50 μL of 0.1% PBST.
9. Apply a drop of 10 μL glycerol onto a glass slide and transfer the discs into the glycerol with a minimum amount of 0.1% PBST (*see* **Note 18**).
10. Place a coverslip onto the discs and wipe off glycerol that comes out from the space between the glass slide and the coverslip with Kimwipes®. Seal the coverslip with a clear nail polish, and dry the nail polish for at least 30 min. Slides covered with aluminum foil can be stored at 4 °C for a few days.
11. Observe the discs under the dissection microscope, and circle the areas where the discs are found.
12. Capture the fluorescent images with an inverted fluorescence microscope in a similar manner as Subheading 3.2, **step 9** (Fig. 3). If additional labeling is done with antibody staining, more lasers on the fluorescence microscope are necessary.

4 Notes

1. pBiFC-VN173 and pBiFC-VC155 developed by Dr. Chang-Deng Hu's lab can be obtained at Addgene (Plasmid #22010 and #22011, respectively) [2]. Dr. Hu's lab also developed pBiFC-VN155(I152L) (Addgene, Plasmid #27097) which shows less self-interaction when co-expressed with pBiFC-VC155 [4]. The pair of pBiFC-bJunVN173 and pBiFC-bFosVC155 can be used as a positive control and can be obtained at Addgene (Plasmid #22012 and #22013, respectively).
2. Sterile coverslips can be prepared by placing the coverslips in a 10-mL glass bottle containing 70% ethanol. In the tissue culture hood, take each coverslip out of the bottle and hold it with a pair of sterile forceps until ethanol evaporates.
3. HEK293T cells can be easily transfected with many commercially available transfection reagents. This cell line can also be transfected with polyethylenimine (PEI). To prepare 1 μg/μL PEI, dissolve PEI in ddH_2O while heating up the ddH_2O up to 80 °C. Once the solution becomes clear, cool down the solution to room temperature and filter the solution with a 0.22 μm syringe filter. Aliquot and store the solution at −20 °C (the

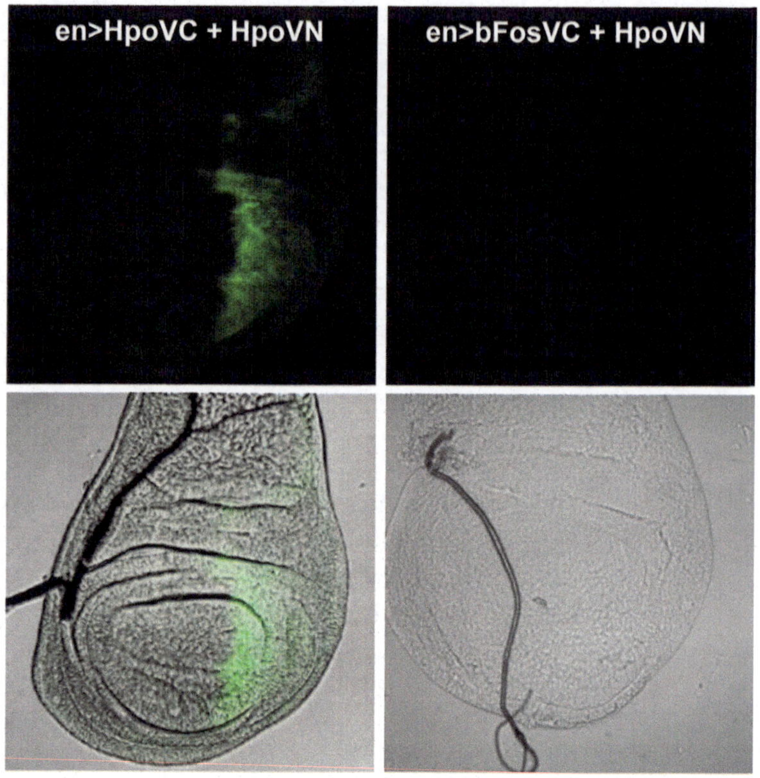

Fig. 3 Homodimerization of Hippo (Hpo) in a *Drosophila* wing imaginal disc of third instar larvae. A transgenic line carrying both VN- and VC-fused *hpo* under UAS was crossed with *engrailed (en)*-Gal4 driver line. The progeny expressed VN- and VC-tagged Hpo in the posterior compartment of the wing. Interaction between VN-tagged Hpo (HpoVN) and VC-tagged Hpo (HpoVC) was observed via BiFC (green; **left panels**). Absence of interaction between VN-tagged Hpo (HpoVN) and VC-tagged bFos (bFosVC) is shown as a negative control (**right panels**) (Reproduced from Deng et al., 2013 with permission from Elsevier) [8]

working solution can be stored at 4 °C). For transfection with PEI, dilute the plasmids (400–600 ng/well of a 24-well plate) with 25 μL of DMEM without FBS. Also, dilute PEI with the reagent-DNA ratio (w/w) of 6:1 (i.e., 2.4–3.6 μL of PEI/well of a 24-well plate) with 25 μL of DMEM without FBS. Mix the diluted plasmids and PEI, and incubate the mixture at room temperature for 15 min before adding to the cells. Incubate the transfected cells for 40–48 h at 37 °C with 5% CO_2 before any further analysis.

4. To prepare 10× PBS, mix 80 g NaCl (1.37 M), 2 g KCl (26.8 mM), 14.4 g Na_2HPO_4 (101 mM), 2.4 g KH_2PO_4 (17.6 mM), and 800 mL ddH_2O. Then, adjust pH to 7.4, add more ddH_2O to make 1 L solution, and autoclave the solution. Dilute the 10× PBS with ddH_2O to prepare 1× PBS.

5. Into a 10-mL glass bottle with a magnetic stir bar, mix 10 mL of 1× PBS, 1 μL of 0.1 M NaOH, and 0.4 g paraformaldehyde. Place the glass bottle in the water bath and increase the temperature of the water bath to 62–64 °C while stirring the mixture. When the mixture becomes clear, take out the glass bottle from the water bath and cool it down to room temperature. Aliquot the fixative (pH should be around 7.3) and store it at −20 °C.
6. In order to make a 5 mg/mL stock solution (10.9 mM of DAPI dilactate), dissolve 10 mg of DAPI dilactate in 2 mL of ddH_2O. Aliquot the stock solution and store it at −20 °C. For the working solution (363.3 nM), dilute the stock solution at the ratio of 1:30 with ddH_2O and store it at 4 °C.
7. In order to make a 200 units/mL stock solution (6.6 μM), dissolve 300 units of lyophilized solids of fluorescent phalloidin in 1.5 mL of methanol. Store the stock solution at −20 °C while protecting it from light.
8. For each coverslip, dilute 0.1 μL of DAPI working solution (final concentration is 1.8 nM) and 0.1 μL of phalloidin stock solution (final concentration is 33 nM) into 20 μL of 0.1% PBST +5% FBS.
9. In order to collect larvae expressing VN- and VC-tagged proteins in the imaginal wing discs, one option is to utilize Gal4/UAS expression system [5]. First, clone the coding sequences of VN- or VC-tagged proteins into an inducible *Drosophila* expression vector, such as pUAST-attB [6, 7]. Microinjection of pUAST-attB constructs into the embryos of a ϕC31 integrase line carrying attP site allows us to target the integration site of our VN- or VC-fused gene when establishing a transgenic line [6, 7]. Finally, set up genetic crosses with a driver line so that both VN- and VC-tagged proteins are simultaneously induced in a developing tissue, such as the wing.
10. pBiFC-VN173 contains FLAG tag at the 5′ end of the multiple cloning site, while pBiFC-VC155 contains HA tag. In order to test the protein expression of the BiFC constructs in the wing imaginal discs, detect the proteins of interest with either gene-specific or FLAG/HA tag primary antibodies. Also, labeling cell adhesion molecules, such as E-cadherin, with their gene-specific primary antibodies can be an alternative way for defining the cell membrane beside phalloidin. If primary antibody staining is done at 4 °C overnight, the primary antibody can be reused. Collect the diluted primary antibody by pipetting from the well of the glass spot plate into a microcentrifuge tube and store it at 4 °C. Collect the antibody while looking through the dissection microscope in order to prevent the stained wing discs from coming along with the antibody solution.

11. If primary antibodies are used for staining genes of interest or defining the cell membrane, choose the secondary antibodies based on which species the primary antibodies were generated from. Also, those secondary antibodies should be conjugated with fluorophores (e.g., Thermo Fisher Scientific Alexa Fluor Secondary Antibodies).
12. The amount of plasmids used for transfection depends on the type of transfection reagents. The amount shown here is for transfection with PEI. Also, in order to minimize false positives, it is recommended to optimize the transfection conditions so that the protein expressions are similar to their endogenous levels.
13. Protein expression of pBiFC constructs needs to be tested by assays such as Western blot. Follow the seeding and transfection conditions as in **steps 2–4** for preparing the cell lysate for Western blot.
14. Transferring of a coverslip becomes easier if there is a trace of liquid in the well. Also, try to pick up the coverslip by its edge in order not to scrape off the cells.
15. For all of the washing steps, remove the solution of the previous step and carefully add 300–500 μL of 0.1% PBST into the well. Leave the solution for 3 min and repeat this for multiple times.
16. Due to the possibility of false positives, it is essential to validate a BiFC-positive result by an independent assay, such as co-immunoprecipitation and pulldown assay. If BiFC result is negative, test the protein expression of both of the pBiFC constructs by an assay, such as Western blot. Also, we recommend trying multiple combinations of differently tagged proteins, i.e., VN at N-terminus (head) with VC at N-terminus (head), VN at N-terminus (head) with VC at C-terminus (tail), or VN at C-terminus (tail) with VC at C-terminus (tail).
17. 0.1% PBST +5% FBS should work fine for detecting proteins in the nucleus and cytoplasm. The efficiency of staining may be enhanced by increasing the concentration of Triton™X-100 for nuclear proteins and decreasing the concentration of Triton™X-100 for membrane proteins.
18. Alternatively, dilute glycerol with the equal amount of 1 × PBS. Add 30–50 μL of this glycerol solution to the well, and transfer the discs onto a glass slide with the minimum amount of solution as possible.

Acknowledgments

We would like to thank Dr. Yaoting Deng for sharing her data on Hpo dimerization. This work was partly supported by the National Science Foundation.

References

1. Yu FX, Guan KL (2013) The Hippo pathway: regulators and regulations. Genes Dev 27:355–371
2. Hu CD, Chinenov Y, Kerppola TK (2002) Visualization of interactions among bZIP and Rel family proteins in living cells using bimolecular fluorescence complementation. Mol Cell 9:789–798
3. Kerppola TK (2006) Design and implementation of bimolecular fluorescence complementation (BiFC) assays for the visualization of protein interactions in living cells. Nat Protoc 1:1278–1286
4. Kodama Y, Hu CD (2010) An improved bimolecular fluorescence complementation assay with a high signal-to-noise ratio. BioTechniques 49:793–803
5. Brand AH, Perrimon N (1993) Targeted gene expression as a means of altering cell fates and generating dominant phenotypes. Development 118:401–415
6. Fish MP, Groth AC, Calos MP, Nusse R (2007) Creating transgenic Drosophila by microinjecting the site-specific ϕC31 integrase mRNA and a transgene-containing donor plasmid. Nat Protoc 2:2325–2331
7. Bischof J, Maeda RK, Hediger M et al (2007) An optimized transgenesis system for Drosophila using germ-line-specific ϕC31 integrases. Proc Natl Acad Sci 104:3312–3317
8. Deng Y, Matsui Y, Zhang Y, Lai ZC (2013) Hippo activation through homodimerization and membrane association for growth inhibition and organ size control. Dev Biol 375:152–159

Part II

Molecular and Cell Biology Studies of the Hippo Pathway

Chapter 7

Immunohistochemistry to Study YAP in Human Tissue Samples

Franziska Haderk, Victor Olivas, and Trever G. Bivona

Abstract

Immunohistochemistry (IHC) analysis of YAP in human tissue samples represents an important means to analyze overall expression levels and subcellular localization of YAP in specimen of interest. As transcriptional coactivator, alterations of YAP levels in the cellular nucleus allow important predictions for YAP activity and transcriptional state of target genes. In the following report, IHC procedures optimized for the detection of YAP in tissue slides of FFPE material are provided. Of note, de-paraffinization and heat-induced antigen retrieval are strictly necessary for successful YAP IHC staining. Further, immunostaining using a labelled polymer-HRP system combined with diaminobenzidine (DAB), as signal-amplifying chromogen, allows strong staining results with minimal unspecific background signal.

Key words YAP, Immunohistochemistry, IHC, YAP nuclear activity, FFPE samples, Hippo signaling, Digital pathology

1 Introduction

YAP is an important transcriptional coactivator which mediates the expression of anti-apoptotic and pro-proliferative genes upon interaction with TEAD transcription factors in the cell nucleus [1–4]. Its nuclear localization and thus functionality are tightly controlled by the inhibitory Hippo signaling cascade [1, 3]. However, amplification and elevated YAP transcriptional activity have been reported in cancer development, progression, and resistance to therapy across different types of neoplasia [2, 5, 6]—illustrating the importance of developing diagnostic tools to study YAP nuclear levels and functionality. The latter can be achieved by IHC analysis for YAP in human tissue samples, providing a combination of spatial as well as quantitative information for YAP levels in the specimen of interest [5, 7, 8].

Franziska Haderk and Victor Olivas contributed equally to this work.

Alexander Hergovich (ed.), *The Hippo Pathway: Methods and Protocols*, Methods in Molecular Biology, vol. 1893, https://doi.org/10.1007/978-1-4939-8910-2_7,

In general, IHC analysis allows the detection of antigens via enzyme-conjugated antibodies. While developed as an enzyme-based assay several decades ago [9], only the description of antigen retrieval methods [10] and signal-amplifying detection methods [11], with DAB-based approaches being one of the most commonly used techniques [12], enabled sensitive antigen detection in fixed tissue specimens and its routine use in clinical diagnostics [13, 14].

In the following report, a standardized protocol for YAP IHC analysis of tissue slides from human FFPE specimens is presented. Key steps for a successful IHC staining include the following: (1) de-paraffinization and antigen retrieval; (2) blocking endogenous enzyme activity and non-specific protein binding sites; (3) immunostaining against YAP—including primary antibody stain, linking with horseradish peroxidase (HRP) enzyme and DAB chromogen—(4) cellular counterstaining by hematoxylin; (5) mounting; and (6) image acquisition and quantitative analysis [15].

2 Materials

2.1 Equipment

1. Staining rack.
2. Staining jars.
3. Slide staining tray.
4. Pressure cooker for IHC antigen retrieval, e.g., Antigen Retriever 2100 (Aptum Biologics Ltd.).
5. Fisherfinest Premium Cover Glasses (Fisher Scientific).

2.2 Solutions

1. PBS: 137 mM NaCl, 2.7 mM KCl, 10 mM Na_2HPO_4, 1.8 mM KH_2PO_4 in ddH_2O, pH 7.4.
2. Ultra-pure/deionized, sterile water (ddH_2O).
3. Xylene.
4. Three different alcohol dilutions in ddH_2O:
   - 100% (v/v) ethanol.
   - 95% (v/v) ethanol.
   - 70% (v/v) ethanol.
5. Target retrieval solution: citrate buffer, pH 6 (10× stock solution, Dako) (*see* **Note 1**).
6. Dual endogenous enzyme block (Dako).
7. Protein blocking buffer: 0.05% (v/v) Tween 20, 15 mM of sodium azide, 1% (w/v) bovine albumin, in PBS.
8. Primary antibody: anti-human YAP mouse monoclonal antibody (e.g., clone H-125, Santa Cruz).

9. EnVision + Dual Link System-HRP (DAB+) (Dako), including labelled polymer-HRP and DAB+ chromogen solution.
10. Hematoxylin.
11. Bluing reagent: 0.1% (w/v) sodium bicarbonate in ddH_2O.
12. Richard-Allan Scientific™ Cytoseal™ 60 (ThermoFisher Scientific).

3 Methods

The following protocol describes the IHC staining procedure for the detection of YAP using tissue slides derived from formalin-fixed, paraffin-embedded (FFPE) specimens.

3.1 Preparation of IHC Solutions

1. Prepare staining jars with solutions required for the IHC procedure in the following order and quantity: xylene (2×), 100% ethanol (2×), 95% ethanol (2×), 70% ethanol (1×), ddH_2O (2×), PBS (1×), hematoxylin (1×), bluing reagent (1×). Solutions can be stored at room temperature and reused calculating a maximum of 50 slides per 50 mL.
2. Prepare primary antibody solution by diluting anti-YAP antibody 1:150 in protein blocking buffer. Primary antibody solution should be prepared fresh and kept at 4 °C during the procedure.

3.2 De-paraffinization and Antigen Retrieval (see Note 2)

Place slides into the staining rack. Perform de-paraffinization and antigen retrieval steps as outlined below, by consecutively transferring the staining rack to jars containing prepared solutions (*see* Subheading 3.1, **step 1**). If not otherwise indicated, incubation steps are performed at room temperature.

3.2.1 De-paraffinization

1. Incubate slides in xylene for 5 min.
2. Move staining rack to second jar containing xylene. Incubate for 5 min.

3.2.2 Hydration

1. Perform a series of incubations in ethanol solutions with descending concentration. Start by incubating in 100% ethanol for 3 min.
2. Move staining rack to second jar containing 100% ethanol. Incubate for 3 min.
3. Transfer slides to 95% ethanol. Incubate for 3 min.
4. Move slides to 95% ethanol. Incubate for 3 min.
5. Move slides to 70% ethanol. Incubate for 3 min.
6. Transfer slides to ddH_2O. Incubate for 3 min.

3.2.3 Antigen Retrieval

1. Place slides in target retrieval solution and incubate for 3 h in pressure cooker (*see* **Note 3**).
2. Remove slides and let them site at room temperature for 10 min.
3. Wash slides three times in PBS, 3 min per wash.

3.3 Blocking Endogenous Enzyme Activity and Non-specific Sites

Incubation with dual endogenous enzyme-blocking reagent and protein blocking buffer are performed on a staining tray, placing slides laying horizontally with the tissue facing upwards. For washing steps in PBS, slides are placed back into the slide rack, and the rack is transferred to respective jar (*see* Subheading 3.1, **step 1**).

1. Apply dual endogenous enzyme-blocking reagent until sample is fully covered. Incubate for 10 min at room temperature.
2. Wash slides three times in PBS, 3 min per wash.
3. Incubate slides in protein blocking buffer for 1 h at room temperature.

3.4 Immunostaining

For primary antibody staining, and incubations with labelled polymer-HRP solution as well as DAB+ chromogen solution, slides are transferred to the horizontal staining tray as described previously. Washing steps in PBS and ddH_2O are performed after slides are placed back into the slide rack and by placing the rack into the respective jar (*see* Subheading 3.1, **step 1**).

1. Stain specimen with primary antibody against YAP at 4 °C overnight (*see* Subheading 3.1, **step 2**).
2. Wash slides three times in PBS, 3 min per wash.
3. Add labelled polymer-HRP solution, until sample is fully covered. Incubate for 25–30 min at room temperature (*see* **Note 4**).
4. Wash slides three times in PBS, 3 min per wash.
5. Incubate slides with DAB+ chromogen solution for 30 s to 3 min at room temperature, until a color change is observed.
6. Transfer slides to ddH_2O. Incubate for 3 min.
7. Move slides to second staining jar containing ddH_2O. Incubate for 3 min.

3.5 Counterstaining and Mounting with Coverslip

Hematoxylin counterstaining and dehydration are performed using the staining rack and by consecutively transferring the staining rack to jars containing prepared solutions (*see* Subheading 3.1, **step 1**). For mounting, slides are transferred to the horizontal staining tray as described previously.

3.5.1 Hematoxylin Counterstain

1. Move slides to staining jar containing hematoxylin. Incubate for 5 min.

2. Wash slides three times in ddH_2O for 3 min each.
3. Dip slides into staining jar containing bluing reagent for a total amount of seven times.
4. Transfer slides to ddH_2O. Incubate for 2 min.

3.5.2 Dehydration

1. Dehydrate slides by a series of incubations in ethanol solutions with increasing concentrations. Start by incubating in 95% ethanol for 3 min.
2. Move staining rack to second staining jar containing 95% ethanol. Incubate for 3 min.
3. Transfer slides to 100% ethanol. Incubate for 3 min.
4. Move to second staining jar containing 100% ethanol. Incubate for 3 min.
5. Place slides into xylene. Incubate for 3 min.
6. Transfer slides to second staining jar containing xylene. Incubate 3 min.

3.5.3 Mounting with Coverslip

1. Transfer slides to staining tray as described previously. Place one drop of Cytoseal 60 on sample (*see* **Note 5**).
2. Cover slides with coverslip by carefully placing coverslip tilted on edge of sample and letting it down slowly.

3.6 Imaging and Data Acquisition

Image analysis and scoring of YAP levels can be done semiquantitatively by manual evaluation of slides under bright-field microscopy or automated and quantitatively using slide scanners and respective software programs (*see* **Note 6**).

4 Notes

1. Sodium citrate buffer (10 mM sodium citrate, 0.05% Tween 20 in ddH_2O, pH 6.0) is commonly used as target retrieval buffer and can easily be prepared if commercial buffers are not available.
2. De-paraffinization and antigen retrieval are essential for successful YAP IHC. Incubation in the organic solvent xylene will allow complete removal of paraffin and accessibility of antigens for subsequent antibody staining. Additional heat-induced epitope retrieval by incubation in citrate buffer (pH 6, target retrieval solution) will remove formaldehyde-induced methylene cross-links.
3. In the following the use of the Antigen Retriever 2100 as pressure cooker for target retrieval is described exemplary. Please see respective manufacturer's instructions if using different instruments in this step.

Using the Antigen Retriever 2100 pressure cooker, slides need to be transferred to slide chambers specific to the machine, and target retrieval solution needs to be added. Chambers need to be placed in the holding basket, and the Antigen Retriever 2100 needs to be filled with 750 mL ddH_2O. The basket can now be placed in the Antigen Retriever 2100. The lid should be closed and the procedure can be started. Further information can be retrieved from vendors, e.g., https://www.emsdiasum.com/microscopy/products/histology/retriever.aspx.

4. The use of a secondary antibody linked with an HRP-labelled polymer (biotin/streptavidin free system) shows reduced unspecific staining due to tissue endogenous biotin-avidin activity (e.g., in the liver or kidney). See also manufacturer's specifications: https://www.agilent.com/en-us/products/immunohistochemistry/visualization-systems/envision-systems/envision-dual-link-kit.
5. Mounting of slides: Prevent introduction of air bubbles since this will interfere with image analysis.
6. Automated, quantitative analysis is strongly recommended for YAP IHC to ensure valid comparison of YAP nuclear levels across different conditions. An example for a scanning and analysis system are Aperio ePathology solutions (Leica), including slide scanners and respective software programs for brightfield IHC image analysis (see also http://www.leicabiosystems.com/digital-pathology/).

Acknowledgments

The authors acknowledge funding from NIH/NCI R01CA211052 and NIH/NCI R01CA204302 (to T.G.B.) as well as from the German Cancer Aid (Mildred Scheel postdoctoral fellowship, to F.H.).

References

1. Huang J, Wu S, Barrera J et al (2005) The Hippo signaling pathway coordinately regulates cell proliferation and apoptosis by inactivating Yorkie, the Drosophila homolog of YAP. Cell 122:421–434
2. Moroishi T, Hansen CG, Guan K-L (2015) The emerging roles of YAP and TAZ in cancer. Nat Rev Cancer 15:73–79
3. Zhao B, Wei X, Li W et al (2007) Inactivation of YAP oncoprotein by the Hippo pathway is involved in cell contact inhibition and tissue growth control. Genes Dev 21:2747–2761
4. Zhao B, Ye X, Yu J et al (2008) TEAD mediates YAP-dependent gene induction and growth control. Genes Dev 22:1962–1971
5. Lin L, Sabnis AJ, Chan E et al (2015) The Hippo effector YAP promotes resistance to RAF- and MEK-targeted cancer therapies. Nat Genet 47:250–256
6. McGowan M, Kleinberg L, Halvorsen AR et al (2017) NSCLC depend upon YAP expression and nuclear localization after acquiring resistance to EGFR inhibitors. Genes Cancer 8:497–504

7. Liu G, Yu F-X, Kim YC et al (2015) Kaposi sarcoma-associated herpesvirus promotes tumorigenesis by modulating the Hippo pathway. Oncogene 34:3536–3546
8. Yang S, Zhang L, Purohit V et al (2015) Active YAP promotes pancreatic cancer cell motility, invasion and tumorigenesis in a mitotic phosphorylation-dependent manner through LPAR3. Oncotarget 6:36019–36031
9. Nakane PK, Pierce GB (1966) Enzyme-labeled antibodies: preparation and application for the localization of antigens. J Histochem Cytochem 14:929–931
10. Huang SN, Minassian H, More JD (1976) Application of immunofluorescent staining on paraffin sections improved by trypsin digestion. Lab Invest 35:383–390
11. Hsu SM, Raine L, Fanger H (1981) Use of avidin-biotin-peroxidase complex (ABC) in immunoperoxidase techniques: a comparison between ABC and unlabeled antibody (PAP) procedures. J Histochem Cytochem 29:577–580
12. Singer SJ (1959) Preparation of an electron-dense antibody conjugate. Nature 183:1523–1524
13. Haines DM, West KH (2005) Immunohistochemistry: forging the links between immunology and pathology. Vet Immunol Immunopathol 108:151–156
14. de Matos LL, Trufelli DC, de Matos MGL et al (2010) Immunohistochemistry as an important tool in biomarkers detection and clinical practice. Biomark Insights 5:9–20
15. Taylor CR, Rudbeck L (2013) Immunohistochemical Staining Methods. Dako Denmark A/S

Chapter 8

Immunofluorescence Study of Endogenous YAP in Mammalian Cells

Valentina Rausch and Carsten G. Hansen

Abstract

Immunocytochemistry enables determination of cellular localization and relative abundance of proteins. This protocol describes a rapid and cost-effective approach to study the cellular localization of YAP (and TAZ), the transcriptional co activators of the Hippo pathway, in mammalian cells. Cells are seeded onto coated cover slips, cultivated and treated as required. Subsequently, they are chemically fixed, and cellular proteins are fluorescently labeled by means of specific antibodies. Multiplexing antibodies enables ascertaining the subcellular localization of YAP and TAZ and thereby also the activation state of the Hippo pathway in various cell types.

Key words Immunofluorescence, Immunocytochemistry, Cellular localization, YAP, Hippo pathway

1 Introduction

The Hippo pathway is central for controlling multiple cellular processes, such as cell survival, proliferation, and differentiation. Consequently, this pathway is a critical regulator of organ regeneration, development as well as carcinogenesis [1]. YAP (yes-associated protein) and TAZ (transcriptional co activator with PDZ-binding motif) are transcriptional co activators of the Hippo pathway [2]. YAP and TAZ shuttle in and out of the nucleus, a process that is predominantly regulated by LATS1-/2-mediated phosphorylation. YAP and TAZ are phosphorylated when the Hippo pathway is active, which results in their cytoplasmic retention and subsequent degradation. Active YAP and TAZ translocate into the nucleus, bind to transcription factors, and thereby regulate gene transcription [1]. Immunofluorescence is used to visualize the cellular localization of YAP and TAZ and provides a way of asserting the activation state of these Hippo pathway co-transcriptional regulators (*see* **Note 1**).

This protocol describes an in vitro approach to immunocytochemically and specifically label YAP and is suitable for most

Alexander Hergovich (ed.), *The Hippo Pathway: Methods and Protocols*, Methods in Molecular Biology, vol. 1893,
https://doi.org/10.1007/978-1-4939-8910-2_8, © Springer Science+Business Media, LLC, part of Springer Nature 2019

mammalian cell types. The cells of interest are seeded onto coated cover slips, cultivated and treated in experimental specific conditions as required. The protocol allows determining the influence of numerous stimuli on YAP activity (*see* **Note 1**), e.g., cell densities, and exposure to or restriction from various chemical stimuli, such as modulators of the cytoskeleton (e.g., latrunculin A, cytochalasins, jasplakinolide, nocodazole, etc. [3–5]), hormones as well as other bioactive components [6, 7]. Subsequent to treatment, cells are chemically fixed, and subcellular proteins such as YAP, other Hippo pathway components, and additional cellular proteins of interest are specifically labeled. Multiplexing antibodies allows simultaneous visualization of various cellular proteins. This is performed using compatible antibodies in combination with fluorophore-conjugated secondary antibodies. Subcellular localization of YAP is detected by confocal microscopy. Due to the low technical expenditure, this method allows the investigation of Hippo pathway activity (*see* **Note 1**) in a short time span in most modern laboratories.

The materials and components used for the coating of cover slips are important features to take into account. The Hippo pathway is a nexus for mechanotransduction and is therefore sensitive to both the hardness of the substrate the cells are grown on and the substrate the cover slips are coated with [8, 9]. Although stiffness of extracellular matrices varies widely between tissue types, glass, in contrast, is a very rigid substrate [10]. Increased stiffness of extracellular matrices activates YAP [11]. Consequently, glass cover slips add an experimental setting which needs to be considered when investigating the Hippo pathway by immunocytochemistry. Yet, due to the absence of autofluorescence, glass is still the preferred material in immunofluorescence studies. Furthermore, the component(s) used to coat cover slips are biological relevant and may interfere with YAP localization and activation state. Fibronectin is widely used in immunocytochemistry to coat surfaces. However, fibronectin activates YAP via integrin signaling, whereas poly-D-lysine and laminin do not appear to interfere with YAP activation state [8]. Finally, this protocol described here is a cost-effective and rapid method of monitoring expression and cellular distribution of the Hippo pathway component YAP.

2 Materials

1. 24-well plate
2. Glass cover slips (12 cm in diameter)
3. Poly-D-lysine (0.1 mg/mL in water; e.g., #P7886 (Sigma))
4. Sterile water

5. Mammalian cells
6. 1× PBS $^{+/+}$ ([+] $CaCl_2$ 100 mg/L, [+] $MgCl_2$ 100 mg/L; e.g., 14,040–091, Gibco)
7. Paraformaldehyde, methanol-free (PFA; e.g., #28908 (Thermo Fisher Scientific), used as 4% PFA in PBS$^{+/+}$ (*see* **Note 2**))
8. Immunofluorescence (IF) buffer (2.5% (v/v) FBS, 0.3% (v/v) Triton-X-100 in PBS$^{+/+}$)
9. Glass slides
10. Tweezers
11. Antibodies (primary, *see* **Notes 3** and **4**), e.g., anti-YAP: ab52771 (Abcam) use at 1:300, #14074 (D8H1X, Cell Signaling Technology) use at 1:200, sc101199 (63.7, Santa Cruz) use at 1:100
12. Antibodies (secondary), such as Alexa Fluor 488 (A-11034, Thermo Fisher Scientific) and Alexa Fluor 594 (A-11032, Thermo Fisher Scientific), use at 1:600
13. Self-curing antifade mounting medium containing DAPI (e.g., #P36962, Invitrogen)
14. Parafilm
15. Petri dish (for overnight incubation)
16. Tissue paper (to wet, for lining the rim of the petri dish)
17. Tin foil or slide box for light-protected storage
18. Pencil (*see* **Note 5**)

3 Methods

The following protocol describes the immunofluorescence study of YAP in mammalian cells cultured in a 24-well plate format. If required, scaling up to a 12- or 6-well plate is possible as long as the size of coverslips and the volumes are adjusted accordingly. Here, poly-D-lysine coating is performed. Importantly, many other substrates are available to coat cover slips. However, careful considerations should be taken into account on which substrate (if any) should be used (*see* **Note 6**). Please comply with local health and safety regulations.

3.1 Preparation of Cover Slips (in the Tissue Culture Hood)

1. Place cover slips into wells of a 24-well plate
2. Cover with sterile poly-D-lysine solution (at least 300 μL per well)
3. Incubate for 10–15 min
4. Aspirate poly-D-lysine, and let cover slips air-dry for 10 min
5. Rinse wells three times with sterile water

6. Let air-dry for at least 1 h exposing the plate to UV light
7. Store plate covered with lid at room temperature or alternatively proceed directly to (8)
8. Seed cells onto cover slips at the desired density
9. Treat cells in an assay-specific manner (as required)

3.2 Fixation of Cells

1. Transfer cells to a class 2 cabinet
2. Aspirate medium from cells (*see* **Note 7**)
3. Rinse cells carefully once with pre-warmed (37 °C) $PBS^{+/+}$
4. Add warm (37 °C) 4% PFA in $PBS^{+/+}$ solution (*see* **Note 2**); around 500 μL/well
5. Incubate for 30 min at 37 °C
6. Aspirate PFA solution
7. Rinse fixed cells briefly with $PBS^{+/+}$
8. Wash fixed cells with $PBS^{+/+}$ twice, 3 min each

3.3 Labeling of Cells (See Note 8)

1. Incubate cells in IF buffer for 20–30 min (for blocking and permeabilization)
2. Prepare moist chambers: line the bottom of a petri dish with parafilm and the rim with wet tissue (as in Fig. 4)
3. Prepare primary antibody solutions in IF buffer
4. Transfer cover slips from plate onto the labeled parafilm using a bent pipette tip and tweezers (Figs. 1 and 2)
5. Cover cells immediately with 30–50 μL of primary antibody solution (Fig. 3)
6. Close petri dish with lid and seal with parafilm (Fig. 4)
7. Incubate overnight at 4 °C
8. (The next day) Prepare secondary antibody solution
9. Aspirate the primary antibody solution
10. Rinse cover slips two times briefly with $PBS^{+/+}$ and then four times for 5 min each
11. Apply secondary antibody solution to cover slips (30–50 μL per cover slip)
12. Incubate for 1 h at room temperature in the dark
13. Aspirate secondary antibody solution
14. Rinse cover slips two times briefly with $PBS^{+/+}$ and then four times for 5 min each
15. Add a small drop of mounting medium (containing DAPI) onto a slide (three to four cover slips fit onto one slide)

Fig. 1 Taking out cover slips from well plate. A cover slip (indicated by dotted red line) is carefully lifted from the 24-well plate by means of tweezers and a bent pipette tip

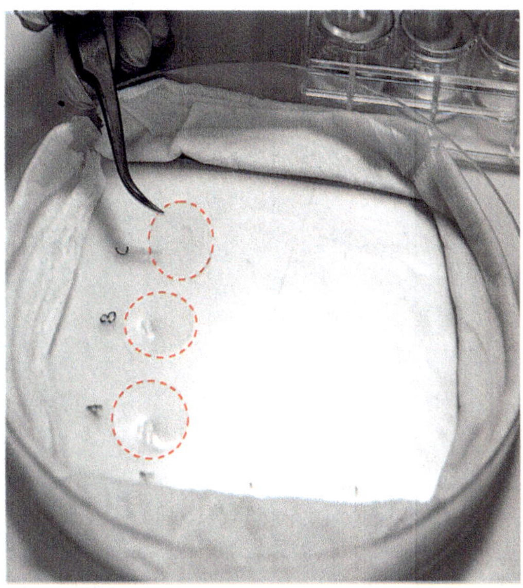

Fig. 2 Placing cover slips into a moist chamber. Cover slips (indicated by dotted red lines) are placed into the moist chamber in order to treat the cells with antibody solution

16. Carefully remove excess liquid from cover slips (Fig. 5)
17. Transfer cover slips from the petri dish onto the slide (avoiding air bubbles), cells facing toward the slide (Fig. 6a, b) (*see* **Note 7**)

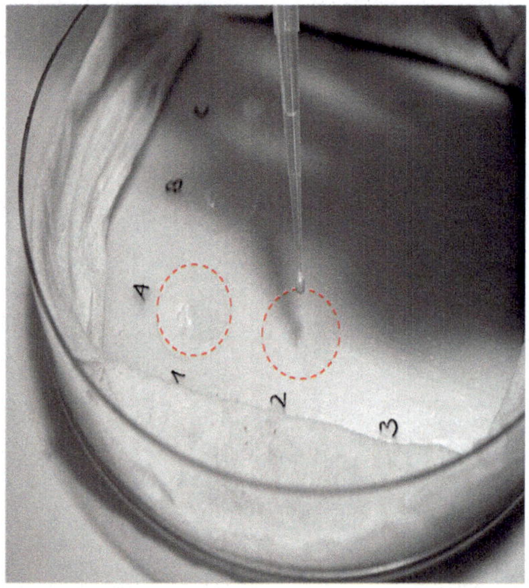

Fig. 3 Adding antibody solution onto fixed cells. Antibody solution is carefully pipetted on top of the fixed cells on cover slips (indicated by dotted red lines)

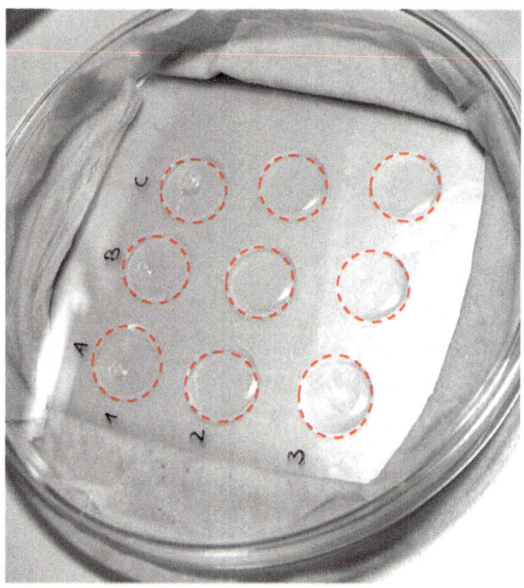

Fig. 4 The moist chamber. The moist chamber is a petri dish lined with parafilm at the bottom and wet tissue paper at the rim. It is sealed with the lid and wrapped with parafilm

18. Let the cover slips, that are now mounted on the glass slide, dry in the dark at room temperature for at least 1 h (or overnight in fridge)
19. Store slides in the dark at 4 °C in a slide container

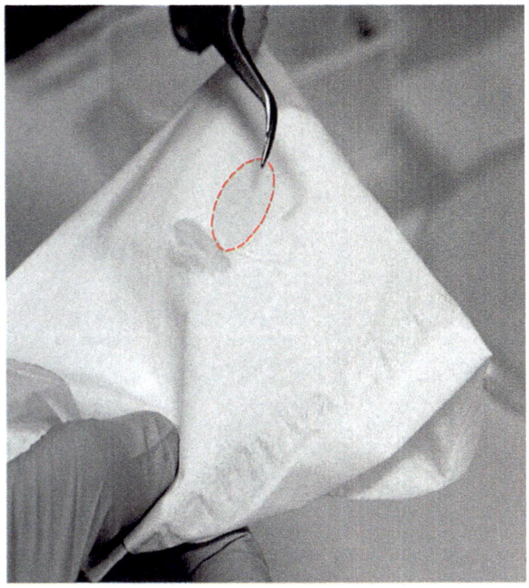

Fig. 5 Preparation of cover slips for mounting. Prior to mounting on a slide, excess liquid on the cover slip (indicated by dotted red line) is carefully removed using tissue paper

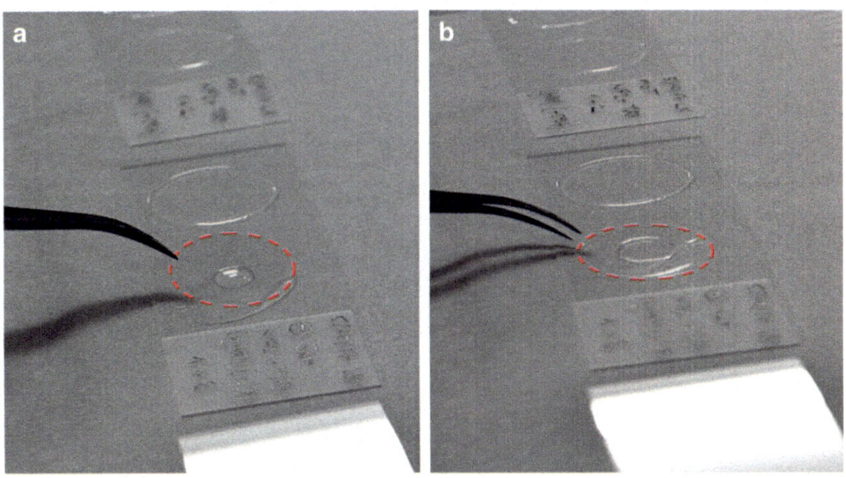

Fig. 6 Mounting of cells on cover slips. (**a**) A drop of DAPI containing mounting medium is placed on a slide, avoiding bubbles, and (**b**) a cover slip (indicated by dotted red line) is placed on top of it, cells facing down, allowing the mounting medium to dispense evenly under the cover slip

3.4 Microscopy

1. Assess localization, and acquire images of YAP (and TAZ) and other proteins of interest by confocal microscopy using optimal fluorophore, and experiment specific microscope settings (Fig. 7)
2. Quantify images using appropriate software

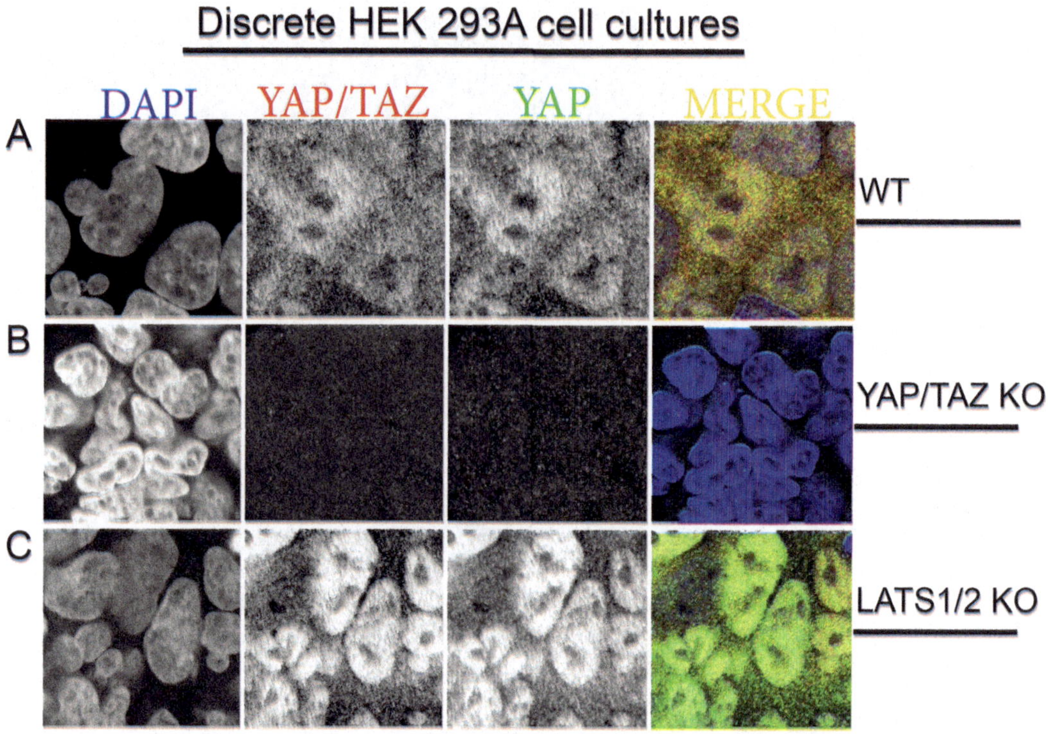

Fig. 7 Example immunofluorescence images of cells. Confocal images of discrete populations of dense (**a**) wild-type (WT), (**b**) YAP/TAZ knockout, and (**c**) LATS1/2 knockout cells labeled with antibodies recognizing YAP/TAZ (red), YAP (green), and DAPI (blue). Cells were prepared side by side using the method described in this chapter using sc101199 (red) and ab52771 (green) antibodies. Images were acquired using the same microscope settings throughout. Note the lack of red and green signal in the YAP/TAZ KO cells and increased nuclear red and green signal in LATS1/2 KO cells compared to WT cells. Additional information on these cells can be found in [15]

4 Notes

1. Additional levels of YAP/TAZ regulation take place in the nucleus [12–14]. The nuclear localization of YAP/TAZ therefore needs to be complemented with techniques such as qPCR to fully justify the use of the term activation of YAP/TAZ.
2. 4% PFA solution is prepared by mixing 16% PFA with 1× PBS$^{+/+}$ (e.g., add 10 mL of 16% PFA to 30 mL of 1× PBS$^{+/+}$). The solution should be stored at −20 °C. If used within 1 month, storage at 4 °C may also be possible.
3. The antibody ab52771 (Abcam) is specific to human YAP, while sc101199 (63.7, Santa Cruz) detects both YAP and TAZ in samples of human, mouse, and rat origin [15] (Fig. 7). To detect murine YAP, we advise using #14074 (D8H1X, Cell Signaling Technology) antibody.

4. The antibodies highlighted in this chapter perform well in our hands (Fig. 7a–c) but are by no means an exhaustive list of antibodies that recognize endogenous YAP by immunofluorescence.
5. Make sure you keep consistent labeling of cover slips with experimental details. Use a pencil as it is resistant to mounting medium and isopropanol.
6. Other substances, such as collagen or laminin, can be used to coat cover slips, but note that these might impact the activation state of YAP [8].
7. Do not allow cover slips to dry out while processing them both before and after fixation, as this will greatly impair the quality of the obtained data.

Acknowledgments

Members of the Gram Hansen lab are thanked for their comments as well as for helping taking the photos for Figs. 1, 2, 3, 4, 5, and 6. Work ongoing in the Gram Hansen lab is supported by a Chancellor's Fellowship start-up fund and by the Wellcome Trust-University of Edinburgh Institutional Strategic Support Fund. We acknowledge the technical support and guidance provided by the Centre for Reproductive Health SuRF Histology and imaging as well as the Centre for Inflammation Research imaging staff.

References

1. Hansen CG, Moroishi T, Guan KL (2015) YAP and TAZ: a nexus for Hippo signaling and beyond. Trends Cell Biol 25(9):499–513. https://doi.org/10.1016/j.tcb.2015.05.002
2. Gomez M, Gomez V, Hergovich A (2014) The Hippo pathway in disease and therapy: cancer and beyond. Clin Transl Med 3(22):1–12. https://doi.org/10.1186/2001-1326-3-22
3. Gaspar P, Tapon N (2014) Sensing the local environment: actin architecture and Hippo signalling. Curr Opin Cell Biol 31:74–83. https://doi.org/10.1016/j.ceb.2014.09.003
4. Wada K, Itoga K, Okano T, Yonemura S, Sasaki H (2011) Hippo pathway regulation by cell morphology and stress fibers. Development 138(18):3907–3914. https://doi.org/10.1242/dev.070987
5. Zhao B, Li L, Wang L, Wang CY, Yu J, Guan KL (2012) Cell detachment activates the Hippo pathway via cytoskeleton reorganization to induce anoikis. Genes Dev 26(1):54–68. https://doi.org/10.1101/gad.173435.111
6. Yu FX, Zhao B, Panupinthu N, Jewell JL, Lian I, Wang LH, Zhao J, Yuan H, Tumaneng K, Li H, Fu XD, Mills GB, Guan KL (2012) Regulation of the Hippo-YAP pathway by G-protein-coupled receptor signaling. Cell 150(4):780–791. https://doi.org/10.1016/j.cell.2012.06.037
7. Park HW, Kim YC, Yu B, Moroishi T, Mo JS, Plouffe SW, Meng Z, Lin KC, Yu FX, Alexander CM, Wang CY, Guan KL (2015) Alternative Wnt signaling activates YAP/TAZ. Cell 162(4):780–794. https://doi.org/10.1016/j.cell.2015.07.013
8. Kim NG, Gumbiner BM (2015) Adhesion to fibronectin regulates Hippo signaling via the FAK-Src-PI3K pathway. J Cell Biol 210(3):503–515. https://doi.org/10.1083/jcb.201501025
9. Dupont S, Morsut L, Aragona M, Enzo E, Giulitti S, Cordenonsi M, Zanconato F, Le Digabel J, Forcato M, Bicciato S, Elvassore N, Piccolo S (2011) Role of YAP/TAZ in

mechanotransduction. Nature 474 (7350):179–183. https://doi.org/10.1038/nature10137

10. Engler AJ, Sen S, Sweeney HL, Discher DE (2006) Matrix elasticity directs stem cell lineage specification. Cell 126(4):677–689. https://doi.org/10.1016/j.cell.2006.06.044

11. Panciera T, Azzolin L, Cordenonsi M, Piccolo S (2017) Mechanobiology of YAP and TAZ in physiology and disease. Nat Rev Mol Cell Biol 18(12):758–770. https://doi.org/10.1038/nrm.2017.87

12. Koontz LM, Liu-Chittenden Y, Yin F, Zheng Y, Yu J, Huang B, Chen Q, Wu S, Pan D (2013) The Hippo effector Yorkie controls normal tissue growth by antagonizing scalloped-mediated default repression. Dev Cell 25(4):388–401. https://doi.org/10.1016/j.devcel.2013.04.021

13. Jiao S, Wang H, Shi Z, Dong A, Zhang W, Song X, He F, Wang Y, Zhang Z, Wang W, Wang X, Guo T, Li P, Zhao Y, Ji H, Zhang L, Zhou Z (2014) A peptide mimicking VGLL4 function acts as a YAP antagonist therapy against gastric cancer. Cancer Cell 25 (2):166–180. https://doi.org/10.1016/j.ccr.2014.01.010

14. Zhang W, Gao Y, Li P, Shi Z, Guo T, Li F, Han X, Feng Y, Zheng C, Wang Z, Li F, Chen H, Zhou Z, Zhang L, Ji H (2014) VGLL4 functions as a new tumor suppressor in lung cancer by negatively regulating the YAP-TEAD transcriptional complex. Cell Res 24(3):331–343. https://doi.org/10.1038/cr.2014.10

15. Hansen CG, Ng YL, Lam WL, Plouffe SW, Guan KL (2015) The Hippo pathway effectors YAP and TAZ promote cell growth by modulating amino acid signaling to mTORC1. Cell Res 25(12):1299–1313. https://doi.org/10.1038/cr.2015.140

Chapter 9

Immunofluorescence Microscopy to Study Endogenous TAZ in Mammalian Cells

Nathan M. Kingston, Andrew M. Tilston-Lunel, Julia Hicks-Berthet, and Xaralabos Varelas

Abstract

The transcriptional coactivator with PDZ-binding motif (TAZ), which is encoded by the *WWTR1* gene, is a key transcriptional effector of the Hippo signaling pathway. TAZ function has been implicated in a variety of developmental processes and diseases, most notably in driving oncogenesis. Given that nuclear-cytoplasmic localization dynamics dictate TAZ activity, techniques for visualizing TAZ localization are critical for its study. Here we describe an immunofluorescence microscopy protocol that allows for the visualization of TAZ subcellular localization in mammalian cells, offering an approach that can aid in the analysis of TAZ regulation and function.

Key words TAZ, *WWTR1*, Hippo pathway, Immunofluorescence microscopy

1 Introduction

The transcriptional coactivator with PDZ-binding motif (TAZ), which is encoded by the *WWTR1* gene, has emerged as a major regulator of cell proliferation, survival, and fate in a variety of organs and tissues [1, 2]. TAZ is a transcriptional regulator that functions as a key effector of Hippo pathway signaling. TAZ directs gene expression by binding to a variety of transcription factors, most notably the TEAD family (TEAD1–4) of proteins [3, 4]. TAZ-regulated transcriptional programs are implicated in a variety of biological processes, which often are redundantly controlled by its paralog Yes-associated protein (YAP) [5–10]. TAZ activity is regulated in large part by alterations in its nuclear-cytoplasmic distribution, which is regulated by a variety of signals [11].

The phosphorylation of TAZ on several conserved serine/threonine residues impacts TAZ localization. For example, phosphorylation of human TAZ on Serine89 by the LATS1 and LATS2

Alexander Hergovich (ed.), *The Hippo Pathway: Methods and Protocols*, Methods in Molecular Biology, vol. 1893, https://doi.org/10.1007/978-1-4939-8910-2_9,

(LATS1/2) kinases promotes binding to 14-3-3 proteins and subsequent cytoplasmic retention [12] and has emerged as a signature mode of TAZ regulation [13]. TAZ stability is also regulated by phosphorylation, contributing to dynamic changes in TAZ subcellular levels [14]. Precise TAZ nuclear-cytoplasmic distribution directs various biological events, including early cell fate determining processes [7]. Further, aberrant nuclear TAZ activity has been implicated in a variety of human cancers [15, 16]. Thus, the ability to examine TAZ localization in cells and tissues has become an important part of the study of TAZ biology.

In this chapter, we describe an immunofluorescence microscopy method for visualizing endogenous TAZ localization in mammalian cells. We demonstrate the efficacy of our method by showing the loss of immunofluorescent signal in primary mouse lung fibroblasts with *Taz* conditionally knocked out.

2 Materials

2.1 Sample Preparation

1. Glass bottom microscopy slides (*see* **Note 1**).
2. Cells of interest (here we used primary mouse lung fibroblasts to validate our method; *see* Fig. 1).
3. Phosphate-buffered saline (PBS) (*see* **Note 2**).
4. 4% (v/v) paraformaldehyde in PBS (*see* **Notes 3** and **4**).
5. Orbital shaker.

2.2 Staining

1. 1% (w/v) SDS in PBS.
2. Blocking solution: 2% (w/v) bovine serum albumin (BSA) in PBS.
3. Primary antibody solution: dilute antibody (BD Biosciences mouse anti-Taz #BD560235) 1:500 in blocking solution (*see* **Notes 4** and **5**).
4. PBS-T: 0.1% (v/v) Tween-20 in PBS.
5. Secondary antibody solution: anti-mouse fluorescent secondary antibody (*see* **Notes 4** and **6**).
6. DNA stain solution: 1 μg/mL solution of DAPI (4′,6-diamidino-2-phenylindole, dihydrochloride) (Thermo Fisher # 62247) in PBS or 1 μg/mL solution of Hoechst 33342 (Thermo Fisher #H1399) in PBS.
7. Prolong Gold anti-fade reagent (Life Technologies P36930).
8. Cover slips.
9. Fluorescent microscope (*see* **Note 7**).

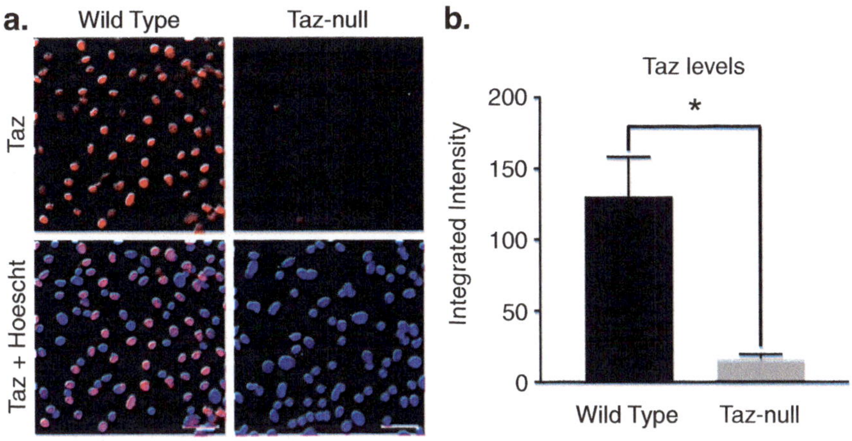

Fig. 1 Validation of TAZ immunofluorescence staining in Taz-null mouse fibroblasts. (**a**) Primary mouse lung fibroblasts were isolated from Taz-loxP/loxP; Col1a1-CreERT2 mice (conditional *Taz* knockout mice, Taz-null) or *Taz*-loxP/loxP mice without any Cre (wild-type). Fibroblasts were grown in the presence of 3 μM 4-hydroxytamoxifen for 5 days to induce CreERT2-mediated knockout and then fixed and stained according to the protocol in this manuscript. Scale bars represent 50 μm. (**b**) Intensity of nuclear Taz staining was quantified for wild-type and Taz-null fibroblasts using CellProfiler as described [20] (±SEM, $^*p < 0.01$)

3 Methods

3.1 Sample Preparation

1. Grow cells to desired confluence in glass bottom chamber slides (*see* **Note 8**).
2. Aspirate the growth media, and add PBS to the well at room temperature (*see* **Note 9**).
3. Aspirate the PBS, and add 4% paraformaldehyde solution to each well at room temperature. Incubate for 15 min at room temperature (*see* **Note 10**).
4. Aspirate the paraformaldehyde, and wash once with PBS (*see* **Note 9**).
5. Add PBS to each well of the chamber slide (*see* **Notes 9** and **11**).

3.2 Staining

1. Aspirate the PBS from the chamber slides, and add 1% SDS/PBS to each well. Incubate at 37 °C for 10 min (*see* **Note 12**).
2. Aspirate the SDS/PBS buffer, and add PBS to each well (*see* **Note 9**). Wash on a shaker at low speed for 5 min at room temperature.
3. Aspirate the PBS. Add blocking solution to each well, and incubate on a shaker at low speed for 1 h at room temperature (*see* **Note 13**).

4. Aspirate the blocking solution. Add primary antibody solution, and incubate on a shaker at low speed at 4 °C for 12–16 h (*see* **Notes 13** and **14**).
5. Aspirate the primary antibody solution. Add PBS-T to each well, and wash on a shaker at low speed for 5 min at room temperature (*see* **Note 9**). Aspirate and repeat the wash for an additional 5 min.
6. Aspirate the PBS-T. Add secondary antibody solution to each well, and incubate on a shaker at low speed for 1 h at room temperature (*see* **Notes 13** and **15**).
7. Aspirate the secondary antibody solution. Add PBS-T to each well, and wash on a shaker at low speed for 5 min at room temperature (*see* **Note 9**). Aspirate and repeat the wash for an additional 5 min.
8. Aspirate the PBS-T. Add DNA stain solution to each well, and incubate on a shaker at low speed for 15 min at room temperature (*see* **Note 16**).
9. Aspirate the DNA stain solution. Add PBS to each well, and wash on a shaker at low speed for 5 min at room temperature (*see* **Note 9**). Aspirate and repeat the wash for an additional 5 min.
10. Aspirate the PBS.
11. Mount the slides for imaging using Prolong Gold anti-fade reagent (Life Technologies P36930) (*see* **Note 17**).
12. Image the slides using a fluorescent microscope (*see* **Notes 7**, **18**, and **19**).

4 Notes

1. We typically use the Thermo Scientific Nunc Lab-Tek II Chamber Slide System (Thermo Fisher 154526) or glass cover slips that fit into a 24-well culture dish (Thermo Fisher 12-545-80).
2. PBS is composed of 137 mM NaCl, 2 mM KCl, 10 mM Na_2HPO_4, sodium phosphate dibasic, and 1.8 mM KH_2PO_4, with the pH adjusted to 7.4 with HCl.
3. We make 4% (v/v) paraformaldehyde by diluting 5 mL 16% paraformaldehyde aqueous solution (Electron Microscopy Sciences # 15710) into 15 mL PBS.
4. The noted solutions should be made on the day that they are used and stored at 4 °C until it is time to use them.
5. Using the protocol described above, as well as variations of this protocol, we have attempted to examine the localization of TAZ using several other commercial antibodies, including

anti-WWTR1 antibody (Sigma #HPA007415), anti-TAZ antibodies (Abcam #ab84927 and #ab110239), and anti-TAZ antibody (Cell signaling #2149). The antibody we report here (BD560235) is the only antibody that we have tested that reliably shows a loss of signal in TAZ depleted or TAZ deleted cells. Some TAZ antibodies, such as the anti-WWTR1 antibody and the YAP/TAZ antibody (Cell signaling #8418), do show a loss of signal when TAZ and YAP are co-depleted or co-deleted, indicating that these antibodies also have an affinity for YAP, which may be appropriate for some experimental conditions. We therefore caution the research community in making conclusions about TAZ levels and localization without proper validation.

6. For our experiment shown in Fig. 1, we made a 1:500 dilution of Invitrogen A-31571 donkey anti-mouse Alexa Fluor 647 in blocking solution.
7. For the images shown in Fig. 1, we used a Zeiss AxioObserver D1 equipped with a X-Cite 120LED System.
8. For the experiment shown in Fig. 1, we used 4-well chamber slides with a culture area of 1.7 cm^2. For larger or smaller chamber slides, adjust all volumes in this protocol accordingly.
9. For 4-well chamber slides with a culture area of 1.7 cm^2, we use 500 μL PBS or PBS-T per well.
10. For 4-well chamber slides with a culture area of 1.7 cm^2, we use 200 μL of 4% PFA solution per well.
11. Chamber slides may be stored in with PBS at 4 °C for several weeks. However, if the slides are not processed right away, wrap them in Parafilm (Sigma #P7793-1) to prevent the wells from drying out.
12. We have found that using 1% SDS for cell permeabilization/antigen retrieval is an important step in our protocol. For 4-well chamber slides with a culture area of 1.7 cm^2, we use 200 μL 1% SDS/PBS per well.
13. For 4-well chamber slides with a culture area of 1.7 cm^2, we use 200 μL of blocking, primary antibody, and secondary antibody solutions per well.
14. This primary antibody incubation step may be done for 2 h at room temperature, but we have found that reducing the time in primary antibody results in a less intense signal.
15. It is important to not incubate in secondary antibody solution for much more than 1 h, as lengthy incubations in secondary antibody will increase the background signal. Also, we recommend covering the chamber slide with foil to protect from light during the washes and incubations for the remainder of the protocol.

16. For 4-well chamber slides with a culture area of 1.7 cm^2, we use 200 μL of DNA stain solution per well.
17. For the chamber slide system, use the directions that accompany the slides, and for cover slips, we carefully pick them up with forceps and place them on a microscopy slide (Thermo Fisher #12-550-15).
18. The highest quality images will be obtained immediately after staining. However, slides can be stored at 4 °C in the dark and retain their signal for several weeks.
19. To show the efficacy of immunofluorescence staining of endogenous TAZ using the BD560235 antibody, we refer the readers to our prior studies [17, 18], as well as Fig. 1, where we show our analysis of primary mouse lung fibroblasts isolated from *Taz*-loxP/loxP [19]; Col1a1CreERT2 (Jax # 027751) mice. Cells isolated from these mice allow for the conditional deletion of *Taz* following treatment with 4-hydroxytamoxifen (Sigma #H7904).

Acknowledgments

We would like to thank Dr. Jeffrey Wrana (Lunenfeld-Tanenbaum Research Institute, Toronto, Canada) for providing the Taz-loxP/loxP mice. X.V. is supported by NIH R01HL124392, the March of Dimes Foundation Grant no. 1-FY17-375, and an American Cancer Society Ellison New England Research Scholar Grant (RSG-17-138-01-CSM). N. M. K. is supported by NIH T32HL007035-40. J.H.B. is supported by NIH F31HL13250601.

References

1. Fu V, Plouffe SW, Guan KL (2018) The Hippo pathway in organ development, homeostasis, and regeneration. Curr Opin Cell Biol 49:99–107. https://doi.org/10.1016/j.ceb.2017.12.012
2. Varelas X (2014) The Hippo pathway effectors TAZ and YAP in development, homeostasis and disease. Development 141 (8):1614–1626. https://doi.org/10.1242/dev.102376
3. Mahoney WM Jr, Hong JH, Yaffe MB, Farrance IK (2005) The transcriptional co-activator TAZ interacts differentially with transcriptional enhancer factor-1 (TEF-1) family members. Biochem J 388(Pt 1):217–225. https://doi.org/10.1042/BJ20041434
4. Zhang H, Liu CY, Zha ZY, Zhao B, Yao J, Zhao S, Xiong Y, Lei QY, Guan KL (2009) TEAD transcription factors mediate the function of TAZ in cell growth and epithelial-mesenchymal transition. J Biol Chem 284 (20):13355–13362. https://doi.org/10.1074/jbc.M900843200
5. Hiemer SE, Szymaniak AD, Varelas X (2014) The transcriptional regulators TAZ and YAP direct transforming growth factor beta-induced tumorigenic phenotypes in breast cancer cells. J Biol Chem 289(19):13461–13474. https://doi.org/10.1074/jbc.M113.529115
6. Levasseur A, St-Jean G, Paquet M, Boerboom D, Boyer A (2017) Targeted disruption of YAP and TAZ impairs the maintenance of the adrenal cortex. Endocrinology 158(11):3738–3753. https://doi.org/10.1210/en.2017-00098
7. Nishioka N, Inoue K, Adachi K, Kiyonari H, Ota M, Ralston A, Yabuta N, Hirahara S, Stephenson RO, Ogonuki N, Makita R,

Kurihara H, Morin-Kensicki EM, Nojima H, Rossant J, Nakao K, Niwa H, Sasaki H (2009) The Hippo signaling pathway components Lats and Yap pattern Tead4 activity to distinguish mouse trophectoderm from inner cell mass. Dev Cell 16(3):398–410. https://doi.org/10.1016/j.devcel.2009.02.003

8. Poitelon Y, Lopez-Anido C, Catignas K, Berti C, Palmisano M, Williamson C, Ameroso D, Abiko K, Hwang Y, Gregorieff A, Wrana JL, Asmani M, Zhao R, Sim FJ, Wrabetz L, Svaren J, Feltri ML (2016) YAP and TAZ control peripheral myelination and the expression of laminin receptors in Schwann cells. Nat Neurosci 19(7):879–887. https://doi.org/10.1038/nn.4316
9. Xin M, Kim Y, Sutherland LB, Murakami M, Qi X, McAnally J, Porrello ER, Mahmoud AI, Tan W, Shelton JM, Richardson JA, Sadek HA, Bassel-Duby R, Olson EN (2013) Hippo pathway effector Yap promotes cardiac regeneration. Proc Natl Acad Sci U S A 110 (34):13839–13844. https://doi.org/10.1073/pnas.1313192110
10. Reginensi A, Hoshi M, Boualia SK, Bouchard M, Jain S, McNeill H (2015) Yap and Taz are required for Ret-dependent urinary tract morphogenesis. Development 142 (15):2696–2703. https://doi.org/10.1242/dev.122044
11. Meng Z, Moroishi T, Guan KL (2016) Mechanisms of Hippo pathway regulation. Genes Dev 30(1):1–17. https://doi.org/10.1101/gad.274027.115
12. Kanai F, Marignani PA, Sarbassova D, Yagi R, Hall RA, Donowitz M, Hisaminato A, Fujiwara T, Ito Y, Cantley LC, Yaffe MB (2000) TAZ: a novel transcriptional co-activator regulated by interactions with 14-3-3 and PDZ domain proteins. EMBO J 19(24):6778–6791. https://doi.org/10.1093/emboj/19.24.6778
13. Lei QY, Zhang H, Zhao B, Zha ZY, Bai F, Pei XH, Zhao S, Xiong Y, Guan KL (2008) TAZ promotes cell proliferation and epithelial-mesenchymal transition and is inhibited by the hippo pathway. Mol Cell Biol 28 (7):2426–2436. https://doi.org/10.1128/MCB.01874-07
14. Liu CY, Zha ZY, Zhou X, Zhang H, Huang W, Zhao D, Li T, Chan SW, Lim CJ, Hong W, Zhao S, Xiong Y, Lei QY, Guan KL (2010) The hippo tumor pathway promotes TAZ degradation by phosphorylating a phosphodegron and recruiting the SCF{beta}-TrCP E3 ligase. J Biol Chem 285(48):37159–37169. https://doi.org/10.1074/jbc.M110.152942
15. Moroishi T, Hansen CG, Guan KL (2015) The emerging roles of YAP and TAZ in cancer. Nat Rev Cancer 15(2):73–79. https://doi.org/10.1038/nrc3876
16. Zanconato F, Cordenonsi M, Piccolo S (2016) YAP/TAZ at the roots of cancer. Cancer Cell 29(6):783–803. https://doi.org/10.1016/j.ccell.2016.05.005
17. Beyer TA, Weiss A, Khomchuk Y, Huang K, Ogunjimi AA, Varelas X, Wrana JL (2013) Switch enhancers interpret TGF-beta and Hippo signaling to control cell fate in human embryonic stem cells. Cell Rep 5 (6):1611–1624. https://doi.org/10.1016/j.celrep.2013.11.021
18. Varelas X, Samavarchi-Tehrani P, Narimatsu M, Weiss A, Cockburn K, Larsen BG, Rossant J, Wrana JL (2010) The Crumbs complex couples cell density sensing to Hippo-dependent control of the TGF-beta-SMAD pathway. Dev Cell 19(6):831–844. https://doi.org/10.1016/j.devcel.2010.11.012
19. Reginensi A, Scott RP, Gregorieff A, Bagherie-Lachidan M, Chung C, Lim DS, Pawson T, Wrana J, McNeill H (2013) Yap- and Cdc42-dependent nephrogenesis and morphogenesis during mouse kidney development. PLoS Genet 9(3):e1003380. https://doi.org/10.1371/journal.pgen.1003380
20. Carpenter AE, Jones TR, Lamprecht MR, Clarke C, Kang IH, Friman O, Guertin DA, Chang JH, Lindquist RA, Moffat J, Golland P, Sabatini DM (2006) CellProfiler: image analysis software for identifying and quantifying cell phenotypes. Genome Biol 7(10):R100. https://doi.org/10.1186/gb-2006-7-10-r100

Chapter 10

Nuclear/Cytoplasmic Fractionation to Study Hippo Effectors

Maria Chatzifrangkeskou and Eric O'Neill

Abstract

The translocation or shuttling of Hippo proteins between the nucleus and cytoplasm is a rapid event following cytoskeletal or mechanical cues as well as stimulation with extracellular growth factors. Here we describe an experimental procedure for a simple and fast separation of nuclear and cytoplasmic fractions which maintains protein integrity and integrity of protein-protein complexes, indicating that it should be applicable to many experimental questions.

Key words Subcellular fractionation, Hippo pathway, Hypotonic buffer, Cytoplasmic compartment, Nuclear compartment

1 Introduction

YAP and TAZ, the major downstream effectors of the Hippo signaling pathway, localize to nucleus as well as cytoplasm. When YAP and TAZ are active, they translocate into the nucleus to bind to transcription factors, most prominently to TEAD family, and induce expression of genes that are involved in cell proliferation, survival, and migration. YAP/TAZ subcellular localization is regulated by a broad range of mechanical and biochemical signals such as cell geometry [1–3], cell density [2, 4, 5], F-actin network status [1], extracellular matrix stiffness [6], and myosin contractility [1]. Cellular fractionation techniques have used to study the location and trafficking of molecular components [7]. To gain a better understanding of the Hippo pathway activity, fractionation of nuclear and cytoplasmic compartments is usually needed.

There are many protocols and commercial kits available that allow separation of the two compartments from cultured cells. However, most require high centrifugation speeds. Our protocol utilizes a small benchtop centrifuge to obtain pure extractions for the cytoplasmic and nuclear compartments in a short period of time. Our protocol for subcellular fractionation is a two-step lysis procedure that provides an efficient preparation of cytoplasmic- and

Alexander Hergovich (ed.), *The Hippo Pathway: Methods and Protocols*, Methods in Molecular Biology, vol. 1893, https://doi.org/10.1007/978-1-4939-8910-2_10,

nuclear-containing fractions from cultured cells. The first step is lysis of the cell membrane and isolation of the cytoplasmic proteins from the cell suspension. This is achieved by the use of a hypotonic extraction buffer which breaks the cell membrane but keeps the nuclear membrane and other compartments intact. With the bulk of the cytoplasmic proteins removed, the nuclei are then lysed in a high-salt nuclear extraction buffer that bursts the nuclear membrane and releases the proteins therein. The protein fractions are compatible with many downstream assays such as Western blot, enzyme activity assays, reporter assays, RNA splicing, and gel shift assay.

2 Materials

All incubations and centrifugation steps should be performed at 4 °C with ice-cold buffers. Protease inhibitors or inhibitor cocktails should be added to the extraction buffers so that the final concentration is 1× (*see* **Note 1**).

1. Phosphate-buffered saline (PBS): Use 10× PBS, pH 7.2 (0.2 M potassium phosphate, 1.5 M NaCl). Dilute appropriate volume to 1× with distilled water.
2. Phenylmethylsulfonyl fluoride (PMSF): Weigh 17.42 g, and add DMSO to a volume of 100 mL. Dissolve completely, and store at −20 °C up to 6 months (*see* **Note 2**).
3. Dithiothreitol (DTT): Dissolve 1.54 g in 10 mL of distilled water to obtain a final concentration of 1 M. Sterilize using 0.22 μm syringe filter, and aliquot in 2 mL tubes. Store at −20 °C (*see* **Note 4**).
4. Cytoplasmic extraction buffer: 10 mM HEPES pH 7.9, 10 mM KCl, 0.1 mM EDTA, 0.1 mM EGTA, 0.5 mM PMSF. Before use, add 1 mM DTT (*see* **Note 3**).
5. Nuclear extraction buffer: 150 mM NaCl, 20 mM HEPES pH 7.5, 0.5 mM EDTA, 1% NP-40.

3 Methods

1. Culture cells on 10 cm culture plate until 70–80% confluent in a 37 °C incubator supplied with 5% CO_2.
2. Discard the culture medium, and wash the cells with room temperature.
3. Trypsinize cells using 1 mL 0.25% Trypsin-EDTA solution. Add 5 mL of culture medium containing 10% fetal bovine serum to inhibit trypsin activity, and collect the cells (*see* **Note 5**).

4. Centrifuge for 5 min at 500 × *g* at room temperature to pellet the cells.
5. Discard the supernatant, and wash the pellet two times with ice-cold 1× PBS and spin again. Use a pipette to carefully remove and discard the supernatant, leaving the cell pellet as dry as possible (*see* **Note 6**).
6. Resuspend the pellet by gentle pipetting in 500 μL of ice-cold cytoplasmic extraction buffer containing protease inhibitors (scale-up or down proportionally).
7. Incubate the cells for 15 min on ice to swell (*see* **Note** 7).
8. Add 10% Nonidet P-40 detergent to a final concentration 0.65% (v/v) (*see* **Note 8**).
9. Vortex the cells vigorously on the highest setting for 10 s, and centrifuge at 1500 × *g* for 5 min, 4 °C. The pellet will contain nuclei, and the supernatant will contain cytoplasm, membrane, and mitochondria. Transfer the supernatant (cytoplasmic fraction) into a clean pre-chilled microcentrifuge tube.
10. The pellet (nuclear fraction) is further lysed in 50 μL ice-cold nuclear extraction buffer supplemented with 1× protease inhibitor cocktail (*see* **Note 9**).
11. Vortex vigorously, and incubate the cell suspension for 30 min on ice with vortexing at 10 min internals (*see* **Note 10**).
12. Centrifuge for 10 min at 14,000 × *g* at 4 °C.
13. Transfer supernatant (nuclear fraction) into a clean pre-chilled microcentrifuge tube. The pellet should, theoretically, just be cell membrane debris which can be discarded.
14. For same-day use, maintain fractions on ice for downstream applications and analysis (*see* **Note 11**). For long-term storage, store fractions at −80 °C.

4 Notes

1. Addition of the protease inhibitor cocktail is required to prevent undesirable proteolysis and maintain protein integrity and function as well as preserve protein-protein interactions. Instead of commercial protease inhibitor cocktails, 4 μg/mL aprotinin and 2 μg/mL pepstatin A could be used, in combination with PMSF.
2. PMSF (targets serine proteases) is very unstable in aqueous solutions and must be added just prior to use. The stability of protease inhibitor-supplemented extraction buffer is 24 h at 4 °C. A common troubleshooting finding is the breakdown products caused by non-specific proteolysis. If necessary, add

phosphatase inhibitors 1 mM sodium vanadate, 50 mM sodium fluoride, 10 mM sodium pyrophosphate) to prevent changes in the target protein's activation state or interactions.

3. HEPES is a zwitterionic biological buffering agent that maintains physiological pH ranging from 6.8 to 8.2.
4. DTT is a strong reducing agent that prevents oxidative damage. At a final 1 mM concentration, DTT is frequently used to disrupt the disulfide bonds of proteins and peptides. Triton X-100 may be substituted for Nonidet P-40.
5. Alternatively, cell scraping could be also used as follows: scrape the cells gently with PBS using a cell scraper, collect the pellet by centrifugation, and resuspend in cytoplasmic extraction buffer.
6. It is very important not to disturb the pellet by pointing the pipette tip toward the opposite side of the tube. It is better to leave some of the supernatant behind.
7. Swollen cells could be disrupted further with the use of Dounce homogenizers, needles, or sonication to aid in cell membrane disruption [8]. Please note that the needle gauge size needs to be chosen so that the cell is disrupted but nuclei can pass through unbroken, such as 21G and 22G. Slowly draw the cell suspension into the syringe, and then eject with a single rapid stroke.
8. The addition of a nonionic non-denaturing detergent (such as NP-40) separates the nuclei from the cytoplasmic fraction allowing leakage of the cytoplasmic proteins. NP-40 is used at a low concentration to allow permeabilization of all the plasma membrane as well as the Golgi, mitochondria membranes, and endoplasmic reticulum while keeping the nuclear membrane intact. SDS is not recommended as it is denaturing, and the extracted proteins will not be in their native form.
9. Glycerol (10%) could be added in the nuclear extraction buffer as it helps to preserve the frozen extracts for functional experiments such as gel shift (EMSA).
10. Vortexing the samples prior to the centrifugation is often used to facilitate lysis. Increasing the vortexing time and incubation time usually results in a higher nuclear protein yield. However, vortexing sometimes isn't enough. It can help to use a fine needle to shear the cellular material.
11. Some applications might require dialysis or desalting to remove detergent and salts. It is recommended to confirm the purity of the separated subcellular compartments using specific marker proteins by immunoblot analysis. For example, beta-actin or alpha-tubulin is used for the cytosolic fraction; histone H3 or Lamin B is used for the nuclear fraction.

References

1. Dupont S, Morsut L, Aragona M et al (2011) Role of YAP/TAZ in mechanotransduction. Nature 474:179–183. https://doi.org/10.1038/nature10137
2. Aragona M, Panciera T, Manfrin A et al (2013) A mechanical checkpoint controls multicellular growth through YAP/TAZ regulation by actin-processing factors. Cell 154:1047–1059. https://doi.org/10.1016/j.cell.2013.07.042
3. Wada K-I, Itoga K, Okano T et al (2011) Hippo pathway regulation by cell morphology and stress fibers. Development 138:3907–3914. https://doi.org/10.1242/dev.070987
4. Zhao B, Wei X, Li W et al (2007) Inactivation of YAP oncoprotein by the Hippo pathway is involved in cell contact inhibition and tissue growth control. Genes Dev 21:2747–2761. https://doi.org/10.1101/gad.1602907
5. Varelas X, Samavarchi-Tehrani P, Narimatsu M et al (2010) The crumbs complex couples cell density sensing to Hippo-dependent control of the TGF-β-SMAD pathway. Dev Cell 19:831–844. https://doi.org/10.1016/j.devcel.2010.11.012
6. Thomasy SM, Morgan JT, Wood JA et al (2013) Substratum stiffness and latrunculin B modulate the gene expression of the mechanotransducers YAP and TAZ in human trabecular meshwork cells. Exp Eye Res 113:66–73. https://doi.org/10.1016/j.exer.2013.05.014
7. Graham JM, Rickwood D (1997) Subcellular fractionation : a practical approach. IRL Press at Oxford University Press, Oxford
8. Das A, Fischer RS, Pan D, Waterman CM (2016) YAP nuclear localization in the absence of cell-cell contact is mediated by a filamentous actin- dependent, myosin II-and phospho-YAP independent pathway during ECM mechanosensing. J Biol Chem 291(12):6096–6110. https://doi.org/10.1074/jbc.M115.708313

Chapter 11

Luciferase Reporter Assays to Determine YAP/TAZ Activity in Mammalian Cells

Sirio Dupont

Abstract

This chapter describes the luciferase assays that are available to monitor YAP/TAZ activity in cell lines and to study their regulation, including the choice for the normalizer, a description of the main YAP-/TAZ-responsive luciferase reporters used so far by the community, and technical notes and experimental considerations on the most appropriate positive controls. Some specific examples are provided to use luciferase assays as the basis to distinguish between Hippo-mediated and phosphorylation-mediated regulatory events and regulatory events that regulate YAP/TAZ independent of these inputs. Finally, typical experimental protocols are outlined briefly for an easier setup of YAP/TAZ luciferase assays.

Key words Luciferase, YAP/TAZ, Hippo, LATS, Phosphorylation, Structure-function analysis

1 Introduction to Monitoring YAP/TAZ Activity by Luciferase Assays

YAP/TAZ were originally identified in mammals as transcriptional coactivators for the TEAD transcription factors [1], an only later rediscovered as downstream effectors of the Hippo pathway, owing to their homology with Drosophila Yorkie [2]. The cooperation between YAP/TAZ and TEADs has been characterized in detail, including the determination of the crystal structure of their interface [3, 4], and underlies the vast majority of the YAP-driven phenotypes so far identified, as indicated by the lack of biological activity of YAP alleles unable to interact with TEADs (S94A mutants). Reflecting this close relationship, YAP and TEAD chromatin immunoprecipitation experiments in multiple cell lines indicate a striking overlap [5–7].

Several systems have been used in the literature to measure YAP/TAZ transcriptional activity. One obvious and important readout is the expression of bona fide endogenous direct YAP/TAZ target genes by real-time PCR, microarray, or RNA sequencing approaches. However, if the analysis is limited to few established

Alexander Hergovich (ed.), *The Hippo Pathway: Methods and Protocols*, Methods in Molecular Biology, vol. 1893, https://doi.org/10.1007/978-1-4939-8910-2_11, © Springer Science+Business Media, LLC, part of Springer Nature 2019

and widely used targets, results can be misleading in some cases, due to the following reasons:

1. Any endogenous target gene is the target of multiple transcription factors.
2. It is hard to define "universal" YAP/TAZ target genes, although some targets (*CTGF*, *CYR61*, *ANKRD1*) have been used and validated across a significant number of cell lines and tissues.
3. With the exception of very few genes for which a TEAD- and YAP-/TAZ-binding element has been mapped in the promoter region and for which mutagenesis experiments indicate this is a functionally relevant regulatory element, experiments formally addressing what are the direct YAP/TAZ targets (i.e., performed in cycloheximide) are still missing.

For this reason, it should be advisable to provide evidence of YAP/TAZ regulation based on multiple endogenous target genes in parallel and by showing coherent regulation of positively and negatively regulated YAP/TAZ target genes (it is worth remembering that microarray analyses indicate that almost half of YAP-/TAZ-regulated genes are inhibited by these factors, although the mechanistic basis for this is still elusive). As alternative or additional evidence, it is possible to monitor YAP/TAZ activity by using luciferase reporter assays.

Luciferase assays provide the advantage of a fast and convenient method to test several experimental conditions in a quantitative manner. Moreover, since luciferase is very sensitive, this assay can be easily scaled-down to 96- or 384-well for large-scale screening purposes [8–12]. Using a luciferase reporter is limited to cell lines in which it is possible to easily transfect plasmid DNA. When establishing a luciferase assay in a new cell line, several transfection reagents should be tested in parallel with several different normalizers to identify the combination giving the strongest signal but also the most consistent results (i.e., similar results are obtained by transfecting the same transfection mix independently). Establishment of stable reporter cell lines can be an option [13], but it requires careful testing and validation of the dynamic range of the randomly inserted transgenic reporter.

2 The Choice of the Normalizer for Your Luciferase Assay

Luciferase assays are based on the comparison of two transfected reporter plasmids, driving expression of two different reporter molecules. Usually, the "experimental" reporter drives luciferase expression, while the "normalizer" drives expression of an enzyme whose activity is orthogonal (i.e., of different enzymatic activity) to luciferase, enabling independent detection. Thus, for any sample of an experiment, the final luciferase value is actually the ratio between the luciferase signal and the signal of the normalizer reporter: in this way it is possible to compensate for sample-to-sample variations in the transfection efficiency and for non-specific effects of the treatments. For example, a treatment leading to general inhibition of the transcriptional machinery or inducing cell death will similarly inhibit both signals, while a specific treatment affecting YAP/TAZ will (preferentially) inhibit luciferase, leaving the other signal unaltered. Clearly, normalization relies on the idea, which should be verified experimentally, that the linearity of the two plasmids is similar (i.e., half DNA transfected = half signal).

The most used normalizers are based on expression of the lacZ (encoding for beta-galactosidase) or Renilla luciferase (Rluc) cDNAs under the control of constitutive and strong promoters such as CMV, EF1a, SV40, or actin. The reasonable assumption is that these promoters will be always expressed at a comparable level, independently of the experimental condition, such that their activity can be used to measure transfection efficiency. Initial testing of different normalizer promoters and/or backbones is recommended because some cell lines, for example, MCF10A that are notoriously considered as hard to transfect, express very poorly cDNAs contained in plasmids of the pCDNA3 family but efficiently from plasmids of the pCS2 or pRK5 family, perhaps owing to the SV40 polyadenylation sequence in the latter.

The use of beta-galactosidase ad normalizer is limited to cell lines in which it is possible to attain enough expression to be faithfully detected by colorimetric methods over endogenous beta-galactosidase activity. Typically, when using CPRG as detection method (see below), it should be desirable to obtain a good signal (i.e., >0.2 OD over the non-transfected samples) within 1–2 h of incubation at 37 °C of the reaction. To obtain a better signal, several strategies can be employed: change normalizer plasmid; increase the amount of normalizer in the mix; increase the total number of cells per sample; and concentrate the lysis buffer or use a higher volume of lysate. Care should be taken to remain in the linear range of the detection reagent (for CPRG, typically up to 2.5OD). The use of Rluc is sometimes preferred because it enables sequential detection of Firefly (Fluc, i.e., standard) and Renilla luciferases with special assay buffers in the same vessel and because it can represent a more sensitive normalizer compared to beta-galactosidase. In any case, since the endogenous or basal signal (i.e., of mock-transfected cells) of the normalization reaction can be different from zero, the basal signal should be subtracted from that of the transfected samples before normalizing data, to avoid introduction of a systematic error.

The choice on the normalizer raises the thorny issue of what is the most appropriate normalizer plasmid and promoter to be used.

Apart from technical considerations and experimental constraints discussed above, in principle one should initially test different normalizers in parallel and use an established treatment known to inhibit YAP/TAZ to choose the best normalizer (*see* Subheading 4). In the case of YAP/TAZ, normalizers based on SV40 promoter should be avoided because this promoter contains TEAD-responsive GTIIC sites [14]. This should also raise general caution with the use of SV40-driven expression plasmids in experiments entailing the modulation of YAP/TAZ activity.

An additional control that we recommend including is the use of an "empty" luciferase plasmid, i.e., a version of the "experimental" YAP-/TAZ-responsive luciferase plasmid devoid of the TEAD-responsive element(s), with the same normalizer. The idea is that in the absence of TEAD-binding elements, not only the normalizer but also the luciferase signal should become insensitive to YAP/-TAZ activity. Several approximations of this control plasmid can be used: a generic luciferase backbone without any promoter (e.g., Promega's pGL3 basic); the very same backbone of the "experimental" plasmid devoid of any promoter; the "experimental" plasmid bearing only the minimal promoter but not the multimeric TEAD-binding sites; and the "experimental" plasmid bearing selected DNA base substitutions or deletions to prevent TEAD recognition. These controls should be made because even an "empty" luciferase backbone in reality harbors multiple potential TF binding sites capable of driving luciferase expression in absence of any promoter. Practically, a good rule should be that the signal of the promoter-less luciferase reporter is at least one or two orders of magnitude (i.e., 10× or 100×) less than the "experimental" reporter and that the effect of the treatment is observed with the "experimental" reporter.

3 The Choice of the YAP/TAZ Reporter Plasmid

Three types of YAP-/TAZ-responsive luciferase reporters have been used so far:

1. Reporters based on a natural promoter region of a YAP/TAZ target gene, spanning some hundreds of base pairs upstream of the transcriptional start site. Examples for these reporters are those based on the promoter elements of *CTGF*, *ANKRD1*, or *RHAMM/HMMR* [5, 13, 15]. A similar kind of reporters has been used in transgenic flies to visualize at a tissue level the activity of the Yorkie YAP/TAZ homolog, based on the promoter of *Expanded* or *DIAP1* driving expression of lacZ. The regulation of these promoter elements is far less complex compared to the endogenous gene, but the existence of multiple transcription factor binding sites other than TEADs

complicates the analysis of the results (i.e., the promoter might be regulated through another transcription factor than YAP/-TAZ). On the other hand, the natural arrangement of TF binding sites enables a precise analysis of cooperating TFs, for example, AP1 and TEAD [5].

2. Reporters bearing synthetic multimerized TEAD-binding elements cloned upstream of a natural minimal promoter. This kind of reporters has been widely used by the signaling community to follow specifically the activity of a pathway, including, among others, TGF-beta, BMP, Notch, Shh, and Wnt. These reporters are in principle more specific for a single TF complex and usually provide much stronger TF-driven luciferase levels compared to natural promoters, likely owing to the more efficient/quantitative recruitment of the TF to the multimerized binding site. One of the most used synthetic reporters for YAP/TAZ is the 8XGTIIC-luciferase [16] (deposited in Addgene #34615), obtained from the 4XGTIIC-luciferase [17]. Other versions have been developed [8, 18].
3. Reporters bearing multimerizedGAL4-binding elements (UAS), to be used in conjunction with GAL4 DNA-binding domain fusions. In the Hippo community, this has been used either based on a YAP/YKI-GAL4 or on a TEAD-GAL4 fusion [9–11, 19, 20]. In this case, the promoter element is completely heterologous to the natural DNA context in which YAP/TAZ binds TEAD, and the assay ultimately measures the strength of the protein-protein interaction (with GAL4-TEAD) and/or the nuclear localization of YAP/TAZ (with GAL4-YAP), plus their ability to recruit the transcriptional machinery to the promoter.
4. Reporters based on alternative DNA-binding platforms for YAP/TAZ. Some reports indicate that YAP/TAZ can bind other TFs than TEADs, for example, RUNX2 or TBX5 [21, 22].

4 Positive Controls for Inhibition of YAP/TAZ in Your Luciferase Assay

A good practice when testing a novel regulatory mechanism is to compare its effects with other established mechanisms; this makes the experimental results more trustable and provides a semiquantitative idea of the overall strength of the new mechanism compared to the others.

Since the endogenous basal activity of YAP/TAZ is high in most cell lines owing to low cell confluence and to the high stiffness of the substrate [16, 23], upon transfection of YAP/TAZ luciferase reporters the signal can quickly become very high. Thus, treatments aimed at inhibiting YAP/TAZ activity should be either started as soon as possible (e.g., by co-transfecting the expression plasmid encoding for the YAP-inhibitory protein) or prolonged for sufficient time to enable degradation of the preexisting luciferase (in our experience, treating cells for 24 h starting on the day after DNA transfection is a good starting point). In alternative, for short treatments it is possible to use either a combination with the doxycycline-inducible system [13] or to modify existing YAP/TAZ reporters with unstable forms of luciferase, by adding a PEST protein degradation signal and/or an mRNA destabilizing sequence.

4.1 Small-Molecule Inhibitors

So far, one of the easiest and universal treatments that leads to fast and quantitative YAP/TAZ inhibition in several cell lines is disassembly of the F-actin cytoskeleton by latrunculin A [9, 16, 24, 25]. Care should be taken to use several dilutions to obtain a quantitative effect (cells should completely round up but remain attached to the dish within 20 min of the treatment), without otherwise compromising cell viability (this should be checked at minimum by visual inspection of floating cells before harvesting). Another widespread YAP-/TAZ-inhibitory treatment is activation of cAMP-dependent protein kinase (PKA) by treatment with forskolin, IBMX, or even better with an equimolar mixture of the two, leading to YAP/TAZ inhibition [26–28]. Other molecules that have been shown to inhibit YAP/TAZ by luciferase assays are statins (inhibitors of HMGCR) and verteporfin [10, 29], although in our hands the inhibition by these compounds is comparatively weaker (unpublished data).

4.2 YAP/TAZ siRNA

Looking for a positive control of YAP/TAZ inhibition, a good approach is the use of cells transfected with YAP/TAZ siRNAs on the day before DNA transfection, which gives very strong down-regulations [13, 16]. In most cell lines tested so far in our hands, it is required to use a combination of both YAP and TAZ siRNAs to achieve a quantitative inhibition, reflecting the redundant role of the two factors for most phenotypes [30]. Cells transfected with an unrelated control siRNA should be used in parallel.

4.3 Overexpression and Activation of Hippo Pathway Components

In principle, activation of the Hippo pathway should be the most established and potent trigger to inhibit YAP/TAZ activity. The overexpression of just one component of the pathway, such as LATS or MST, has weak effects of YAP/TAZ luciferase activity in many cell lines, such that sometimes a combination of LATS and MST has been used. A better YAP/TAZ inhibition can be achieved in cell lines genetically deficient of NF2/merlin, such as schwannoma cell lines or MDA-MB-231 cells, by re-expression of NF2. As discussed below, this system based on NF2 expression is one of the few faithful LATS-dependent assay available in mammalian cell cultures so far, while MST1-/MST2-dependent assays are still lacking.

5 Positive Controls for Activation of YAP/TAZ in Luciferase Assays

If one desires positive controls for YAP/TAZ activation, to gauge the entire dynamic range of the luciferase assay, the easiest option is to co-transfect a plasmid encoding for a constitutive-activated form of YAP (YAP 5SA) or TAZ (TAZ 4SA). These cDNAs bear point mutations in the established LATS phosphorylation sites [23, 31], which enables the encoded protein to accumulate in the nucleus upon overexpression, at difference with wild-type YAP or TAZ that when overexpressed are mainly localized in the cytoplasm (personal observation). An alternative, depending on the cell type, can be the use of cells transfected with LATS1/LATS2 siRNAs or the expression of the PAR2/PAR3/PAR4 kinases, also known as MARK2/MARK3/MARK4) [12, 32] (and our unpublished observations), although the mechanism by which these control YAP/TAZ activity is not entirely understood [8].

6 A Cautionary Note on the Use of siRNAs in Luciferase Assays

RNA interference is an extremely powerful tool to study gene function. It is however well known that transfection of siRNAs or shRNAs can have both specific and non-specific effects, such that their use in combination with luciferase assays requires a set of dedicated experimental controls. One issue is how to define a good negative control siRNA sequence, because it is common experience that different "nontargeting" siRNAs can have different effects on luciferase expression. This can be ascribed mainly to two effects: (a) if transfection of siRNAs is done before DNA transfection (and with a separate reagent), which is convenient because the reporter will be transfected in cells already depleted of the YAP regulatory protein, it sometimes happens that some siRNAs affect overall DNA transfection efficiency and/or the expression of the normalizer plasmid, which complicates the analysis; (b) siRNAs, including controls, can have multiple off-target effects. Thus, when setting up such type of experiments, it would be advisable to use different control siRNA sequences and to include positive controls (e.g., YAP/TAZ and LATS1/LATS2 siRNAs) to experimentally determine what is the best setup. Of course any knockdown of a new gene should be presented by using two single independent siRNAs of known sequence, and/or data based on one single siRNA should be rescued by expression of a siRNA-insensitive isoform of the gene cDNA.

7 Reconstitution Experiments to Assay the Mechanism Underlying YAP/TAZ Regulation(s)

The number of upstream YAP/TAZ regulatory inputs is ever expanding and ranges from the Hippo pathway to mechanical and cytoskeletal cues, from Wnt to GPCR signaling, from PKA to the mevalonate and glycolysis pathways, and counting. For several of these inputs, it remains unclear whether regulation of YAP/TAZ works uniquely through phosphorylation by LATS1/LATS2 kinases or whether other mechanisms are involved. Exclusion of YAP/TAZ from the nucleus, rescued by LATS1/LATS2 depletion, is a strong argument for LATS involvement, but since nuclear YAP can be inactive, this should be always accompanied by evidence for rescued YAP/TAZ co-transcriptional activity. Curiously, despite genetic systems are available including LATS1/LATS2 KO MEFs or LATS1/LATS2 CRISPR cell lines [26, 33], this latter evidence is still lacking for many supposed LATS-dependent inputs.

7.1 Reconstitution with Mutant Isoforms of YAP/TAZ

To answer this specific question, we developed an experimental system based on depletion of endogenous YAP/TAZ followed by reconstitution with plasmids encoding for siRNA-insensitive YAP or TAZ cDNAs, co-transfected with the luciferase reporter plasmid. This has been done by direct mutagenesis of the human YAP cDNA or by using a mouse TAZ cDNA that is naturally insensitive to human TAZ siRNAs. In this way it is possible to compare selectively the transfected isoforms of YAP or TAZ, without confounding effects due to the parallel regulation of endogenous YAP/TAZ. For example, we used this system to carefully dissect the relevance of LATS phosphorylation sites for the functional inhibition of YAP/TAZ: while both NF2 overexpression and RHO inhibition are unable to inhibit reconstituted YAP 5SA or TAZ 4SA, inhibition of F-actin by latrunculin A or culturing cells on a soft ECM is still able to inhibit them [16, 29, 34]. Care should be taken when using the YAP 5SA available in Addgene #27371 because it contains three additional serine/threonine mutations other than the LATS1/LATS2 sites.

In principle, this system can be easily used to test the importance of other posttranslational modifications of YAP/TAZ or of their protein domains (SH3-binding, WW). For example, the PDZ-binding mutant is completely inactive in reconstitution assays, in keeping with early evidence [20].

7.2 Reconstitution with NF2/Merlin

A similar experimental setup can be used to test NF2-/LATS-specific regulation of YAP/TAZ based on functional assays. The basis for this assay was the serendipitous discovery that MDA-MB-231 cells bear inactivating mutations in the NF2/merlin gene, such that co-transfection of NF2-expressing plasmids (e.g., Addgene #19701) very efficiently inhibits YAP/TAZ activity in luciferase assays [16, 34]. This inhibition is completely dependent on endogenous LATS1/LATS2 activity, because transfection of two different siRNA mixes targeting both LATS1 and LATS2 abrogated NF2-induced 8XGTIIC-luc inhibition [29, 34]. Moreover, this also completely depends on LATS phosphorylation sites on YAP/-TAZ, because reconstitution of cells with YAP 5SA or TAZ 4SA (Subheading 7.1) makes 8XGTIIC-luc activity insensitive to NF2 expression. Thus, NF2 expression in MDA-MB-231 cells can be used as a bona fide LATS-specific regulation of YAP/TAZ activity, which may be used in the future to gauge the activity and specificity of other LATS regulatory inputs.

8 Outline of the Setup of Luciferase Experiments

8.1 Luciferase Experiment Involving Treatments with Small Molecules

Day 0: plate the cells in 24-well format. Seed a proper number of cells to avoid overcrowding during the final phases of the experiment, as this may inhibit YAP/TAZ and/or introduce sample-to-sample variability (e.g., one sample reaches confluence, while another remains sparse).

Day 1: transfect DNA. Change medium if/when necessary.

Day 2: start small-molecule treatments.

Day 3: harvest cells (24 h treatment).

In this regard, it is important to note that with several transfection reagents, it is recommended to use a standard ratio between cells/DNA/reagent. In this case, we usually keep the amount of reporter and normalizer equal for each sample, add other expression plasmids where needed, and finally use a filler DNA (pBluescript) to obtain the same total amount of DNA per sample.

A typical DNA transfection mix for MDA-MB-231 cells performed with LT1 transfection reagent (MirusBio) is:

8XGTIIC-luciferase	50 ng/cm^2
CMV-lacZ	125 ng/cm^2
pBluescript	Fill to 375 ng/cm^2

The amount of CMV-lacZ can be adjusted depending on how strongly it is expressed (e.g., in HEK293T cells, it is sufficient to use 25 ng/cm^2). It is also possible to save the remaining extracts and to use them for Western blotting to check for expression of the co-transfected epitope-tagged proteins. In this case, load according to normalizer expression and not to total protein content to account for sample-to-sample differences in transfection efficiency.

8.2 Luciferase Experiment with siRNA Transfection

Day 0: plate the cells in 24-well format. Seed cells at the ideal confluence to obtain an efficient siRNA transfection (in our hands, a low confluence is best).

Day 1: transfect siRNA.

Day 2: renew medium and transfect DNA. Change medium if/when necessary.

Day 3/4: harvest cells.

It is also possible, depending on the siRNA transfection reagent/efficiency, to perform siRNA transfection on the morning of Day 1 and DNA transfection on the evening of the same day. This will save one day. In alternative, if siRNA transfection requires high cell confluence, it is also possible to replate cells (Subheading 8.3).

8.3 Luciferase Experiment with Replating

This approach is used to transfect cells in batch, which are then replated to perform large-scale screenings (e.g., small-molecule treatments) or to study the effects of cell crowding and of the mechanical properties of the ECM (by using fibronectin-functionalized polyacrylamide hydrogels of defined stiffness). In our experience, confluence-mediated inhibition of YAP occurs in very packed cell monolayers, more concentrated than what is usually considered a uniform confluent culture [34]. Thus, when replating cells to reach high confluence, we recommend to seed cells at 120–150% compared to the number of cells forming a monolayer (100%) and to harvest after 48 h [35]. Since not all cell lines undergo efficient contact inhibition of YAP, including many cancer cells that underwent EMT [36], a parallel experiment should be run to verify efficient nuclear exclusion of YAP by immunofluorescence.

Day 0: plate cells in a large dish. Seed cells to reach the optimal density for DNA transfection on Day 1.

Day 1: transfect DNA. Change medium if/when necessary.

Day 2: trypsinize and replate cells.

Day 3/4: harvest cells.

It is also possible to perform DNA transfection on the morning of Day 1 and to replate on the evening of the same day. This will save one day, for example, to start a small-molecule treatment on Day 2, before luciferase accumulation.

9 Analysis of Data from Luciferase Assays

The output of any luciferase experiment is calculated as the ratio between the luciferase and the normalizer signals. Care should be taken when some samples display normalization levels strongly reduced or increased compared to the average of the other samples, because this may artificially boost or reduce, respectively, luciferase normalization. Reduction of the normalizer might indicate some toxic or non-specific effect due to some component of the transfection mix; enhancement might complicate the analysis because the sample could be outside of the supposed range for linearity. If co-transfecting cDNA expression plasmids, the issue is even more complicated (e.g., is the cDNA expressed more or less compared to other samples? Is the altered normalizer expression an effect of the cDNA transfection itself?). If possible, samples displaying normalization levels too low/high should be excluded from further analysis a priori.

Within a single experiment developed at once, it is trivial to compare samples. However, when repeating the whole experiment a second time at a distance, the absolute numbers might not be directly comparable, due, for example, to variations in the time of incubation of the beta-galactosidase assay reaction or to different transfection efficiencies. Still, the comparison of different experiments is important to perform statistical significance analyses on the observed differences. To overcome this problem, one possible solution is to calculate, within each experiment, the ratio between the biological replicates of each condition and the mean of the biological replicates of the control sample(s) (e.g., wells transfected with only the experimental and normalizer plasmids). In this manner, the biological replicates of the control sample(s) will always have their mean = 1, which will enable the direct comparison between different experiments. This will allow to calculate the experimental error (SEM or SD) by taking in account all the data of different experiments together.

10 Recommended Buffers and Their Use for YAP/TAZ Luciferase Assays

10.1 Lysis Buffer

25 mM Tris-HCl pH 7.5.

2.5 mM EDTA.

10% (v/v) glycerol.

1% (v/v) NP40.

2 mM DTT (freshly added just before lysis).

Briefly, our standard procedure using this lysis buffer is as follows:

1. Aspirate medium with a Pasteur pipette.
2. Add ice-cold lysis buffer on cells (100 μL/cm^2).
3. Freeze cells at −80 °C.
4. Thaw cells on ice.
5. Flush lysis buffer on cells three to four times to ensure uniform recollection of cell lysates.

6. Transfer lysates to fresh Eppendorf tubes on ice.
7. Vortex 20 s.
8. Spin 10 min >16000 rcf at 4 °C.
9. Keep tubes on ice while aliquoting the lysates in 96-well plates for luciferase and beta-galactosidase assays.

10.2 Luciferase Assay Buffer

This buffer is freshly prepared as a mix of:

½ volume 2× buffer (40 mM tricine, 2 mM $(MgCO_3)_4 Mg(OH)_2$, 5.4 mM $MgSO_4$, 0.2 mM EDTA).

½ volume 2× D(−)-luciferin (1 mM stock solution in water, stored in aliquots at −80 °C; this solution will be cloudy until it is mixed with the 2× buffer).

1/100 volume ATP (53 mM stock solution in water, stored at −20 °C).

1/100 volume coenzyme A (27 mM stock solution in water, stored at −20 °C).

1/30 volume DTT (1 M stock solution in water, stored at −20 °C).

This buffer is then added to cell lysates directly to read luciferase activity (typically, 60 μL of assay buffer to 10–40 μL of lysate). If reading luminescence in a multi-well reader, it is better to use white plastic plates to enhance the signal and to avoid interference from neighboring wells. Background luminescence in non-transfected cells is usually very low and thus negligible.

10.3 Beta-Galactosidase Assay Buffer

This buffer is freshly prepared as a mix of:

1/5 volume Z buffer (stock solution: 0.5 M sodium phosphate NaH_2PO_4/Na_2HPO_4 buffer pH = 8.0, 50 mM KCl, 5 mM $MgSO_4$).

1/40 volume CPRG (stock solution: 30 mg/mL Chlorophenol Red-β-D-galactopyranoside in water, stored at −20 °C).

Remaining volume is made of cell lysate and mQ water, depending on the transfection efficiency/amount of cells lysed (in general, we use 10–40 μL lysate and add assay buffer to a final volume of 200 μL).

We usually add ice-cold beta-galactosidase assay buffer (yellow) with a multichannel pipettor to transparent 96-well plates in which we previously aliquoted cell lysates. The reaction is then incubated at room temperature or at 37 °C until the color (red) develops within the linear range of detection. Reaction can be quantified with a spectrophotometer at 570–590 nm. Non-transfected lysates should be included to subtract background activity. If the reaction develops too fast, it can be stopped by adding a small volume of 1 M Na_2CO_3.

10.4 Renilla Luciferase Assay Buffer

1/500 volume of coelenterazine (Fluka, 2 μg/μL stock in methanol stored at −80 °C) in TBS (final concentration: 50 mM Tris-HCl pH 7.5, 150 mM NaCl).

This buffer is then added to cell lysates directly to read luciferase activity (typically, 50 μL of assay buffer to 10 μL of lysate). This buffer is not suitable for sequential Firefly/Renilla luciferase assay in the same well and thus requires two separate aliquots of the cell lysate.

References

1. Vassilev A, Kaneko KJ, Shu H, Zhao Y, DePamphilis ML (2001) TEAD/TEF transcription factors utilize the activation domain of YAP65, a Src/Yes-associated protein localized in the cytoplasm. Genes Dev 15:1229–1241. https://doi.org/10.1101/gad.888601
2. Dong J, Feldmann G, Huang J, Wu S, Zhang N, Comerford SA et al (2007) Elucidation of a universal size-control mechanism in Drosophila and mammals. Cell 130:1120–1133. https://doi.org/10.1016/j.cell.2007.07.019
3. Chen L, Chan SW, Zhang X, Walsh M, Lim CJ, Hong W et al (2010) Structural basis of YAP recognition by TEAD4 in the hippo pathway. Genes Dev 24:290–300. https://doi.org/10.1101/gad.1865310
4. Mesrouze Y, Bokhovchuk F, Meyerhofer M, Fontana P, Zimmermann C, Martin T, et al (2017) Dissection of the interaction between the intrinsically disordered YAP protein and the transcription factor TEAD. eLife 6. https://doi.org/10.7554/eLife.25068
5. Zanconato F, Forcato M, Battilana G, Azzolin L, Quaranta E, Bodega B et al (2015) Genome-wide association between YAP/TAZ/TEAD and AP-1 at enhancers drives oncogenic growth. Nat Cell Biol 17:1218–1227. https://doi.org/10.1038/ncb3216
6. Galli GG, Carrara M, Yuan W-C, Valdes-Quezada C, Gurung B, Pepe-Mooney B et al (2015) YAP drives growth by controlling transcriptional pause release from dynamic enhancers. Mol Cell 60(2):328–337. https://doi.org/10.1016/j.molcel.2015.09.001
7. Stein C, Bardet AF, Roma G, Bergling S, Clay I, Ruchti A et al (2015) YAP1 exerts its transcriptional control via TEAD-mediated activation of enhancers. PLoS Genet 11:e1005465. https://doi.org/10.1371/journal.pgen.1005465
8. Mohseni M, Sun J, Lau A, Curtis S, Goldsmith J, Fox VL et al (2014) A genetic screen identifies an LKB1-MARK signalling axis controlling the Hippo-YAP pathway. Nat Cell Biol 16:108–117. https://doi.org/10.1038/ncb2884
9. Sansores-Garcia L, Bossuyt W, Wada K-I, Yonemura S, Tao C, Sasaki H et al (2011) Modulating F-actin organization induces organ growth by affecting the Hippo pathway. EMBO J 30:2325–2335. https://doi.org/10.1038/emboj.2011.157
10. Liu-Chittenden Y, Huang B, Shim JS, Chen Q, Lee S-J, Anders RA et al (2012) Genetic and pharmacological disruption of the TEAD-YAP complex suppresses the oncogenic activity of YAP. Genes Dev 26:1300–1305. https://doi.org/10.1101/gad.192856.112
11. Ribeiro PS, Josué F, Wepf A, Wehr MC, Rinner O, Kelly G et al (2010) Combined functional genomic and proteomic approaches identify a PP2A complex as a negative regulator of Hippo signaling. Mol Cell 39:521–534. https://doi.org/10.1016/j.molcel.2010.08.002
12. Heidary Arash E, Shiban A, Song S, Attisano L (2017) MARK4 inhibits Hippo signaling to promote proliferation and migration of breast cancer cells. EMBO Rep 18:420–436. https://doi.org/10.15252/embr.201642455
13. Enzo E, Santinon G, Pocaterra A, Aragona M, Bresolin S, Forcato M et al (2015) Aerobic glycolysis tunes YAP/TAZ transcriptional activity. EMBO J 34:1349–1370. https://doi.org/10.15252/embj.201490379
14. McLean BG, Lee KS, Simpson PC, Farrance IKG (2003) Basal and alpha1-adrenergic-induced activity of minimal rat betaMHC promoters in cardiac myocytes requires multiple TEF-1 but not NFAT binding sites. J Mol Cell Cardiol 35:461–471
15. Wang Z, Wu Y, Wang H, Zhang Y, Mei L, Fang X et al (2014) Interplay of mevalonate and Hippo pathways regulates RHAMM transcription via YAP to modulate breast cancer cell motility. Proc Natl Acad Sci 111:E89–E98. https://doi.org/10.1073/pnas.1319190110

16. Dupont S, Morsut L, Aragona M, Enzo E, Giulitti S, Cordenonsi M et al (2011) Role of YAP/TAZ in mechanotransduction. Nature 474:179–183. https://doi.org/10.1038/nature10137
17. Mahoney WM, Hong J-H, Yaffe MB, Farrance IKG (2005) The transcriptional co-activator TAZ interacts differentially with transcriptional enhancer factor-1 (TEF-1) family members. Biochem J 388:217–225. https://doi.org/10.1042/BJ20041434
18. Couzens AL, Knight JDR, Kean MJ, Teo G, Weiss A, Dunham WH et al (2013) Protein interaction network of the mammalian Hippo pathway reveals mechanisms of kinase-phosphatase interactions. Sci Signal 6: rs15–rs15. https://doi.org/10.1126/scisignal.2004712
19. Zhao B, Ye X, Yu J, Li L, Li W, Li S et al (2008) TEAD mediates YAP-dependent gene induction and growth control. Genes Dev 22:1962–1971. https://doi.org/10.1101/gad.1664408
20. Kanai F, Marignani PA, Sarbassova D, Yagi R, Hall RA, Donowitz M et al (2000) TAZ: a novel transcriptional co-activator regulated by interactions with 14-3-3 and PDZ domain proteins. EMBO J 19:6778–6791. https://doi.org/10.1093/emboj/19.24.6778
21. Hong J-H, Hwang ES, McManus MT, Amsterdam A, Tian Y, Kalmukova R et al (2005) TAZ, a transcriptional modulator of mesenchymal stem cell differentiation. Science 309:1074–1078. https://doi.org/10.1126/science.1110955
22. Murakami M, Nakagawa M, Olson EN, Nakagawa O (2005) A WW domain protein TAZ is a critical coactivator for TBX5, a transcription factor implicated in Holt-Oram syndrome. Proc Natl Acad Sci U S A 102:18034–18039. https://doi.org/10.1073/pnas.0509109102
23. Zhao B, Wei X, Li W, Udan RS, Yang Q, Kim J et al (2007) Inactivation of YAP oncoprotein by the Hippo pathway is involved in cell contact inhibition and tissue growth control. Genes Dev 21:2747–2761. https://doi.org/10.1101/gad.1602907
24. Wada K-I, Itoga K, Okano T, Yonemura S, Sasaki H (2011) Hippo pathway regulation by cell morphology and stress fibers. Development 138:3907–3914. https://doi.org/10.1242/dev.070987
25. Zhao B, Li L, Wang L, Wang C-Y, Yu J, Guan K-L (2012) Cell detachment activates the Hippo pathway via cytoskeleton reorganization to induce anoikis. Genes Dev 26:54–68. https://doi.org/10.1101/gad.173435.111
26. Kim M, Kim M, Lee S, Kuninaka S, Saya H, Lee H et al (2013) cAMP/PKA signalling reinforces the LATS-YAP pathway to fully suppress YAP in response to actin cytoskeletal changes. EMBO J 32:1543–1555. https://doi.org/10.1038/emboj.2013.102
27. Yu F-X, Zhang Y, Park HW, Jewell JL, Chen Q, Deng Y et al (2013) Protein kinase A activates the Hippo pathway to modulate cell proliferation and differentiation. Genes Dev 27:1223–1232. https://doi.org/10.1101/gad.219402.113
28. Iglesias-Bartolome R, Torres D, Marone R, Feng X, Martin D, Simaan M et al (2015) Inactivation of a Gα(s)-PKA tumour suppressor pathway in skin stem cells initiates basal-cell carcinogenesis. Nat Cell Biol 17:793–803. https://doi.org/10.1038/ncb3164
29. Sorrentino G, Ruggeri N, Specchia V, Cordenonsi M, Mano M, Dupont S et al (2014) Metabolic control of YAP and TAZ by the mevalonate pathway. Nat Cell Biol 16:357–366. https://doi.org/10.1038/ncb2936
30. Piccolo S, Dupont S, Cordenonsi M (2014) The biology of YAP/TAZ: hippo signaling and beyond. Physiol Rev 94:1287–1312. https://doi.org/10.1152/physrev.00005.2014
31. Liu C-Y, Zha Z-Y, Zhou X, Zhang H, Huang W, Zhao D et al (2010) The hippo tumor pathway promotes TAZ degradation by phosphorylating a phosphodegron and recruiting the SCF{beta}-TrCP E3 ligase. J Biol Chem 285:37159–37169. https://doi.org/10.1074/jbc.M110.152942
32. Kwan J, Sczaniecka A, Arash EH, Nguyen L, Chen C-C, Ratkovic S et al (2016) DLG5 connects cell polarity and Hippo signaling protein networks by linking PAR-1 with MST1/2. Genes Dev 30:2696–2709. https://doi.org/10.1101/gad.284539.116
33. Meng Z, Moroishi T, Mottier-Pavie V, Plouffe SW, Hansen CG, Hong AW et al (2015) MAP4K family kinases act in parallel to MST1/2 to activate LATS1/2 in the Hippo pathway. Nat Commun 6:8357. https://doi.org/10.1038/ncomms9357
34. Aragona M, Panciera T, Manfrin A, Giulitti S, Michielin F, Elvassore N et al (2013) A mechanical checkpoint controls multicellular growth through YAP/TAZ regulation by actin-processing factors. Cell 154:1047–1059. https://doi.org/10.1016/j.cell.2013.07.042

35. Azzolin L, Zanconato F, Bresolin S, Forcato M, Basso G, Bicciato S et al (2012) Role of TAZ as mediator of Wnt signaling. Cell 151:1443–1456. https://doi.org/10.1016/j.cell.2012.11.027

36. Cordenonsi M, Zanconato F, Azzolin L, Forcato M, Rosato A, Frasson C et al (2011) The Hippo transducer TAZ confers cancer stem cell-related traits on breast cancer cells. Cell 147:759–772. https://doi.org/10.1016/j.cell.2011.09.048

Chapter 12

Quantitative Real-Time PCR to Measure YAP/TAZ Activity in Human Cells

Xiaolei Cao and Bin Zhao

Abstract

Transcription coactivators Yes-associated protein (YAP) and transcriptional coactivator with PDZ-binding motif (TAZ, also known as WWTR1) are homologs of the *Drosophila* Yorkie (Yki) protein and are major downstream effectors of the evolutionarily conserved Hippo pathway. YAP/TAZ play critical roles in regulation of cell proliferation, apoptosis, and stemness, thus mediate functions of the Hippo pathway in organ size control and tumorigenesis. The Hippo pathway inhibits YAP/TAZ through phosphorylation, which leads to YAP/TAZ cytoplasmic retention and degradation. Dephosphorylated and nuclear-localized YAP/TAZ bind to transcription factors, especially the TEAD family proteins, thus transactivate the expression of specific genes. Therefore, measuring the expression level of YAP/TAZ target genes is a critical approach to assess Hippo pathway activity. Through gene expression profiling in different tissues and cells using techniques such as microarray and RNA-seq, many target genes of YAP/TAZ have been identified. Some of these genes were confirmed to be direct YAP/TAZ targets by chromatin immunoprecipitation (ChIP)-PCR or ChIP-seq. These works made it possible to quickly determine YAP/TAZ activity by measuring the mRNA levels of several YAP/TAZ target genes, such as *CTGF*, *CYR61*, and *miR-130a* by quantitative real-time PCR (qPCR). In this chapter, we demonstrate the use of qPCR to measure YAP/-TAZ activity in MCF10A cells.

Key words YAP, TAZ, Hippo pathway, Quantitative real-time PCR, CTGF, CYR61, miR-130a

1 Introduction

YAP cDNA was first cloned by screening expression library for proteins that could associate with SH3 domains of Yes and Src tyrosine kinases [1]. Besides the SH3-binding motif, YAP contains WW domains (one in the YAP1 isoform and two in the YAP2 isoform), a PDZ domain-binding motif, an N-terminal TEAD-binding domain (TBD), and a C-terminal transcriptional activation domain (TAD) [2]. Through a yeast two-hybrid screen, the Pan lab found that the *Drosophila* YAP/TAZ homolog Yki associates with the Hippo pathway kinase Wts (homolog of mammalian LATS1/LATS2) [3]. YAP/TAZ/Yki were then demonstrated to be critical effectors of the Hippo pathway [4, 5]. Activation of the Hippo

Alexander Hergovich (ed.), *The Hippo Pathway: Methods and Protocols*, Methods in Molecular Biology, vol. 1893, https://doi.org/10.1007/978-1-4939-8910-2_12, © Springer Science+Business Media, LLC, part of Springer Nature 2019

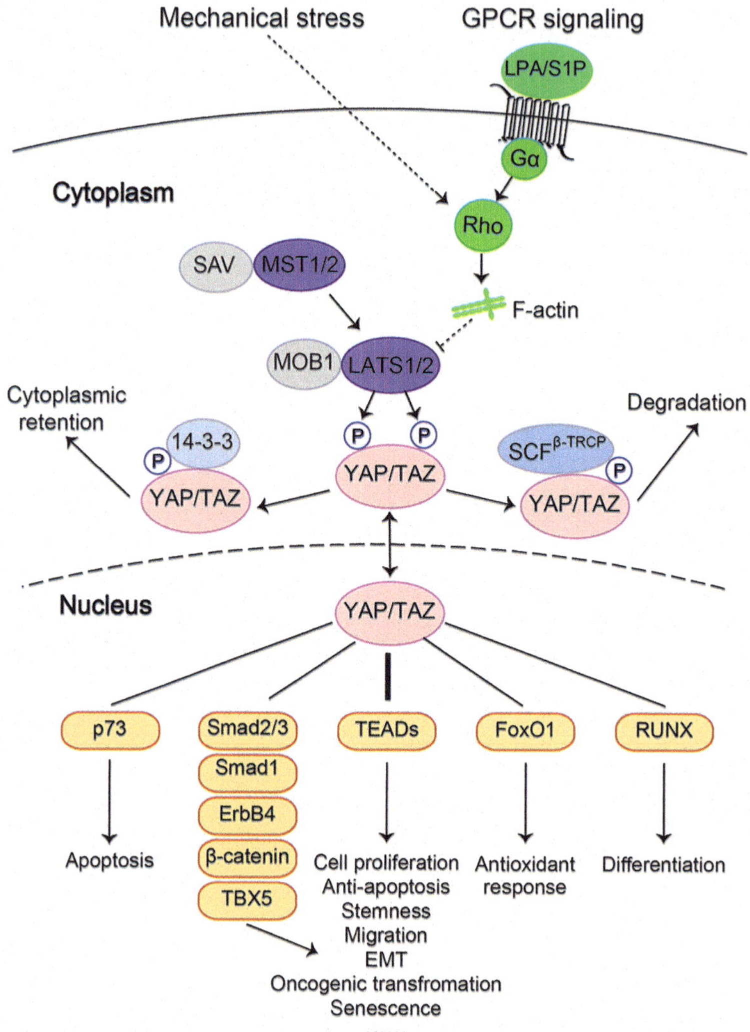

Fig. 1 The Hippo pathway regulates gene transcription through inhibition of YAP/TAZ. Mechanisms of YAP/TAZ regulation by the Hippo pathway as well as mechanisms of YAP/TAZ in regulation of various cellular activities through specific transcription factors are illustrated

pathway leads to YAP phosphorylation by LATS1/LATS2 on multiple residues [4, 5], including serine 127 (S127) and S381, which mediates interaction with 14-3-3 and $SCF^{\beta\text{-TRCP}}$, respectively [6]. As a result, YAP is sequestered in the cytoplasm and degraded (Fig. 1). TAZ is the paralog of YAP sharing similar structure, regulation, and functions with YAP [7–9]. YAP/TAZ play an evolutionarily conserved role in organ size control [10]. For example, transgenic expression of *YAP* or knockout of the Hippo pathway kinases *MST1/MST2* in mouse liver leads to drastic and reversible hepatomegaly [4, 11, 12]. However, sustained YAP activation leads to massive liver tumorigenesis [4]. Importantly, deregulation of YAP/TAZ and the Hippo pathway is also widely observed in different types of human cancers [10]. In consistent with the potent activity of YAP/TAZ in promoting stemness and dedifferentiation, they were also found playing critical roles in regeneration of tissues such as the myocardium and the intestinal epithelium [10].

YAP/TAZ are transcription coactivators without DNA-binding ability. They bind to specific genomic regions through interaction with transcription factors. Using affinity purification, transcription factor library screening, and *Drosophila* genetic approaches, TEAD family transcription factors (TEAD1, TEAD2, TEAD3, TEAD4, homologs of the *Drosophila* scalloped) were identified as key partners of YAP/TAZ in mediating tissue overgrowth and organ size enlargement [13–17]. Importantly, keratinocyte- or cardiomyocyte-specific knock-in of a point mutation in YAP that abolishes interaction with TEAD phenocopies YAP knockout. This result highlights a major role of TEAD in YAP-induced tissue growth and stemness. Nevertheless, YAP/TAZ could also interact with other transcription factors such as Smad1, RUNX, ErbB4 cytoplasmic domain, p73, TBX5, β-catenin, and FoxO1 [2, 18–23], mainly through their WW domains to activate or repress target gene expression (Fig. 1). YAP binds phosphorylated Smad1 and promotes its transcriptional activity, which is required for BMP-dependent suppression of neural differentiation of mouse embryonic stem cells [18]. Additionally, it has also been reported that TAZ interacts with Smad2/Smad3 through the coiled-coil domain and this interaction is believed to dictate the subcellular localization of Smad2/Smad3 [24, 25]. Through interaction with various transcription factors, YAP/TAZ regulate gene expression, thus controlling cell proliferation, apoptosis, stemness, antioxidant response, and other cellular activities related to organ development and cancer.

Gene expression profiling studies using microarray and RNA-seq have revealed many target genes of the Hippo pathway [4, 16]. Genes directly induced by YAP/TAZ have also been demonstrated using ChIP-seq [26–28]. In *Drosophila*, Yki induces *diap1* [14, 15], which inhibits apoptosis, and *cycE* [29] and *E2F1* [13], two genes that are involved in cell-autonomous regulation of cell proliferation. Yki also induces transcription of *dMyc* [30, 31], a potent promoter of ribosome biogenesis and cell growth. Furthermore, Yki induces EGFR ligands *Vein*, *Keren*, and *Spitz* [32, 33] and Jak-Stat pathway ligands *Unpaired1/Unpaired2/Unpaired3* [32, 34–36], which may mediate non-cell-autonomous functions of the Hippo pathway. In addition, Yki also induces expression of Hippo pathway components such as *Ex*, *Kibra*, *Crb*, and *Fj* [37–40] and thus may establish a feedback loop. Mammalian

YAP/TAZ activate a largely different set of target genes, although a few of them were conserved from *Drosophila*. For example, *BIRC5*, a mammalian homolog of *diap1*, and *Myc* are also induced by YAP/TAZ [4]. YAP also induces genes encoding secreted factors such as *CTGF*, *CYR61*, and *AREG* [16, 33, 41], which are important for YAP-induced cell proliferation and anchorage-independent growth. In addition, it was recently reported that YAP induces *Ccl2* and *Csf1* [42], which mediate recruitment of type 2 macrophages by tumor-initiating cells in mouse liver. YAP also induces a negative feedback loop through induction of upstream negative regulators *LATS2*, *AMOTL2*, and *NF2* in mammalian cells [43, 44]. Noteworthy, these genes are different from those in the *Drosophila* negative feedback loop induced by Yki.

Besides protein-coding genes, Yki/YAP/TAZ also induces expression of microRNAs (miRNAs). *bantam* is a *Drosophila* miRNA induced by Yki and partially mediates functions of the Hippo pathway in growth control [45, 46]. However, *bantam* is not conserved in mammalian cells. YAP induces miR-130a, which represses translation of VGLL4, a cofactor competing with YAP for TEAD binding, thus establishing a positive feedback loop [47]. Interestingly, *bantam* functionally mimics miR-130a by inhibiting SdBP/Tgi, the *Drosophila* homolog of VGLL4, hence confirms an evolutionarily conserved positive feedback mechanism [47, 48]. Furthermore, YAP induces miR-29 to inhibit translation of tumor suppressor PTEN, a negative regulator of mTOR, thereby activates mTOR and promotes cell growth [48]. In addition, TAZ promotes proliferation and migration of osteosarcoma cells through inducing miR-224 that targets TGF-β effector Smad4 [49]. Table 1 summarizes YAP/TAZ target genes including miRNAs.

Noteworthy, target genes induced by YAP/TAZ vary in a tissue/cell type-dependent manner (Table 1). For example, while *CTGF* and *CYR61* were found induced by YAP/TAZ in multiple tissues and cells, *Nanog* and *Oct4* [50, 51] are induced specifically in mouse embryonic stem cells. Therefore, it is critical to make sure that target genes used as a readout of YAP/TAZ activity are relevant in the tissue/cell context under investigation.

Quantitative real-time PCR (qPCR) is a convenient and sensitive method for routine examination of YAP/TAZ activity in mammalian cells. In this chapter we use well-established YAP/TAZ-TEAD target genes *CTGF* and *CYR61* as well as miR-130a as examples (*see* **Note 1**) to demonstrate how qPCR could be used to examine YAP/TAZ activity in MCF10A cells. This protocol falls into two parts covering qPCR examination of regular and miRNA genes, respectively.

Table 1
YAP/TAZ target transcription factors and target genes

Transcription factors	Target genes	Tissues/cells	Functions	References
TEADs	*CTGF*	Multiple tissues and mammalian cell lines	Cell proliferation, migration and metastasis	[16, 17]
	CYR61	Multiple tissues and mammalian cell lines	Cell proliferation	[52]
	c-Myc	Mouse liver and multiple human renal cell carcinoma (RCC) cell lines	Cell proliferation, migration and metastasis	[4, 53]
	EDN1/EDN2	Multiple human RCC cell lines and A549, HepG2 cell lines	Cell proliferation, migration and metastasis	[54]
	RHAMM	The human breast cancer cell line MDA-MB-231	Cell proliferation, migration and metastasis	[55]
	Axl	The immortalized non-tumorigenic hepatocyte cell line MIHA and the primary HCC cell line PLC/PRF/5	Cell proliferation, metastasis	[56]
	CDK6	The human fibroblast cells cell line IMR90	Senescence	[57]
	BCL2, BIRC5	Mouse liver and HPNE cell line	Anti-apoptosis	[4]
	Oct4, Nanog	Mouse embryonic stem cells	Stemness	[50, 51]
	Cdx2, Gata3	Mouse embryonic stem cells	Trophoblast differentiation	[58–60]
	Pax3	Embryonic neural cells	Embryogenesis	[8, 61, 62]
	SOX9	Non-transformed cell types of gastrointestinal origin, including primary esophageal epithelium cells, immortalized embryonic liver cells as well as esophageal cancer cells	Cancer stem cell maintenance	[63]
	ZEB2	Mouse lung adenocarcinoma	Barrier for lung cancer cell fate conversion and regulation of cancer plasticity	[64]
	LATS2, AMOTL2, NF2	Multiple mammalian cell lines	Negative feedback	[43, 44]
	Ccl2, Csf1	Mouse liver	Recruitment of type 2 macrophages	[42]
	miR-130a	Mouse liver and MCF10A, HMLE and HepG2 cell lines	Positive feedback	[47]
	miR-224	Osteosarcoma cells	Osteosarcoma cell proliferation, migration	[49]
	miR-29	Mouse intestine and MCF10A cell line	Cell proliferation and cell growth	[48]
Unknown	*miR-206*	Skeletal muscle and cardiomyocyte	Cardiomyocyte survival and hypertrophy	[65]
Smad1	*Id1/Id2/Id3*	Mouse embryonic stem cells	Stemness	[18]
Smad2/Smad3-TEADs	*Snail, Twist1, Slug, NEGR1, UCA1, CTGF*	Mouse atrioventricular valves and the human breast cancer cell line MDA-MB-231	Cell proliferation and endothelial to mesenchymal transition	[24, 25, 66, 67]
Cytoplasmic domain of ErbB4	*CTGF*	MCF10A and human breast cancer cell line T47D	Cell proliferation	[19, 68]

(continued)

Table 1
(continued)

Transcription factors	Target genes	Tissues/cells	Functions	References
Unknown	*AREG*	MCF10A cell line	Cell proliferation, cell migration	[33]
RUNX	*osteocalcin*	MSCs	Osteoblast differentiation	[2, 69]
p73	*PML, Bax, puma*	Human colon carcinoma cell line HCT116, mouse embryo fibroblasts MEFs, human breast cancer cell line SKBR3 and MCF7	Apoptosis	[20, 70, 71]
TBX5	*ANF, BCL2L1, BIRC5*	Mouse cardiomyocytes	Anti-apoptosis and cardiac development	[21, 72]
β-catenin	*Sox2, Snail2*	Mouse heart	Cardiomyocyte proliferation	[22]
FoxO1	*catalase, MnSOD*	Mouse heart	Antioxidant	[23]

2 Materials

1. Human mammary epithelial cell line MCF10A.
2. MCF10A growth medium (DMEM/F12 supplemented with 5% horse serum, 20 ng/mL EGF, 0.5 μg/mL hydrocortisone, 10 μg/mL insulin, 100 ng/mL cholera toxin, and 50 μg/mL penicillin/streptomycin).
3. Fetal bovine serum (FBS).
4. TRIzol (Thermo Fisher Scientific).
5. Chloroform (100%).
6. Ethanol (70% solution in nuclease-free water).
7. Isopropanol (100%).
8. Nuclease-free water.
9. cDNA synthesis kit (e.g., from Takara).
10. Primer stock solutions (1.25 μM solutions of primers in nuclease-free water) (Table 2).
11. SYBR green qPCR mix (e.g., from KAPA Biosystems).
12. mirVana miRNA isolation kit (e.g., from Thermo Fisher Scientific).
13. TaqMan MicroRNA Reverse Transcription Kit (e.g., from Thermo Fisher Scientific).
14. Two TaqMan MicroRNA Assays (*see* **Note 2**) for Hsa-miR-130a and U6 snRNA including miRNA-specific RT primers (Table 2).

Table 2
miRNA, control U6 snRNA, and primer sequences used for qPCR

Name	Sequence
Human *HPRT* forward	AGCCCTGGCGTCGTGATTA
Human *HPRT* reverse	ACAATGTGATGGCCTCCCA
Human *CTGF* forward	CCAATGACAACGCCTCCTG
Human *CTGF* reverse	TGGTGCAGCCAGAAAGCTC
Human *CYR61* forward	AGCCTCGCATCCTATACAACC
Human *CYR61* reverse	TTCTTTCACAAGGCGGCACTC
Hsa-miR-130a	CAGUGCAAUGUUAAAAGGGCAU
Human U6 snRNA	GTGCTCGCTTCGGCAGCACATATACTAAAATTGGAACGA TACAGAGAAGATTAGCATGGCCCCTGCGCAAGGATGACACGCAAA TTCGTGAAGCGTTCCATATTTT

15. TaqMan Universal Master Mix II (*see* **Note 3**) (e.g., from Thermo Fisher Scientific).
16. qPCR reaction plates (e.g., from Bio-Rad).
17. Adhesive films.
18. NanoDrop spectrometer.
19. Thermo cycler.
20. Real-time PCR instrument (e.g., CFX96 from Bio-Rad).

3 Methods

3.1 Cell Culture and Stimulation

MCF10A cells are maintained in MCF10A growth medium at 37 °C. Early passage MCF10A cells are split into 6-well plates with 3×10^5 cells/well. To activate the Hippo pathway, cells are serum-starved overnight (*see* **Note 4**). After serum starvation, serum is added back for 1 h to inactivate the Hippo pathway (*see* **Note 5**). To overexpress YAP, MCF10A cells are infected by YAP-expressing lentiviral vectors. Cells were then selected in medium containing 2 μg/mL of puromycin for 3 days [47].

3.2 RNA Isolation for Regular mRNA

1. Discard the culture medium and rinse cells once with PBS. Add 1 mL of TRIzol to cells in each well of 6-well plates, pipette up and down several times (*see* **Note 6**).

2. Incubate for 5 min at room temperature (RT).
3. Transfer cell lysate to a new 1.5 mL microcentrifuge tube, and then add 200 μL chloroform (*see* **Note 7**).
4. Vortex vigorously for 15 s, and incubate at RT for 2 min.
5. Centrifuge at 12,000 × *g* for 10 min at 4 °C to separate the aqueous and organic phases.
6. Transfer the upper aqueous layer to a new 1.5 mL microcentrifuge tube (*see* **Note 8**).
7. Add 500 μL isopropanol to the sample, vortex for 15 s, and incubate at RT for 10 min.
8. Centrifuge at 12,000 × *g* for 10 min at 4 °C.
9. Discard the supernatant and wash the pellet with 500 μL 70% ethanol.
10. Centrifuge at 7500 × *g* for 5 min at 4 °C.
11. Discard the supernatant and air-dry the pellet for 5–10 min (*see* **Note 9**).
12. Dissolve the total RNA pellet with 50 μL nuclease-free water.
13. Measure the concentration of RNA sample with NanoDrop (*see* **Note 10**).

3.3 RT for Regular mRNA (see Note 11)

1. Dilute total RNA sample with nuclease-free water to 0.5 μg/μL (*see* **Note 12**).
2. Prepare the RT reaction mix. For each 10 μL reaction, mix 7 μL of nuclease-free water with 2 μL 5 × PrimeScript RT Master Mix from cDNA synthesis kit and 1 μL RNA sample (0.5 μg/μL). Remaining RNA samples should be stored at −80 °C.
3. Mix well and spin down with microcentrifuge.
4. Perform the RT reaction. Incubate the reaction mix for 15 min at 37 °C, followed by 5 s at 85 °C in the thermo cycler.

3.4 Quantitative Real-Time PCR for Regular mRNA

1. Thaw SYBR green qPCR mix on ice and protect it from light.
2. Prepare PCR reaction mix. For each 10 μL reaction, PCR reaction mix consists of 5 μL of 2× SYBR green qPCR master mix, 0.5 μL of cDNA from RT reaction product, 1 μL of 1.25 μM forward and reverse primers each, and 2.5 μL of nuclease-free water (*see* **Note 13**).
3. Mix well and spin down with microcentrifuge.
4. Transfer 10 μL of mixture with multichannel pipettes into the qPCR reaction plate (*see* **Note 14**). Reactions should be triplicated.
5. Seal the qPCR reaction plate with an optical adhesive film and centrifuge the plate at 1000 × *g* for 1 min at 4 °C to spin down contents and eliminate bubbles.

6. Perform qPCR following the instruction of instrument. Parameters are as follows. 95 °C for 3 min, followed by 40 interval cycles of 95 °C for 10 s and 60 °C for 30 s.

3.5 RNA Isolation for miRNA (see Note 15)

1. Discard the culture medium and rinse cells once with cold PBS.
2. Remove PBS and add 600 μL Lysis/Binding Solution to cells in each well of 6-well plate (*see* **Note 16**).
3. Transfer cell lysate to a new 1.5 mL microcentrifuge tube, and add 60 μL (*see* **Note 17**) of miRNA Homogenate Additive.
4. Vortex for 15 s, and incubate on ice for 10 min.
5. Add 600 μL (*see* **Note 18**) of Acid-Phenol:Chloroform and vortex for 1 min.
6. Centrifuge at 10,000 × *g* for 5 min at RT to separate the aqueous and organic phases (*see* **Note 19**).
7. Transfer the upper aqueous layer to a new 1.5 mL microcentrifuge tube (*see* **Note 20**).
8. Add 750 μL of 100% ethanol at RT (*see* **Note 21**) and vortex for 15 s.
9. Place a filter cartridge into a collection tube for each sample.
10. Transfer 700 μL of the mixture (from **step 8**) into the filter cartridge (*see* **Note 22**).
11. Centrifuge at 10,000 × *g* for 15 s at RT.
12. Discard the waste in collection tube.
13. Add 700 μL miRNA Wash Solution 1 (*see* **Note 23**) to the filter cartridge, and centrifuge at 10,000 × *g* for 15 s at RT.
14. Discard the waste in collection tube.
15. Add 500 μL Wash Solution 2/3 (*see* **Note 24**) to the filter cartridge, and centrifuge at 10,000 × *g* for 15 s at RT.
16. Repeat **steps 14** and **15** once more.
17. Centrifuge at 10,000 × *g* for 1 min at RT to remove residual fluid from the filter.
18. During centrifugation, heat up enough nuclease-free water to 95 °C.
19. Transfer the filter cartridge to a new collection tube. Add 100 μL of pre-heated nuclease-free water to the center of the filter, and close the cap.
20. Incubate for 2 min at RT.
21. Centrifuge at 10,000 × *g* for 1 min at RT to collect RNA.
22. Measure the concentration of RNA samples with NanoDrop.

3.6 RT for miRNA (see Note 25)

1. Dilute total RNA sample with nuclease-free water to 10 ng/μL.
2. Prepare the RT master mix. For each 15 μL RT reaction, add 8.16 μL of nuclease-free water, 1.5 μL of 10 × reverse transcription buffer, 1 μL of MultiScribe Reverse Transcriptase (50 U/μL), 0.15 μL of 100 mM dNTPs, and 0.19 μL of RNase inhibitor (20 U/μL) (*see* **Note 13**).
3. Add 3 μL of 5 × RT primer and 1 uL RNA sample (10 ng/μL) into the master mix (from **step 2**) to make a final 15 μL RT reaction. Store the remaining RNA samples at −80 °C.
4. Mix well and spin down the contents with microcentrifuge.
5. Perform the RT reaction. Incubate the reaction mix for 30 min at 16 °C, followed by 30 min at 42 °C and 5 min at 85 °C in a thermo cycler.

3.7 Quantitative Real-Time PCR for miRNA

1. Prepare PCR reaction mix. For each 20 μL reaction, PCR reaction mix consists of 1 μL of the specific miRNA PCR primer from TaqMan MicroRNA Assay, 1.33 μL of cDNA from RT reaction (from Subheading 3.6), 10 μL of TaqMan Universal Master Mix II and 7.67 μL of nuclease-free water (*see* **Note 13**).
2. Mix well and spin down the contents with microcentrifuge.
3. Transfer 20 μL of mixture with multichannel pipettes into a qPCR reaction plate (*see* **Note 14**).
4. Seal the qPCR reaction plate with an optical adhesive film and centrifuge the plate at 1000 × *g* for 1 min at 4 °C to spin down the contents and eliminate bubbles.
5. Perform qPCR following the instruction of instrument. Parameters are as follows: 95 °C for 10 min, followed by 40 interval cycles of 95 °C for 15 s and 60 °C for 1 min.

3.8 Data Analysis for Both Regular mRNA and miRNA

Data analysis is performed using the comparative C_T ($\Delta\Delta C_T$) method in the accompanying CFX Maestro Software (Bio-Rad). The threshold cycle (C_T) is defined as the fractional cycle number at which the fluorescence passes the fixed threshold. Values represent means ± SD from three technical repeats for one specific gene in each sample (Fig. 2). For regular mRNA, relative expression levels are normalized to human hypoxanthine phosphoribosyltransferase 1 (HPRT) mRNA. To normalize relative expression levels of miRNAs, human U6 snRNA is used.

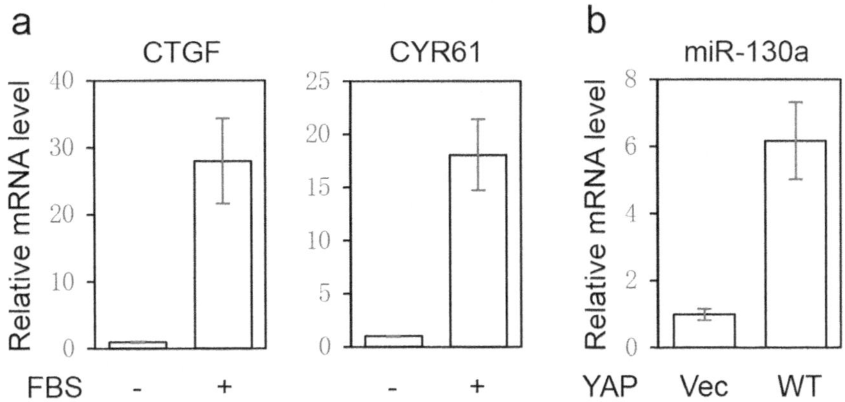

Fig. 2 Regulation of *CTGF*, *CYR61*, and *miR-130a* expression by the Hippo pathway. (**a**) Serum stimulation increases *CTGF* and *CYR61* mRNA levels. MCF10A cells were serum-starved overnight and were then treated with 10% FBS for 1 h. The expression levels of *CTGF* and *CYR61* were determined by qPCR. (**b**) Overexpression of YAP increases *miR-130a* level. MCF10A cells were infected with lentivirus expressing YAP. The expression level of *miR-130a* was determined by qPCR

4 Notes

1. It is important to choose YAP/TAZ target genes according to the specific tissue/cell context under investigation. While *CTGF* and *CYR61* are examples of more general YAP/TAZ target genes, some other genes were induced only in specific tissues/cells.
2. Each TaqMan MicroRNA Assay contains one tube of small RNA-specific RT primer for RT reaction and one tube of mix of small RNA-specific PCR primers and TaqMan minor groove binder (MGB) probe for qPCR.
3. TaqMan Universal Master Mix II contains AmpliTaq Gold DNA Polymerase, UP (Ultra Pure) dNTPs (with dUTP), ROX passive reference, and qPCR reagent buffer.
4. Resin cells twice with PBS, then add back serum-free MCF10A growth medium for 12 h.
5. After serum starvation, FBS is added back to a final concentration of 10% (volume/volume) into MCF10A culture medium for 1 h.
6. Use RNase ZAP spray, filter tip, and nuclease-free water throughout all the experimental procedures to avoid RNase contamination.
7. Do not add chloroform to the 6-well plate. It will dissolve the plate.

8. Volume of the aqueous layer is about 600 μL. It is enough to take 450 μL thus avoids touching the interphase, which contains genomic DNA.
9. Carefully avoid touching the pellet. The pellet should be transparent when it dries. Do not let the RNA dry completely, which will be difficult to dissolve.
10. The level of protein contamination in the sample is indicated by A260/A280 ratio. The ideal RNA sample has an A260/A280 ratio of 2.1; values between 1.8 and 2.0 are considered acceptable. The level of guanidine salts and phenol contamination are indicated by A230, a high peak at which indicates contamination with either of these. The ideal A260/A230 ratio is greater than 1.5.
11. RT reactions should include negative controls without addition of RNAs. All PCR reactions should include samples that are not treated with RT as negative controls.
12. From now on, all procedures should to be done on ice.
13. 10–20% excess of volume is recommended to compensate for losses during pipetting.
14. Multichannel pipette should be carefully used to avoid technical variation. Visually double check volumes in the pipette tips.
15. All reagents are from mirVana miRNA isolation kit.
16. No need to wait. Cells will lyse immediately upon exposure to the Lysis/Binding Solution.
17. 1/10 volume of initial Lysis/Binding Solution.
18. The volume of Acid-Phenol:Chloroform should be equal to the initial Lysis/Binding Solution.
19. Note that the interphase would be compact after centrifugation; if not, repeat the centrifugation.
20. Volume of the aqueous layer is about 500 μL. It is enough to take 400 μL thus avoids touching the interphase.
21. The volume of 100% ethanol should be 1.25 times of the initial Lysis/Binding Solution. Ethanol at RT should be used.
22. If the volume of lysate/ethanol mixture is larger than 700 μL, add maximum 700 μL at a time, and then repeat the procedure to let all solution pass the filter cartridge.
23. Before use, 30 mL of miRNA Wash Solution 1 should be mixed with 21 mL of 100% ethanol.
24. Before use, 50 mL of Wash Solution 2/3 should be mixed with 40 mL of 100% ethanol.
25. The specific RT primers of miR-130a and U6 snRNA from respective TaqMan MicroRNA Assay and other reagents from TaqMan MicroRNA Reverse Transcription Kit should be thawed on ice.

Acknowledgments

This work was supported by grants to B. Zhao from the National Natural Science Foundation of China Key Project (81730069), General Project (31471316) and the National Key R&D Program of China (2017YFA0504502), the National Natural Science Foundation of China International Collaboration Project (31661130150), the Fundamental Research Funds for the Central Universities, and the Qianjiang Scholar Plan of Hangzhou, the Thousand Young Talents Plan of China, and the Newton Advanced Fellowship from the Academy of Medical Sciences, UK.

References

1. Sudol M (1994) Yes-associated protein (YAP65) is a proline-rich phosphoprotein that binds to the SH3 domain of the Yes proto-oncogene product. Oncogene 9 (8):2145–2152
2. Yagi R, Chen LF, Shigesada K, Murakami Y, Ito Y (1999) A WW domain-containing yes-associated protein (YAP) is a novel transcriptional co-activator. EMBO J 18(9):2551–2562
3. Huang J, Wu S, Barrera J, Matthews K, Pan D (2005) The Hippo signaling pathway coordinately regulates cell proliferation and apoptosis by inactivating Yorkie, the Drosophila Homolog of YAP. Cell 122(3):421–434
4. Dong J, Feldmann G, Huang J, Wu S, Zhang N, Comerford SA, Gayyed MF, Anders RA, Maitra A, Pan D (2007) Elucidation of a universal size-control mechanism in Drosophila and mammals. Cell 130(6):1120–1133. https://doi.org/10.1016/j.cell.2007.07.019
5. Zhao B, Wei X, Li W, Udan RS, Yang Q, Kim J, Xie J, Ikenoue T, Yu J, Li L, Zheng P, Ye K, Chinnaiyan A, Halder G, Lai ZC, Guan KL (2007) Inactivation of YAP oncoprotein by the Hippo pathway is involved in cell contact inhibition and tissue growth control. Genes Dev 21(21):2747–2761. https://doi.org/10.1101/gad.1602907
6. Zhao B, Li L, Tumaneng K, Wang CY, Guan KL (2010) A coordinated phosphorylation by Lats and CK1 regulates YAP stability through SCF(beta-TRCP). Genes Dev 24(1):72–85. https://doi.org/10.1101/gad.1843810
7. Kanai F, Marignani PA, Sarbassova D, Yagi R, Hall RA, Donowitz M, Hisaminato A, Fujiwara T, Ito Y, Cantley LC, Yaffe MB (2000) TAZ: a novel transcriptional co-activator regulated by interactions with 14-3-3 and PDZ domain proteins. EMBO J 19(24):6778–6791
8. Murakami M, Tominaga J, Makita R, Uchijima Y, Kurihara Y, Nakagawa O, Asano T, Kurihara H (2006) Transcriptional activity of Pax3 is co-activated by TAZ. Biochem Biophys Res Commun 339(2):533–539
9. Mahoney WM Jr, Hong JH, Yaffe MB, Farrance IK (2005) The transcriptional co-activator TAZ interacts differentially with transcriptional enhancer factor-1 (TEF-1) family members. Biochem J 388(Pt 1):217–225. https://doi.org/10.1042/BJ20041434
10. Yu FX, Zhao B, Guan KL (2015) Hippo pathway in organ size control, tissue homeostasis, and cancer. Cell 163(4):811–828. https://doi.org/10.1016/j.cell.2015.10.044
11. Zhou D, Conrad C, Xia F, Park JS, Payer B, Yin Y, Lauwers GY, Thasler W, Lee JT, Avruch J, Bardeesy N (2009) Mst1 and Mst2 maintain hepatocyte quiescence and suppress hepatocellular carcinoma development through inactivation of the Yap1 oncogene. Cancer Cell 16(5):425–438. https://doi.org/10.1016/j.ccr.2009.09.026
12. Camargo FD, Gokhale S, Johnnidis JB, Fu D, Bell GW, Jaenisch R, Brummelkamp TR (2007) YAP1 increases organ size and expands undifferentiated progenitor cells. Curr Biol 17 (23):2054–2060. https://doi.org/10.1016/j.cub.2007.10.039
13. Goulev Y, Fauny JD, Gonzalez-Marti B, Flagiello D, Silber J, Zider A (2008) SCALLOPED interacts with YORKIE, the nuclear effector of the hippo tumor-suppressor pathway in Drosophila. Curr Biol 18(6):435–441. https://doi.org/10.1016/j.cub.2008.02.034
14. Wu S, Liu Y, Zheng Y, Dong J, Pan D (2008) The TEAD/TEF family protein Scalloped mediates transcriptional output of the Hippo growth-regulatory pathway. Dev Cell 14

(3):388–398. https://doi.org/10.1016/j.devcel.2008.01.007
15. Zhang L, Ren F, Zhang Q, Chen Y, Wang B, Jiang J (2008) The TEAD/TEF family of transcription factor Scalloped mediates Hippo signaling in organ size control. Dev Cell 14 (3):377–387. https://doi.org/10.1016/j.devcel.2008.01.006
16. Zhao B, Ye X, Yu J, Li L, Li W, Li S, Lin JD, Wang CY, Chinnaiyan AM, Lai ZC, Guan KL (2008) TEAD mediates YAP-dependent gene induction and growth control. Genes Dev 22 (14):1962–1971. https://doi.org/10.1101/gad.1664408
17. Vassilev A, Kaneko KJ, Shu H, Zhao Y, DePamphilis ML (2001) TEAD/TEF transcription factors utilize the activation domain of YAP65, a Src/Yes-associated protein localized in the cytoplasm. Genes Dev 15 (10):1229–1241
18. Alarcon C, Zaromytidou AI, Xi Q, Gao S, Yu J, Fujisawa S, Barlas A, Miller AN, Manova-Todorova K, Macias MJ, Sapkota G, Pan D, Massague J (2009) Nuclear CDKs drive Smad transcriptional activation and turnover in BMP and TGF-beta pathways. Cell 139 (4):757–769. https://doi.org/10.1016/j.cell.2009.09.035
19. Komuro A, Nagai M, Navin NE, Sudol M (2003) WW domain-containing protein YAP associates with ErbB-4 and acts as a co-transcriptional activator for the carboxyl-terminal fragment of ErbB-4 that translocates to the nucleus. J Biol Chem 278 (35):33334–33341
20. Strano S, Munarriz E, Rossi M, Castagnoli L, Shaul Y, Sacchi A, Oren M, Sudol M, Cesareni G, Blandino G (2001) Physical interaction with yes-associated protein enhances p73 transcriptional activity. Eur J Cancer 37 (18):15164–15173
21. Rosenbluh J, Nijhawan D, Cox AG, Li X, Neal JT, Schafer EJ, Zack TI, Wang X, Tsherniak A, Schinzel AC (2012) β-Catenin-driven cancers require a YAP1 transcriptional complex for survival and tumorigenesis. Cell 151 (7):1457–1473
22. Heallen T, Zhang M, Wang J, Bonillaclaudio M, Klysik E, Johnson RL, Martin JF (2011) Hippo pathway inhibits Wnt signaling to restrain cardiomyocyte proliferation and heart size. Science 332(6028):458
23. Shao D, Zhai P, Del Re DP, Sciarretta S, Yabuta N, Nojima H, Lim DS, Pan D, Sadoshima J (2014) A functional interaction between Hippo-YAP signalling and FoxO1 mediates the oxidative stress response. Nat Commun 5:3315
24. Varelas X, Samavarchi-Tehrani P, Narimatsu M, Weiss A, Cockburn K, Larsen BG, Rossant J, Wrana JL (2010) The Crumbs complex couples cell density sensing to Hippo-dependent control of the TGF-β-SMAD pathway. Dev Cell 19(6):831–844
25. Varelas X, Sakuma R, Samavarchi-Tehrani P, Peerani R, Rao BM, Dembowy J, Yaffe MB, Zandstra PW, Wrana JL (2008) TAZ controls Smad nucleocytoplasmic shuttling and regulates human embryonic stem-cell self-renewal. Nat Cell Biol 10(7):837
26. Stein C, Bardet AF, Roma G, Bergling S, Clay I, Ruchti A, Agarinis C, Schmelzle T, Bouwmeester T, Schübeler D (2015) YAP1 exerts its transcriptional control via TEAD-mediated activation of enhancers. PLoS Genet 11(8):e1005465
27. Galli GG, Carrara M, Yuan WC, Valdes-Quezada C, Gurung B, Pepe-Mooney B, Zhang T, Geeven G, Gray NS, De LW (2015) YAP drives growth by controlling transcriptional pause release from dynamic enhancers. Mol Cell 60(2):328
28. Zanconato F, Forcato M, Battilana G, Azzolin L, Quaranta E, Bodega B, Rosato A, Bicciato S, Cordenonsi M, Piccolo S (2015) Genome-wide association between YAP/-TAZ/TEAD and AP-1 at enhancers drives oncogenic growth. Nat Cell Biol 17 (9):1218–1227
29. Tapon N, Harvey KF, Bell DW, Wahrer DC, Schiripo TA, Haber DA, Hariharan IK (2002) Salvador promotes both cell cycle exit and apoptosis in Drosophila and is mutated in human cancer cell lines. Cell 110(4):467–478
30. Neto-Silva RM, de Beco S, Johnston LA (2010) Evidence for a growth-stabilizing regulatory feedback mechanism between Myc and Yorkie, the Drosophila homolog of Yap. Dev Cell 19(4):507–520. https://doi.org/10.1016/j.devcel.2010.09.009
31. Ziosi M, Baena-Lopez LA, Grifoni D, Froldi F, Pession A, Garoia F, Trotta V, Bellosta P, Cavicchi S (2010) dMyc functions downstream of Yorkie to promote the supercompetitive behavior of hippo pathway mutant cells. PLoS Genet 6(9):e1001140. https://doi.org/10.1371/journal.pgen.1001140
32. Ren F, Wang B, Yue T, Yun EY, Ip YT, Jiang J (2010) Hippo signaling regulates Drosophila intestine stem cell proliferation through multiple pathways. Proc Natl Acad Sci U S A 107 (49):21064–21069. https://doi.org/10.1073/pnas.1012759107
33. Zhang J, Ji JY, Yu M, Overholtzer M, Smolen GA, Wang R, Brugge JS, Dyson NJ, Haber DA (2009) YAP-dependent induction of

amphiregulin identifies a non-cell-autonomous component of the Hippo pathway. Nat Cell Biol 11(12):1444–1450

34. Staley BK, Irvine KD (2010) Warts and Yorkie mediate intestinal regeneration by influencing stem cell proliferation. Curr Biol 20 (17):1580–1587. https://doi.org/10.1016/j.cub.2010.07.041

35. Shaw RL, Kohlmaier A, Polesello C, Veelken C, Edgar BA, Tapon N (2010) The Hippo pathway regulates intestinal stem cell proliferation during Drosophila adult midgut regeneration. Development 137(24):4147–4158. https://doi.org/10.1242/dev.052506

36. Karpowicz P, Perez J, Perrimon N (2010) The Hippo tumor suppressor pathway regulates intestinal stem cell regeneration. Development 137(24):4135–4145. https://doi.org/10.1242/dev.060483

37. Hamaratoglu F, Willecke M, Kango-Singh M, Nolo R, Hyun E, Tao C, Jafar-Nejad H, Halder G (2006) The tumour-suppressor genes NF2/Merlin and Expanded act through Hippo signalling to regulate cell proliferation and apoptosis. Nat Cell Biol 8(1):27–36

38. Genevet A, Wehr MC, Brain R, Thompson BJ, Tapon N (2010) Kibra is a regulator of the salvador/warts/hippo signaling network. Dev Cell 18(2):300–308. https://doi.org/10.1016/j.devcel.2009.12.011

39. Cho E, Feng Y, Rauskolb C, Maitra S, Fehon R, Irvine KD (2006) Delineation of a Fat tumor suppressor pathway. Nat Genet 38 (10):1142–1150

40. Genevet A, Polesello C, Blight K, Robertson F, Collinson LM, Pichaud F, Tapon N (2009) The Hippo pathway regulates apical-domain size independently of its growth-control function. J Cell Sci 122(Pt 14):2360–2370. https://doi.org/10.1242/jcs.041806

41. Lai D, Ho KC, Hao Y, Yang X (2011) Taxol resistance in breast cancer cells is mediated by the hippo pathway component TAZ and its downstream transcriptional targets Cyr61 and CTGF. Cancer Res 71(7):2728–2738. https://doi.org/10.1158/0008-5472.CAN-10-2711

42. Guo X, Zhao Y, Yan H, Yang Y, Shen S, Dai X, Ji X, Ji F, Gong XG, Li L (2017) Single tumor-initiating cells evade immune clearance by recruiting type II macrophages. Genes Dev 31 (3):247–259

43. Moroishi T, Park HW, Qin B, Chen Q, Meng Z, Plouffe SW, Taniguchi K, Yu FX, Karin M, Pan D (2015) A YAP/TAZ-induced feedback mechanism regulates Hippo pathway homeostasis. Genes Dev 29(12):1271–1284

44. Dai X, Liu H, Shen S, Guo X, Yan H, Ji X, Li L, Huang J, Feng XH, Zhao B (2017) YAP activates the Hippo pathway in a negative feedback loop. Cell Res 25(10):1175

45. Thompson BJ, Cohen SM (2006) The Hippo pathway regulates the bantam microRNA to control cell proliferation and apoptosis in Drosophila. Cell 126(4):767–774

46. Nolo R, Morrison CM, Tao C, Zhang X, Halder G (2006) The bantam microRNA is a target of the hippo tumor-suppressor pathway. Curr Biol 16(19):1895–1904. https://doi.org/10.1016/j.cub.2006.08.057

47. Shen S, Guo X, Yan H, Lu Y, Ji X, Li L, Liang T, Zhou D, Feng XH, Zhao JC, Yu J, Gong XG, Zhang L, Zhao B (2015) A miR-130a-YAP positive feedback loop promotes organ size and tumorigenesis. Cell Res 25(9):997–1012. https://doi.org/10.1038/cr.2015.98

48. Tumaneng K, Schlegelmilch K, Russell RC, Yimlamai D, Basnet H, Mahadevan N, Fitamant J, Bardeesy N, Camargo FD, Guan KL (2012) YAP mediates crosstalk between the Hippo and PI3K-TOR pathways by suppressing PTEN via miR-29. Nat Cell Biol 14 (12):1322

49. Ma J, Huang K, Ma Y, Zhou M, Fan S (2017) The TAZ-miR-224-SMAD4 axis promotes tumorigenesis in osteosarcoma. Cell Death Dis 8(1):e2539

50. Tamm C, Bower N, Anneren C (2011) Regulation of mouse embryonic stem cell self-renewal by a Yes-YAP-TEAD2 signaling pathway downstream of LIF. J Cell Sci 124 (Pt 7):1136–1144. https://doi.org/10.1242/jcs.075796

51. Lian I, Kim J, Okazawa H, Zhao J, Zhao B, Yu J, Chinnaiyan A, Israel MA, Goldstein LS, Abujarour R, Ding S, Guan KL (2010) The role of YAP transcription coactivator in regulating stem cell self-renewal and differentiation. Genes Dev 24(11):1106–1118. https://doi.org/10.1101/gad.1903310

52. Zhang H, Pasolli HA, Fuchs E (2011) Yes-associated protein (YAP) transcriptional coactivator functions in balancing growth and differentiation in skin. Proc Natl Acad Sci U S A 108(6):2270–2275. https://doi.org/10.1073/pnas.1019603108

53. Schutte U, Bisht S, Heukamp LC, Kebschull M, Florin A, Haarmann J, Hoffmann P, Bendas G, Buettner R, Brossart P, Feldmann G (2014) Hippo signaling mediates proliferation, invasiveness, and metastatic potential of clear cell renal cell carcinoma. Transl Oncol 7(2):309–321. https://doi.org/10.1016/j.tranon.2014.02.005

54. Zhang Y, Xia H, Ge X, Chen Q, Yuan D, Qi C, Leng W, Liang C, Tang Q, Feng B (2014) CD44 acts through RhoA to regulate YAP signaling. Cell Signal 26(11):2504–2513
55. Wang Z, Wu Y, Wang H, Zhang Y, Mei L, Fang X, Zhang X, Zhang F, Chen H, Liu Y (2014) Interplay of mevalonate and Hippo pathways regulates RHAMM transcription via YAP to modulate breast cancer cell motility. Proc Natl Acad Sci U S A 111(1):E89
56. Xu MZ, Chan SW, Liu AM, Wong KF, Fan ST, Chen J, Poon RT, Zender L, Lowe SW, Hong W (2011) AXL receptor kinase is a mediator of YAP-dependent oncogenic functions in hepatocellular carcinoma. Oncogene 30 (10):1229–1240
57. Xie Q, Chen J, Feng H, Peng S, Adams U, Bai Y, Huang L, Li J, Huang J, Meng S (2013) YAP/TEAD-mediated transcription controls cellular senescence. Cancer Res 73 (12):3615–3624
58. Ralston A, Cox BJ, Nishioka N, Sasaki H, Chea E, Rugg-Gunn P, Guo G, Robson P, Draper JS, Rossant J (2010) Gata3 regulates trophoblast development downstream of Tead4 and in parallel to Cdx2. Development 137(3):395
59. Nishioka N, Yamamoto S, Kiyonari H, Sato H, Sawada A, Ota M, Nakao K, Sasaki H (2008) Tead4 is required for specification of trophectoderm in pre-implantation mouse embryos. Mech Dev 125(3–4):270–283
60. Nishioka N, Inoue K, Adachi K, Kiyonari H, Ota M, Ralston A, Yabuta N, Hirahara S, Stephenson RO, Ogonuki N (2009) The Hippo signaling pathway components Lats and Yap pattern Tead4 activity to distinguish mouse trophectoderm from inner cell mass. Dev Cell 16(3):398–410
61. Milewski RC, Chi NC, Li J, Brown C, Lu MM, Epstein JA (2004) Identification of minimal enhancer elements sufficient for Pax3 expression in neural crest and implication of Tead2 as a regulator of Pax3. Development 131(4):829
62. Gee ST, Milgram SL, Kramer KL, Conlon FL, Moody SA (2011) Yes-associated protein 65 (YAP) expands neural progenitors and regulates Pax3 expression in the neural plate border zone. PLoS One 6(6):e20309
63. Song S, Ajani JA, Honjo S, Maru DM, Chen Q, Scott AW, Heallen TR, Xiao L, Hofstetter WL, Weston B (2014) Hippo coactivator YAP1 upregulates SOX9 and endows esophageal cancer cells with stem-like properties. Cancer Res 74(15):4170–4182
64. Gao Y, Zhang W, Han X, Li F, Wang X, Wang R, Fang Z, Tong X, Yao S, Li F (2015) YAP inhibits squamous transdifferentiation of Lkb1-deficient lung adenocarcinoma through ZEB2-dependent DNp63 repression. Nat Commun 5(1):4629
65. Yang Y, Del Re DP, Nakano N, Sciarretta S, Zhai P, Park J, Sayed D, Shirakabe A, Matsushima S, Park Y (2015) miR-206 mediates YAP-induced cardiac hypertrophy and survival. Circ Res 117(10):891
66. Zhang H, Gise AV, Liu Q, Hu T, Tian X, He L, Pu W, Huang X, He L, Cai CL (2014) Yap1 is required for endothelial to mesenchymal transition of the atrioventricular cushion. J Biol Chem 289(27):18681
67. Hiemer SE, Szymaniak AD, Varelas X (2014) The transcriptional regulators TAZ and YAP direct transforming growth factor β-induced tumorigenic phenotypes in breast cancer cells. J Biol Chem 289(19):13461–13474
68. Haskins JW, Nguyen DX, Stern DF (2014) Neuregulin 1-activated ERBB4 interacts with YAP to induce Hippo pathway target genes and promote cell migration. Sci Signal 7(355): ra116
69. Zaidi SK, Sullivan AJ, Medina R, Ito Y, van Wijnen AJ, Stein JL, Lian JB, Stein GS (2004) Tyrosine phosphorylation controls Runx2-mediated subnuclear targeting of YAP to repress transcription. EMBO J 23(4):790–799
70. Basu S, Totty NF, Irwin MS, Sudol M, Downward J (2003) Akt Phosphorylates the Yes-Associated Protein, YAP, to Induce Interaction with 14-3-3 and Attenuation of p73-Mediated Apoptosis. Mol Cell 11(1):11
71. Lapi E, Di AS, Donzelli S, Gal H, Domany E, Rechavi G, Pandolfi PP, Givol D, Strano S, Lu X (2008) PML, YAP, and p73 are components of a proapoptotic autoregulatory feedback loop. Mol Cell 32(6):803
72. Murakami M, Nakagawa M, Olson EN, Nakagawa O (2005) A WW domain protein TAZ is a critical coactivator for TBX5, a transcription factor implicated in Holt-Oram syndrome. Proc Natl Acad Sci U S A 102 (50):18034–18039

Chapter 13

HTRF® Total and Phospho-YAP (Ser127) Cellular Assays

Diana Zindel, Claire Vol, Odile Lecha, Isabelle Bequignon, Merve Bilgic, Marion Vereecke, Fabienne Charrier-Savournin, Maïté Romier, Eric Trinquet, Jean-Philippe Pin, Julie Pannequin, Thomas Roux, Elodie Dupuis, and Laurent Prézeau

Abstract

The YAP protein is a co-transcription factor increasing the expression of genes involved in cell proliferation and repressing the expression of genes important for cell differentiation and apoptosis. It is regulated by several inputs, like the Hippo pathway, through the action of kinases that phosphorylate YAP on several residues. The level of phosphorylation of the residues serine 127 (S127) of YAP is generally assessed in cellular models, native tissues, and organs, as a marker of YAP activity and location, and is regulated by numerous partners. This phosphorylation event is classically detected using a western blot technical approach. Here, we describe a novel approach to detect both the relative amount of total YAP (T-YAP assay) and the phosphorylation of the residue S127 of YAP (S127-P-YAP assay) using a HTRF®-based method. This easy-to-run method can easily be miniaturized and allows for a high-throughput analysis in 96/384-well plate format, requiring less cellular material and being more rapid than other approaches.

Key words YAP phosphorylation, YAP-S127-phosphorylation-directed antibodies, Antibody-based assay, Resonance energy transfer, HTRF®

1 Introduction

The Hippo-YAP-TAZ pathway is a potent regulator of physiological processes controlling organ size as well as cellular proliferation and differentiation. Deregulation of this pathway is frequently associated with anomalous cellular growth leading to cancer [1]. Thus, accurate assessing of the activity of this pathway is crucial to better understanding of its function and to perform screening of compounds and setup strategies to control it. The Hippo pathway comprises a kinase module and a transcriptional module. The kinase module is the major regulator of YAP/TAZ transcriptional module activity [2–4]. By phosphorylation mostly through LATS1/LATS2 kinases, YAP and TAZ co-transcription factors are retained in the cytosol and degraded, and thereby the transcriptional module is

Alexander Hergovich (ed.), *The Hippo Pathway: Methods and Protocols*, Methods in Molecular Biology, vol. 1893, https://doi.org/10.1007/978-1-4939-8910-2_13,

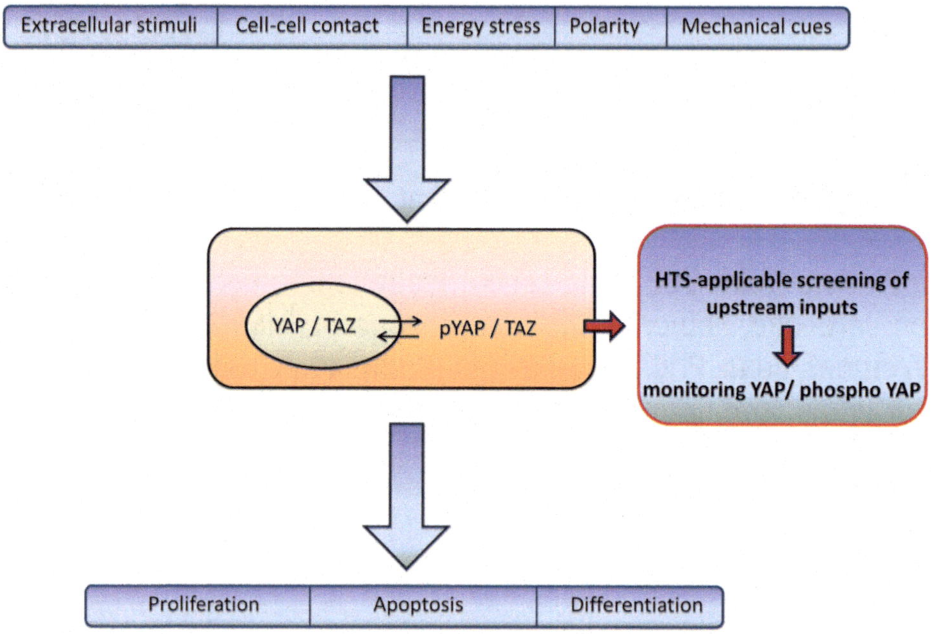

Fig. 1 Multiple upstream inputs regulate the Hippo-YAP-TAZ axis through phosphorylation. When phosphorylated, YAP/TAZ are retained in the cytosol. Non-phosphorylated YAP/TAZ translocate into the nucleus, act as transcriptional coactivators, and induce target gene expression to control cellular proliferation, differentiation, and apoptosis

inactivated (Fig. 1). Thus, there is a crucial need to identify inputs and screen bioactive molecules that could modulate the Hippo-YAP/TAZ pathway in pathological processes like cancer, leading to generating innovative therapeutic antitumor strategies. In the present chapter, we present an innovative HTS-applicable HTRF® approach to determine YAP expression and YAP activity by its phosphorylation status of the residues S127 (Fig. 1) [5], generally assessed in cellular models, native tissues, and organs, as a marker of YAP activity, location, and regulation by numerous partners.

Resonance energy transfer (RET)-based biophysical methods like FRET and BRET (for Förster and bioluminescence resonance energy transfer, respectively) have been used by many laboratories to generate signaling assays. The physical principle states that an excited donor fluorophore that displays compatible spectra with another fluorophore (called acceptor) can transfer its energy to the latter in a non-radiative manner by dipole-dipole interaction. FRET requires (I) that the integral of the excited donor's emission spectrum overlaps with the integral of the acceptors absorption spectrum, (II) that both fluorophores are in close distance to one another in the range of a few nm, and (III) that the transition dipole orientation of the donor and the acceptor is approximately parallel to each other [6]. Thus, signaling assays are built on variation of distance or orientation of two fluorophore attached to the signaling proteins of interest (e.g., G protein BRET assays, EPAC cAMP FRET assays, etc. [7–9]). Although well adapted for specific experimental conditions, mostly in recombinant systems, these methods display several limitations. Indeed, the BRET or FRET donor and acceptor fluorophore are generally luminescent and fluorescent proteins (luciferase, green fluorescent protein (GFP) variants, etc.) that need to be fused to the proteins of interest and then expressed by the cells after transfection of the encoding plasmids. Second, the absorption and emission spectra of the donor and acceptor fluorophore are not very well separated, meaning that excitation of the donor will lead to a background excitation of the acceptor (cross excitation). Moreover, the emission spectrum of the acceptor is contaminated by the emission of the donor (bleed through).

To overcome these issues, we used the Homogeneous Time-Resolved FRET (HTRF®) (Fig. 2), based on RET between a rare earth europium [10] or terbium cryptate as a donor fluorophore and a Cy5-like dye (D2) as an acceptor fluorophore [11–13]. As an advantage, this type of donor displays a long fluorescence emission lifetime in the millisecond range (in comparison to the nanosecond range of standard fluorophores) (Fig. 2), allowing a recording of FRET signal in a time-gated manner (Time-Resolved FRET or TR-FRET). Indeed, recording the TR-FRET signal 50 μs after the excitation of the donor allows the removal of the background arising from both the autofluorescence of the cells and the fluorescence of the acceptor, as their fluorescence is rapidly negligible. As such, TR-FRET signal can be measured without washing out the reagents, then in a homogenous way (HTRF®). In addition, considering that this pair of donor-acceptor displays optimal compatible spectral properties that the D2 acceptor emits light at wavelengths where the donor emission is barely detectable, TR-FRET provides a good signal-to-noise ratio compared to most other RET methods.

Thus, HTRF® is used to generate signaling assays suitable for HTS (https://www.cisbio.com/drug-discovery/htrf-technology) [14, 15]. Four main advantages can be pointed out: first, HTRF® displays a good signal-to-noise ratio due to its properties explained above; second, HTRF® assays are easy-to-run assays, avoiding purification or washing steps, allowing miniaturizing, and thus are ideal for HTS; third, for most of the assays, great specificity is given by a coincidence detector principle; and fourth, such assays are ideally suited to record endogenous signaling events in native systems. Furthermore, classical FRET assays are limited by the relative orientation of the fluorophores, whereas there is minimal orientation constraint as terbium has multiple electronic transitions [16]. For many assays, the coincidence detection is due to the use of two different antibodies, one recognizing a general state of the signaling protein of interest and the other one recognizing an active state

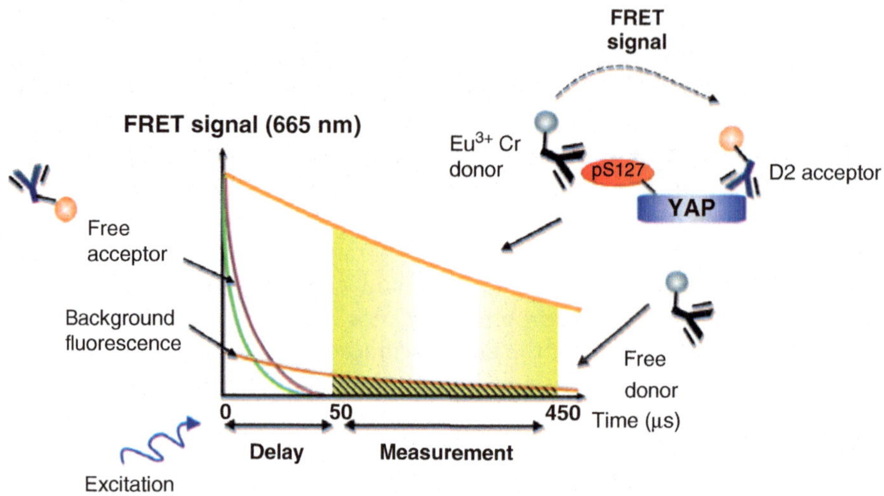

Fig. 2 Features of TR-FRET techniques. HTRF® (*Homogeneous Time-Resolved FRET*) donor, e.g., Eu^{3+}cryptate, is excited at 337 nm and shows long luminescence lifetime in the range of several hundred µs. After initial excitation, the signal is cleared from unspecific short-lived emissions through a delay (50–100 µs) between donor excitation and recording of acceptor fluorescence emission. During the measurement, only the stably emitting donor fluorophore and the subsequent activated acceptor fluorophore will be fluorescent. The donor emission signal peaking at 620 nm and acceptor emission signal peaking at 665 nm are monitored simultaneously. HTRF® values are calculated from 665 nm/620 nm ratios enabling the data to be normalized with respect to between-assay variations

(e.g., phosphorylated state) of the signaling protein of interest. By using these antibodies being labeled by either donor or acceptor TR-FRET fluorophore, the TR-FRET signal is recorded only when both antibodies are bound to the signaling protein of interest, thereby highly reducing any non-specific signal due to binding to non-specific targets of each antibody. This is typically the case for the S127-P-YAP assay. Note that, for T-YAP assay, the coincidence detector is due to the binding of both antibodies to the same protein whatever its active-inactive state (see below). In such sandwich assays [15], normalized TR-FRET signal is generated through energy transfer between two labeled antibodies, and the larger the amount of target produced, the higher the normalized FRET signal. The TR-FRET signal measured ratiometrically is then proportional to the quantity of target produced.

In the present chapter, we describe the HTRF®protocol for two assays, one detecting the total amount of YAP protein (T-YAP assay) and one detecting the serine 127 phosphorylation state of the protein YAP (S127-P-YAP assay) (Fig. 3). T-YAP assay will then be used (1) to determine the relative amount of YAP in different samples and its regulation over time and (2) to help normalizing the level of phosphorylation of YAP in samples, especially when regulated by drugs or other experimental conditions.

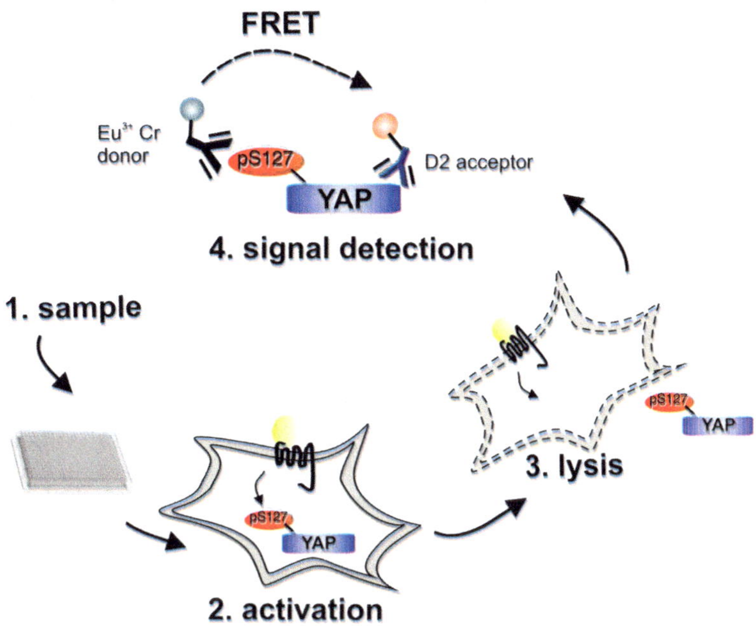

Fig. 3 Overview of the HTRF®assay to detect YAP-S127 phosphorylation in a 96-well or 384-well format based on Förster resonance energy transfer (FRET). For total YAP detection with the T-YAP assay, the antibody recognizing the S127 phosphorylated form of YAP in the S127-P-YAP assay is replaced by another antibody recognizing all forms of YAP. Cells in suspension or in adherence are exposed to an appropriate stimulus, lysed, and incubated with Eu^{3+}Cr-donor-labeled and d2-acceptor-labeled antibodies raised against total YAP and S127 phosphorylated YAP. After 2 h (up to 24 h) incubation with the antibodies, HTRF signal is detected using a plate reader

2 Materials

Some of the materials are common for both T-YAP and S127-P-YAP assays except for the antibodies.

2.1 Sample Preparation (HEK293 Cells)

1. Phosphate-buffered saline (PBS), containing KH_2PO_4 (144 mg/L), NaCl (9000 mg/L), and Na_2HPO_4 (795 mg/L).
2. 96-well black plate with black bottom (CellStar®; Greiner). Note that 384-well plates could also be used depending on the type of cells and the level of YAP expression (*see* **Note 1**).
3. HEK293 cells (from American Type Culture collection, ATCC). In the present chapter/section, the protocol will be described for HEK293 and can be adapted to any other cell types, heterologous or primary cells (cultured or freshly isolated), and even pieces of tissue (*see* **Note 1**).
4. Culture medium for HEK293 cells: complete Dulbecco's Modified Eagle's Medium (DMEM) – DMEM supplemented with 10% (v/v) fetal bovine serum, 1% (v/v) penicillin/

streptomycin, and 1% (v/v) nonessential amino acids. All the products used for cell culture are purchased from Gibco/BRL-Life Technologies/Thermo Fisher Scientific.

2.2 YAP Detection

1. Stimulation agents, agonists, and inhibitors as required by the scientific project.
2. Deionized water.
3. Low volume 384-well plates (*see* **Note 2**).

The following materials are included in the Phospho-YAP (Ser127) & Total YAP Cellular Assay HTRF®kits (*see* **Note 3**):

4. Antibodies raised against YAP and labeled with donor or acceptor HTRF®fluorophore. Two antibodies recognizing all forms of YAP but with non-overlapping epitopes for the T-YAP assay. For the S127-P-YAP assay, one antibody recognizing all forms of YAP and one antibody recognizing the S127-phosphorylated form of YAP.
5. HTRF® detection buffer.
6. Lysis buffer 4× concentrated.
7. Blocking reagent.
8. Control cell lysate.

2.3 FRET Measurements

1. HTRF® compatible reader (*see* **Note 4**).

3 Methods

In this chapter, we will focus on the two-plate protocol for adherent cells (Fig. 4). Few steps of the protocol for HTRF® YAP assays will differ when cells are grown in suspension, as only one plate will be used (one-plate protocol), and they will be briefly described at the end of the methods section in Subheading 3.7. For accurate total YAP and phosphorylated YAP detection, it is crucial to determine the optimal amount of sample to use (*see* **Notes 1** and **5**), and the process to follow will be described in Subheading 3.6 below, after the description of the different steps to run the protocol for the assays. T-YAP and S127-P-YAP assays can be run separately. Indeed, T-YAP is dedicated to the assessment of the relative amount of total YAP in samples and its regulation over time or upon treatment, on a time scale different than that of YAP phosphorylation (*see* **Note 6**). On the other side, S127-P-YAP assay by itself detects the phosphorylation state of YAP and its modulations by drugs and experimental conditions. But both T-YAP and S127-P-YAP assays can also be run in parallel, and this will be used to normalize the data obtained with the S127-P-YAP assay with that obtained with the T-YAP assay (*see* **Note 7**). The protocols are the same for both assays except for the required antibodies.

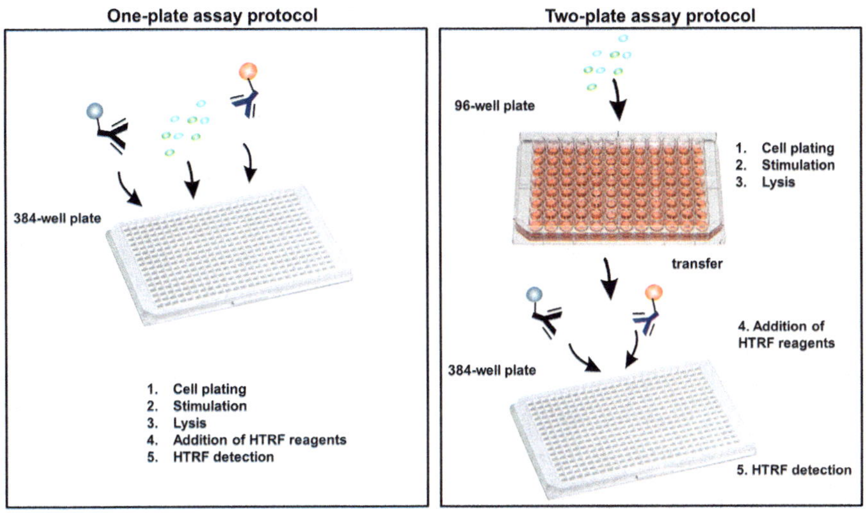

Fig. 4 Overview of both one-plate and two-plate assay protocols for non-adherent and adherent cells, respectively

3.1 Preparing and Dissolving the Reagents

Allow all reagents to thaw before use and centrifuge the vials gently before pipetting the stock solutions. Prepare the working solutions from stock solutions by following the instructions below:

1. Dilute the stock solution of phospho-YAP-d2-antibody/phospho-YAP-europium cryptate antibody (20-fold concentrated antibody stock solution) 1:20 with detection buffer, e.g., dilute 0.05 mL of d2-antibody or phospho-YAP-cryptate antibody solution in 0.95 mL of detection buffer.
2. Dilute the 4× concentrated lysis buffer 1:4 in deionized water, e.g., add 1.25 mL of 4× lysis buffer to 3.75 mL deionized water, and mix gently.
3. The blocking reagent is provided as 100× concentrated. Dilute the 100× blocking reagent 1:100 in 1× lysis buffer, e.g., add 0.05 mL of blocking reagent to 4.95 mL of 1× lysis buffer. Mix gently.
4. The control lysate is ready to use. It is provided as an internal positive control to determine the assay quality.

3.2 Cell Preparation in 96-Well Plates

1. Cells are plated in 96-well plates (but could also be plated in 384-well plates; *see* Subheading 3.7 on non-adherent cells), but the detection and reading of the HTRF® signal will be performed in low-volume 384-well plates, to minimize the volume of reagents to be used (donor- and acceptor- fluorophore bearing antibodies).

2. Seed 5–8 × 10^4 HEK293 cells/well in appropriate (usually black plate with black bottom) 96-well plates, and incubate them in DMEM supplied with 10% FCS at 37 °C and 5% CO_2 for 6–8 h until they are well attached. For each cell line, optimization of cell seeding densities is recommended (*see* **Notes 1** and **5** and Subheading 3.6).
3. Depending on the signaling input you intend to study, a starvation step with serum-free medium can be included (*see* **Note 8**). Wash the cells one to two times carefully with serum-free DMEM, and starve them several hours or overnight in serum-free DMEM, 50 μL/well.

3.3 Cell Stimulation and Lysis

Steps 1 and **2** are required when one wants to assess the effect of exogenous added compounds acting on receptors, enzymes, etc., on YAP phosphorylation and level of expression. Otherwise, directly start at **step 4**.

1. Add 50 μL/well containing the compounds to be tested diluted at the desired concentration in serum free DMEM, or 50 μL/well of buffer for control conditions.
2. Incubate at 37°C and 5% CO_2. A time-course study to determine the optimal stimulation time is recommended for each compound or target stimulated. Note that the stimulation incubation time is different when running the T-YAP assay (4 h to overnight) separately from the S127-P-YAP assay (usually 30 min, but ranges from 10 min to 2 h) (*see* **Note 6** and Fig. 5).
3. Remove the stimulation buffer from the cells, and immediately add 50 μL of lysis buffer/well. Lysis buffer volume can be decreased down to 25 μL if higher volume of lysate is required.

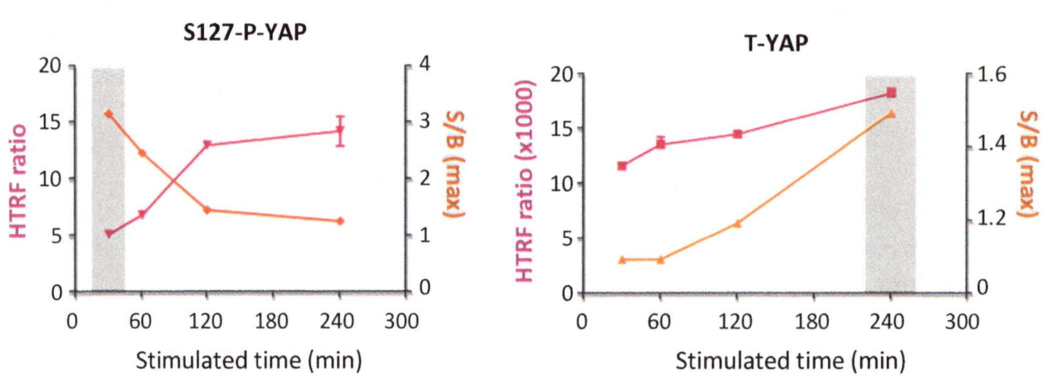

Fig. 5 Time course for S127 phosphorylated and total YAP detection. HEK 293 cells stably expressing PAR1 were starved and treated with 100 μM of TRAP-6 over a large range of time at 37 °C. Cells were lysed for 30 min at room temperature and assessed for YAP phosphorylation in position S127 and total YAP by HTRF®. Results are displayed as the mean ± S.E.M. performed in triplicates

4. Incubate for 30 min at room temperature under gentle shaking (*see* **Note 9**).
5. Homogenize the lysate by gently pipetting it up and down, avoiding bubbles.
6. Transfer 16 μL of each test condition well from the 96-well plates, onto a low-volume 384-well white plate. Depending on the cell line used, it may be necessary to dilute the cell lysate to ensure samples are within the assay linear detection range (*see* **Notes 1** and **5** and Subheading 3.6).
7. Transfer 16 μL of the control lysate for S127-P-YAP or total T-YAP assays, for positive control conditions.

3.4 Conjugate-Antibody Addition and Signal Detection

1. For T-YAP detection, mix 2 μL of diluted anti-total YAP-d2 antibody with 2 μL of diluted anti-total YAP-europium cryptate antibody. Add 4 μL of the mix to the 16 μL of lysate for a final volume of 20 μL/well for the detection of the total endogenous YAP proteins in the test condition or in the positive/negative control condition, respectively.
2. For S127-P-YAP detection, mix 2 μL of anti-phospho-YAP-d2 antibody with 2 μL anti-total YAP-europium cryptate antibody. Add 4 μL of the mix to the 16 μL of lysate for a final volume of 20 μL/well to detect the fraction of phosphorylated YAP proteins in the test condition or in the positive/negative control condition, respectively.
3. Incubate at least 2 h (up to overnight) at room temperature in the dark.
4. Read the fluorescence emission at 665 nm and 620 nm with an appropriate plate reader displaying an HTRF®detection module (e.g., PHERAstar FS, BMG Labtech, Ortenberg, Germany).

3.5 Data Evaluation and Normalization

HTRF® ratio must be calculated for each individual well. The mean and standard deviation can then be worked out from ratio replicates. Normalization of the data obtained with S127-P-YAP vs. T-YAP assays is most of the time required for proper analysis (*see* **Note** 7 and Fig. 6).

1. Calculate the ratio between the emission at 665 nm and the emission at 620 nm from raw data obtained with the plate reader:

$$\text{HTRF}^{\circledR}\ \text{ratio} = [(\text{Signal 665 nm}) / (\text{Signal 620 nm})] \times 10^4$$

2. Normalize the HTRF® signal for phospho-YAP to the HTRF® signal of total YAP (*see* **Note** 7).

The normalization value can be used to monitor changes in YAP expression versus changes in YAP phosphorylation. This relative value represents the proportion of phosphorylated YAP

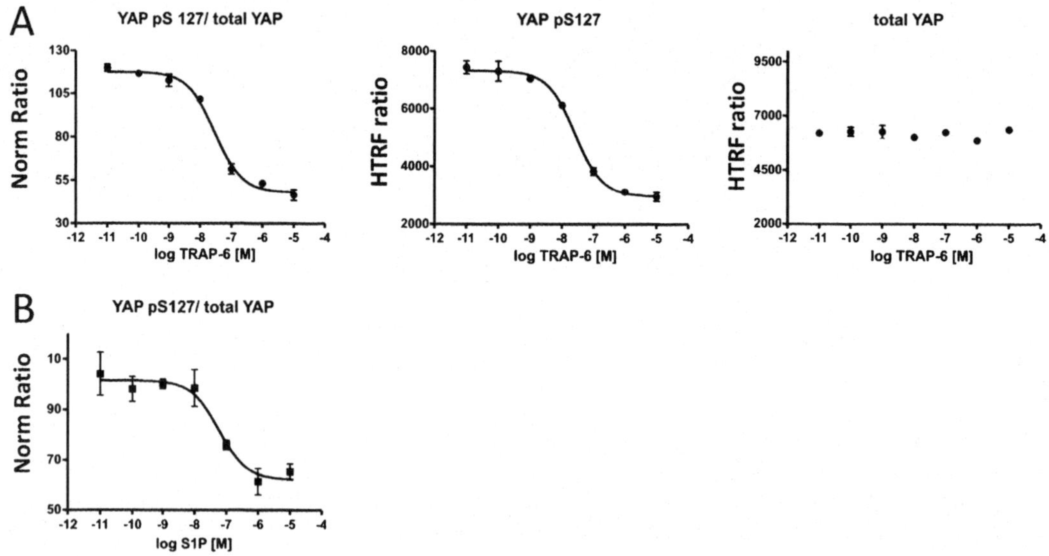

Fig. 6 YAP activation monitored by HTRF. HEK 293 cells stably expressing PAR1 (**a**) or endogenously expressing S1P receptors (**b**) were starved overnight and treated with the indicated concentration of TRAP-6 or S1P for 30 min at 37 °C. Cells were lysed for 30 min at room temperature and assessed for YAP phosphorylated in position S127 and total YAP by HTRF®. Results are displayed as the mean ± S.E.M. performed in triplicates. In (**a**), all data from S127-P-YAP and T-YAP assays and the normalized data are shown

compared to the total YAP present within the sample and is then calculated for each well as follows:

$$\text{Normalization value } (P/T) = [(\text{Phospho YAP HTRF}^{\circledR} \text{ ratio})/(\text{Total YAP HTRF}^{\circledR}\text{ratio})] \times 100$$

3.6 Determination of the Amount of Lysate Required for Optimal Detection with T-YAP and S127-P-YAP

When starting with new cellular model, samples, conditions, etc., it is recommended to determine the amount of lysate necessary to obtain a HTRF ratio lying within the linear range of the assay (*see* **Notes 1** and **5**).

1. Transfer a range of 1 to 16 μL of cell lysate (out of the 50 μL in 96-well plate) into a 384-well plate; add the respective amount of lysis buffer to a volume of 16 μL lysate/well. Add the antibody conjugates (Subheading 3.4 and Table 1) to a total volume of 20 μL/well.
2. Determine the HTRF values for the negative control by pipetting 16 μL of supplemented lysis buffer (1×) per well, add 2 μL of d2 antibody and 2 μL of europium cryptate antibody to the final volume of 20 μL per well (*see* Table 1).
3. Determine the HTRF values for the positive control by pipetting 16 μL of control lysate; add 2 μL of d2 antibody and 2 μL of europium cryptate antibody to the final volume of 20 μL per

Table 1
Composition of reaction buffers to optimize HTRF ratios

	Non-treated cell lysate	Treated cell lysate	Control lysate	Blank control	Negative control
Non-treated cell lysate	16 μL	–	–	16 μL	–
Treated cell lysate	–	16 μL	–	–	–
Control lysate	–		16 μL	–	–
Supplemented lysis buffer (1×)	–	–	–	–	16 μL
Detection buffer	–	–	–	2 μL	–
S127-P-YAP-d2 antibody	2 μL	2 μL	2 μL		2 μL
S127-P-YAP cryptate antibody	2 μL	2 μL	2 μL	2 μL	2 μL
Total volume	20 μL	20 μL	20 μL	20 μL	20 μL

well (*see* Table 1). The ratio of positive control lysate signal/non-specific signal from the negative control should be ≥2 (*see* **step 3.5** *for data evaluation*).

3.7 One-Plate Protocol for Cells Grown in Suspension

1. Transfer 8 μL of cells in small volume white 384 well plates in appropriate medium (culture medium, for example, depending on the cells) 3.3 × 10^5 cells are generally sufficient but optimal density needs to be determined experimentally (*see* **Note 1**).
2. Dispense 4 μL of compound (3×) in appropriate medium (e.g., culture medium, depending of the cells) and incubate between 10 and 30 min at 37 °C. In order to determine the optimal stimulation time, perform a time-course study.
3. Add 4 μL of supplemented lysis buffer (4×) and incubate for 30 min at room temperature. The optimal lysis time may vary depending on the cell line (*see* **Note 9**).
4. Add 4 μL of premixed antibodies prepared in detection buffer. Cover the plate with a sealer and incubate overnight at room temperature.
5. Read the fluorescence emission on a compatible HTRF reader as described in Subheadings 3.4 and 3.5.

4 Notes

1. It is generally required to probe different cellular densities to determine the amount or volume of cell lysate to use. Indeed, cell lines express various levels of YAP proteins and then will generate various levels of HTRF® signal. Moreover, the confluency greatly influences the expression and activity of Hippo-YAP/TAZ pathway. High cell density induces YAP phosphorylation in position S127 [17] via contact inhibition [5], resulting in cytosolic sequestration of YAP whereas low cellular density results in nuclear YAP accumulation [18]. This is correlated to the fact that YAP/TAZ inhibition is the key mechanism of contact inhibition through Hippo kinase activation that in turn inhibits and induces degradation of YAP [5, 19]. Thus, there is more YAP protein in proliferating cells. That is the reason why the cells types, the conditions of culture and treatment, the confluence, meaning the cell seeding density, have to be taken into account for optimal conditions of detection. The determination of the amount or volume of cell lysate to be used is described in Subheading 3.6.
2. The low-volume 384-well plates we have tested and we can recommend are the white low-volume 384-well microplates, Ref. 784,075, from Greiner Bio-One GmbH.
3. The HTRF® assay kits are described on the Cisbio website: https://www.cisbio.com/drug-discovery/phospho-yap-ser127-total-yap-cellular-assay-kits.
4. Several devices on the market possess an HTRF® module that was approved by Cisbio. But we would recommend the PHERAStar FS® plate reader (BMG Labtechnologies).
5. It is crucial to obtained values in the linear range of each of both assays to be able to compare data. This is crucial to precisely detect any regulation of YAP expression or YAP phosphorylation upon stimulation by drugs, upon conditions of culture, etc., as the variation of signal has to stay in the optimal linear range detection of the assay. The linear range values vary depending on the biological models, the amount of endogenous YAP, the way the cells or samples are treated, etc. Thus, in order to work properly and get data in the linear range of detection, it is recommended to check the linearity of the data obtained with a range of diluted samples, in addition to the negative controls. This will give the range of variation of signal optimally detected for a given sample model and help adjusting the volume of samples to use for detection.
6. Modulation of total YAP amount and of S127 phosphorylated YAP displays a different time course as illustrated in Fig. 5. Indeed, T-YAP is mostly dedicated to assess the relative

expression of YAP itself, in different samples, in different conditions, or upon drug actions. In the case of a modulation of YAP expression, the time course needed to record this variation is in the range of 4 h to OVN. By comparison, the time course for S127 phosphorylation of YAP is in the 10 min to 2 h range. Time course has to be taken into account when assessing the dynamics of either the YAP protein amount or the YAP phosphorylation regulation.

7. YAP phosphorylation in position S127 and S381 has been shown to activate a phosphodegron resulting in polyubiquitination and degradation [20]. Compounds inducing YAP phosphorylation consequently can induce YAP degradation. Hence, measuring S127-P-YAP before and after compound incubation may possibly result in an apparent decrease in phosphorylation, due to a reduction of the total protein level. In order to avoid false-negative results for detection of S127-P-YAP, it is advantageous to normalize the HTRF® result of S127-P-YAP to the HTRF® result determined for T-YAP. See also the website: https://www.cisbio.com/drug-discovery/guideline-htrfr-cell-based-phospho-protein-data-normalization.
8. Serum-derived factors such as the lipids S1P or LPA are known to be potent small molecule activators of YAP [21]. It has been demonstrated that serum deprivation for several hours induces YAP cytoplasmic retention accompanied by an increase of YAP phosphorylation in position S127. The addition of serum to starved cells leads to a concentration-dependent decrease in pS127 YAP phosphorylation resulting in nuclear YAP localization and target gene expression [21, 22]. It is therefore recommended to test different serum concentrations or to apply a serum starvation step before probing for YAP activity in cellular systems.
9. Optimal lysis time may be cell line dependent, and it is recommended to carefully determine it experimentally by performing a time-course experiment.

Acknowledgment

The work described was made possible through Cisbio Bioassays and the technological pharmacology facilities ARPEGE (Pharmacology Screening Interactome) of Biocampus (IGF). This work has been supported by CisBio (cooperative research team Eidos, CNRS N°039293/CBB DRD-09-04 avenant 4), the Labex EpiGenMed, an "Investissements d'avenir" program (ANR-10-LABX-12-01), the Fondation pour la Recherche Médicale (DEQ20170336747), CNRS, Inserm, and the University of Montpellier.

References

1. Cordenonsi M, Zanconato F, Azzolin L, Forcato M, Rosato A, Frasson C, Inui M, Montagner M, Parenti AR, Poletti A et al (2011) The hippo transducer TAZ confers cancer stem cell-related traits on breast cancer cells. Cell 147:759–772
2. Dupont S, Morsut L, Aragona M, Enzo E, Giulitti S, Cordenonsi M, Zanconato F, Le Digabel J, Forcato M, Bicciato S et al (2011) Role of YAP/TAZ in mechanotransduction. Nature 474:179–183
3. Azzolin L, Panciera T, Soligo S, Enzo E, Bicciato S, Dupont S, Bresolin S, Frasson C, Basso G, Guzzardo V et al (2014) YAP/TAZ incorporation in the b -catenin destruction complex orchestrates the Wnt response. Cell 158:157–170
4. Hansen CG, Moroishi T, Guan K-L (2015) YAP and TAZ: a nexus for Hippo signaling and beyond. Trends Cell Biol 9:499–513
5. Zhao B, Wei X, Li W, Udan RS, Yang Q, Kim J, Xie J, Ikenoue T, Yu J, Li L et al (2007) Inactivation of YAP oncoprotein by the Hippo pathway is involved in cell contact inhibition and tissue growth control. Genes Dev 21:2747–2761
6. Shrestha D, Jenei A, Nagy P, Vereb G, Szöllősi J (2015) Understanding FRET as a research tool for cellular studies. Int J Mol Sci 16:6718–6756
7. Jares-Erijman EA, Jovin TM (2003) FRET imaging. Nat Biotechnol 21:1387–1395
8. Comps-Agrar L, Maurel D, Rondard P, Pin J-P, Trinquet E, Prézeau L (2011) Cell-surface protein—protein interaction analysis with time-resolved FRET and snap-tag technologies: application to G protein-coupled receptor oligomerization. In: Luttrell LM, Ferguson SSG (eds) Signal Transduction Protocols, p 201–214
9. Zindel D, Engel S, Bottrill AR, Pin J-P, Prézeau L, Tobin AB, Bünemann M, Krasel C, Butcher AJ (2016) Identification of key phosphorylation sites in PTH1R that determine arrestin3 binding and fine-tune receptor signaling. Biochem J 473:4173–4192
10. Lehn J-M, Roth CO (1991) Synthesis and properties of sodium and europium(III) cryptates incorporating the 2,2′-bipyridine 1,1-′-dioxide and 3,3′-biisoquinoline 2,2′-dioxide units. Helv Chim Acta 74:572–578
11. Bazin H, Trinquet E, Mathis G (2002) Time resolved amplification of cryptate emission: a versatile technology to trace biomolecular interactions. Rev Mol Biotechnol 82:233–250
12. Maurel D, Kniazeff J, Mathis G, Trinquet E, Pin J-P, Ansanay H (2004) Cell surface detection of membrane protein interaction with homogeneous time-resolved fluorescence resonance energy transfer technology. Anal Biochem 329:253–262
13. Mathis G (1995) Probing molecular interactions with homogeneous techniques based on rare earth cryptates and fluorescence energy transfer. Clin Chem 41:1391–1397
14. Ayoub MA, Trebaux J, Vallaghe J, Charrier-savournin F, Al-hosaini K, Moya AG, Pin J, Pfleger KDG, Trinquet E (2014) Homogeneous time-resolved fluorescence-based assay to monitor extracellular signal-regulated kinase signaling in a high-throughput format. Front Endocrinol 5:1–11
15. Scholler P, Zwier JM, Trinquet E, Rondard P, Pin JP, Prézeau L, Kniazeff J (2013) Chapter seven - time-resolved Förster resonance energy transfer-based technologies to investigate G protein-coupled receptor machinery: high-throughput screening assays and future development. In: Prog Mol Biol Transl Sci, p 275–312
16. Selvin P, Hearst J (1994) Luminescence energy transfer using a terbium chelate : Improvements on fluorescence energy transfer. PNAS 91:10024–10028
17. Meng Z, Moroishi T, Mottier-Pavie V, Plouffe SW, Hansen CG, Hong AW, Park HW, Mo J-S, Lu W, Lu S et al (2015) MAP4K family kinases act in parallel to MST1/2 to activate LATS1/2 in the Hippo pathway. Nat Commun 6:8357
18. Juan WC, Hong W (2016) Targeting the Hippo signaling pathway for tissue regeneration and cancer therapy. Genes (Basel) 7:1–25
19. Aragona M, Panciera T, Manfrin A, Giulitti S, Michielin F, Elvassore N, Dupont S, Piccolo S (2018) A mechanical checkpoint controls multicellular growth through YAP/TAZ regulation by actin-processing factors. Cell 154:1047–1059
20. Zhao B, Li L, Lei Q, Guan K (2010) The Hippo – YAP pathway in organ size control and tumorigenesis : an updated version, p 862–874
21. Miller E, Yang J, Deran M, Wu C, Su AI, Bonamy GMC, Liu J, Peters EC, Wu X (2012) Identification of serum-derived sphingosine-1-phosphate as a small molecule regulator of YAP. Chem Biol 19:955–962
22. Yu FX, Zhao B, Panupinthu N, Jewell JL, Lian I, Wang LH, Zhao J, Yuan H, Tumaneng K, Li H et al (2012) Regulation of the Hippo-YAP pathway by G-protein-coupled receptor signaling. Cell 150:780–791

Chapter 14

Studying YAP-Mediated 3D Morphogenesis Using Fish Embryos and Human Spheroids

Yoichi Asaoka, Hitoshi Morita, Hiroko Furumoto, Carl-Philipp Heisenberg, and Makoto Furutani-Seiki

Abstract

The transcription coactivator, Yes-associated protein (YAP), which is a nuclear effector of the Hippo signaling pathway, has been shown to be a mechano-transducer. By using mutant fish and human 3D spheroids, we have recently demonstrated that YAP is also a mechano-effector. YAP functions in three-dimensional (3D) morphogenesis of organ and global body shape by controlling actomyosin-mediated tissue tension. In this chapter, we present a platform that links the findings in fish embryos with human cells. The protocols for analyzing tissue tension-mediated global body shape/organ morphogenesis in vivo and ex vivo using medaka fish embryos and in vitro using human cell spheroids represent useful tools for unraveling the molecular mechanisms by which YAP functions in regulating global body/organ morphogenesis.

Key words Tissue tension, 3D morphogenesis, YAP, Medaka, Zebrafish, Mutant, Human spheroid

1 Introduction

Animals display elaborate and complex body shapes. To genetically dissect vertebrate morphogenesis, we carried out the genome-wide mutagenesis screen using medaka fish (*Oryzias latipes*) that complement zebrafish to identify new gene functions by virtue of rapid independent evolution of duplicated genes between the two species [2, 3]. This screen identified the *hirame* (*hir*) mutant displaying a unique phenotype throughout medaka, zebrafish, and mouse mutants: a flat body affecting 3D organogenesis and global body shape [4]. The combined analysis of *hir* mutants in which YAP is mutated, YAP/TAZ double knockdown zebrafish embryos, and YAP knockdown human 3D spheroids revealed that YAP is essential for the body and its internal organs to withstand external forces [1]

Yoichi Asaoka and Hitoshi Morita contributed equally to this work.

Alexander Hergovich (ed.), *The Hippo Pathway: Methods and Protocols*, Methods in Molecular Biology, vol. 1893, https://doi.org/10.1007/978-1-4939-8910-2_14,

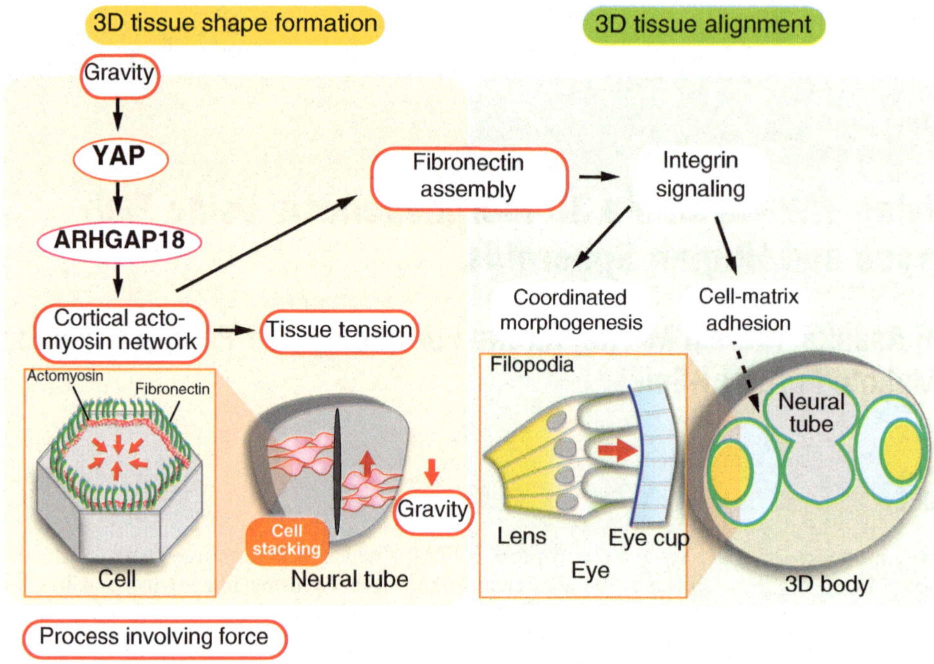

Fig. 1 Summary of how YAP-dependent actomyosin network contraction controls tissue shape and alignment

(Fig. 1); (1) the analysis of *hir* mutants showed that 3D tissue formation and their alignment underlie the generation of 3D organ/body shape. (2) The in vivo analysis of YAP/TAZ knockdown zebrafish gastrula and ex vivo analysis of *hir* mutant tissues showed that actomyosin contractility and tissue surface tension are reduced. (3) Live imaging of neurula medaka embryos revealed defects in cell stacking causing tissue neural tube flattening. (4) Immunohistochemical analysis showed that force-mediated fibronectin fibrillogenesis is affected in *hir* mutants. (5) Knocking down YAP in 3D spheroids using human RPE-1 cell line mimicked the *hir* tissue flattening and fibronectin phenotype and allowed the identification of the YAP target gene ARHGAP18 controlling F-actin turnover.

Here, we will present a simple platform for analyzing 3D morphogenesis in medaka embryos and human spheroids (Fig. 2). We will describe how those two assay systems can be used to elucidate the function of Hippo-YAP signaling at various scales of resolution ranging from the gene and protein level to the level of the entire organism.

1.1 Why 3D Analysis Is Needed?

In an animal body, all cells exist within a complex 3D environment and have their own physical characteristics, such as the cell shape, cell geometry, and tension either intrinsically generated or extrinsically applied by neighboring cells and/or the extracellular matrix [5]. Recently, it has become clear that the physical properties of

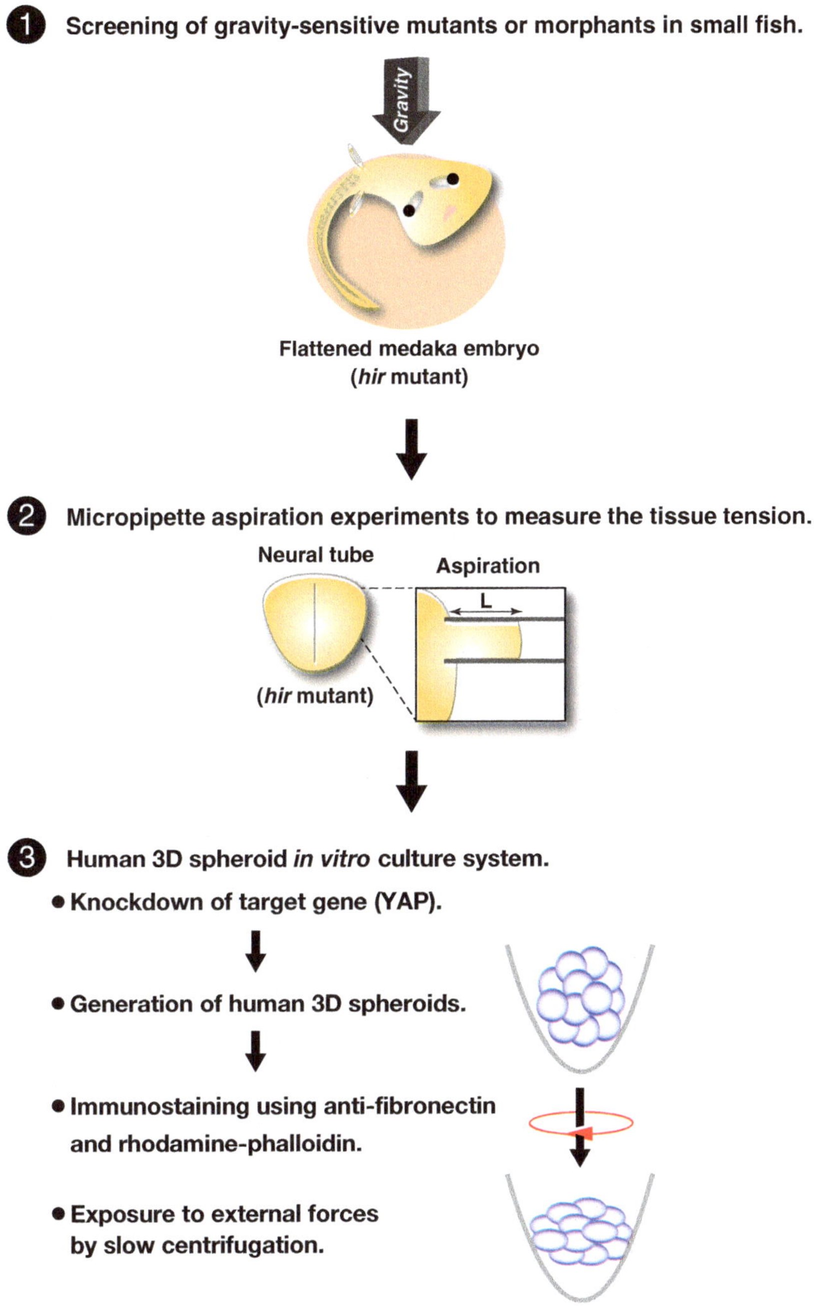

Fig. 2 A platform for analyzing gravity-resisting genes using small fish and spheroids

these 3D tissue environments play a pivotal role in organ physiology and pathogenesis [6]. In particular, the Hippo-YAP signaling pathway that controls organ size precisely responds to the mechanical properties of tissues and regulates cell proliferation and differentiation [7]. In fact, biophysical signals such as extracellular matrix stiffness are linked to YAP activity in cancer [8]. However, conventional 2D cell culture experimental systems do not accurately reflect the in vivo physical tissue conditions. For example, the stiffness of 2D cell culture dishes made from glass or plastic (GPa range) is several orders of magnitude higher than typical in vivo substrates (KPa range) [9], leading to ectopic activation of Hippo-YAP signaling in 2D culture system to abnormally promote cell proliferation. Therefore, it is necessary to carefully interpret the results from such an artificial 2D experimental system. Notably, in 2D cultures drugs are ten times more effective than in 3D cultures, so that 2D systems do not faithfully reflect in vivo drug efficacy [10].

1.2 The Merit of Combined Use of Fish and Human 3D Spheroids

It is important to select an appropriate animal model in order to investigate the fundamental questions in 3D morphogenesis. The fish models, zebrafish and medaka, are powerful models for studying the in vivo mechanism of 3D body formation [3]. First, transparent embryos and rapid development allow observation of the biological processes in the intact 3D environment. They also allow genetic dissection of biophysical processes in vivo [11]. Actually, the *hir* mutant can serve as a useful tool for the study of how 3D organs are generated and how they align to give rise to the global body shape [1]. To gain mechanistic insights into YAP regulation of tissue tension and its potential conservation in humans, a human 3D spheroid culture is an ideal system, because it can serve as an accessible in vitro system for testing human gene function in a controllable 3D environment. The human telomerase reverse transcriptase (hTERT) immortalized retinal pigmented epithelial cell line, hTERT-RPE1, was selected to generate 3D spheroids, since it is a normal cell line (non-cancerous) and displayed a relatively mild proliferation defect upon YAP knockdown. The human cell line also allowed advanced molecular biology tools such as YAP knockdown using siRNA and differential gene expression analysis using microarrays, which led to the identification of the Rho GTPase-activating protein ARHGAP18 as an effector of YAP in controlling tissue tension. Together, these analyses using both fish model organisms and human spheroids suggest a new evolutionarily conserved function of YAP in regulating tissue shape and alignment required for proper 3D morphogenesis [1].

2 Materials

2.1 Analysis of Medaka Mutants Using the Micropipette Aspiration Method

2.1.1 Micropipette Preparation

1. Glass capillary: length 76 mm, outer diameter 1.0 mm, inner diameter 0.75 mm.
2. Micropipette puller (e.g., Sutter Instrument, P97).
3. Microforge (e.g., Microdata 20 instrument, MFG5).
4. Fetal bovine serum (FBS).

2.1.2 Micropipette Aspiration Assay

1. Inverted confocal microscope (e.g., Leica, SP5).
2. Microfluidic flow control system (e.g., Fluigent, MFCS).
3. Micromanipulator (e.g., Eppendorf, TransferMan NK2).
4. Glass bottom dish: dish diameter 50 mm, glass area diameter 14 mm.
5. Phosphate buffered saline (PBS): 137 mM NaCl, 2.7 mM KCl, 8.1 mM Na_2HPO_4, 1.47 mM KH_2PO_4.
6. 1× Balanced salt solution (BSS): 20× BSS: 130 g NaCl, 8 g KCl, 4 g $MgSO_4{\cdot}7H_2O$, 4 g $CaCl_2{\cdot}2H_2O$, and 10 mg phenol red in 1 l Milli-Q water and autoclave; 500 mM HEPES in Milli-Q water autoclaved. Add 25 ml 20× BSS and 15 ml 500 mM HEPES, pH 7.0, fill up to 500 ml, and filter sterilize before use [12].

2.1.3 Data Analysis

1. Fiji software (https://fiji.sc/).

2.2 Analysis of Human 3D Spheroid

1. hTERT-RPE1 cells (American Type Culture Collection; CRL-4000).
2. Dulbecco's Modified Eagle's Medium (DMEM) high glucose with L-glutamine.
3. Fetal bovine serum (FBS).
4. Penicillin-Streptomycin Solution (100×): Streptomycin sulfate 10,000 mg/l (approximately 13.7 mM), penicillin G 10,000,000 units/l (approximately 18.7 mM).
5. 0.25% (w/v) trypsin/1 mM EDTA solution.
6. D-Phosphate buffered saline (PBS).
7. Low adherent cell culture plates, such as HydroCell™ (CellSeed Inc.), Cellstar® Cell-Repellent, 96-well plate (Greiner bio-one #650970), and Elplasia Micro-Space Cell Culture Plate Non-adherent surface (Kuraray Co., Ltd.). The characteristics of each culture plate are summarized in Table 1.
8. Wide bore tip (BMT-1000 W, BM).
9. Opti-Mem medium (Life Technologies).

Table 1
Characteristics of each cell culture plate for spheroid formation

Name of plate	HydroCell™ 12-well plate (#CS2002, CellSeed Inc.)	Cellstar® Cell-Repellent, 96-well plate (#650970, Greiner bio-one)	Elplasia® Micro-Space Cell Culture Plate, 24-well plate, Non-adherent surface (#RB 500400 NA 24, Kuraray Co., Ltd.)
Quantity of spheroid	Large quantities of spheroid	Small number of spheroid	Large number of spheroid
Shape of spheroid	Irregular shaped	Uniform	Uniform
Number of spheroid per well	NA	One spheroid/well	600 spheroids/well
Number of cells to be seeded per well	$1–2 \times 10^5$ cells/well	250–2000 cells/well	$0.5–1.2 \times 10^6$ cells/well
Conveniency	Good	Fine work is required	Good
Suitable experiment	Western blotting/ RNA isolation (RT-PCR)	The first condition setting, immunohistochemistry, drug response studies	Western blotting/ RNA isolation (RT-PCR)/ immunohistochemistry

10. Lipofectamine RNAiMax (Life Technologies).
11. Stealth RNA (100 pmol).
12. 6-well plate.
13. 12-well plate.
14. 440 mM [4% (w/v)] paraformaldehyde solution.
15. Triton X-100 (polyoxyethylene octylphenyl ether).
16. Image-iT™ Fx Signal Enhancer (Thermo Fisher Scientific Inc.).
17. Rhodamine-Phalloidin solution: Stock solution: Add 1.5 ml of methanol to the vial to dissolve the contents to yield a final concentration of 200 units/ml (approximately 6.6 μM). Working solution: Add 1 μl Rhodamine-Phalloidin stock solution and 200 μl PBS containing 1% (w/v) bovine serum albumin (BSA).
18. BSA.
19. Tween-20 (polyoxyethylene sorbitan monolaurate).
20. Normal goat serum.
21. Anti-human fibronectin antibody (rabbit) (Novotec).
22. Alexa Fluor® 488 goat anti-rabbit IgG H&L (Thermo Fisher Scientific Inc., ab150077).

23. DAPI solution: Stock solution: 10 mg DAPI dihydrochloride in 2 ml Milli-Q water (approximately 14.3 mM). Working solution: Add 1–2 μl DAPI stock solution and 1 ml PBS.
24. Anti-fade reagent (e.g., ProLong® Gold or SlowFade® Gold reagent).

3 Methods

3.1 Analysis of Medaka Mutants Using the Micropipette Aspiration Method

3.1.1 Preparing Micropipettes

1. Pull glass capillaries using a micropipette puller to make a 1 cm long taper [13].
2. Using a microforge, cut the tip of the taper of the capillary so that internal radius of the opening becomes 30–35 μm (*see* **Note 1**), and bend it to 40–45° angle at the point about 2–3 mm away from the tip (Fig. 3).
3. To coat the inner surface of the capillary, immerse its tip in FBS for 7 min.
4. Wash the inside of the capillary by PBS and store it with its tip kept in PBS (*see* **Note 2**).

3.1.2 Setting up Micropipette Aspiration System

1. Fill in the microfluidic flow control system with PBS.
2. Install micropipette filled with PBS to a pipette holder which is situated on a micromanipulator (*see* **Note 3**).
3. Set a glass bottom dish filled with BSS on the stage of the microscope, and put the tip of the micropipette into the medium. BSS in the dish should be heated at 28 °C.
4. Adjust the system to an initial state in which no flow occurs between the dish medium and inside of the micropipette.
5. Set a software of the microscope for time-lapse imaging with 500 msec time intervals.

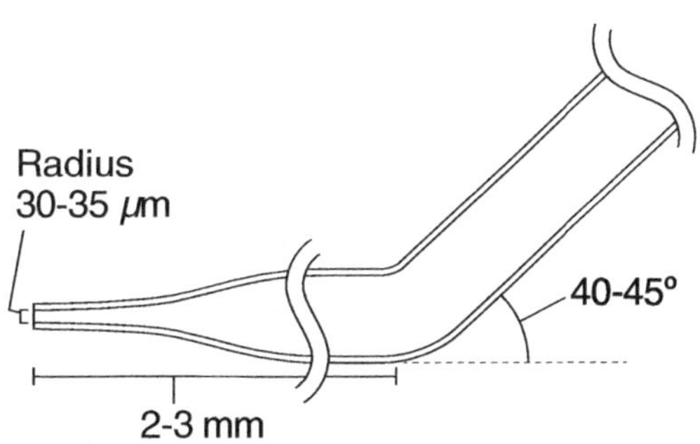

Fig. 3 Dimension of micropipette. The tip of taper is cut perpendicularly to the flow path at the point where the inner radius is 30–35 μm, followed by bending the body of the pipette to 40–45° about 2–3 mm away from the tip

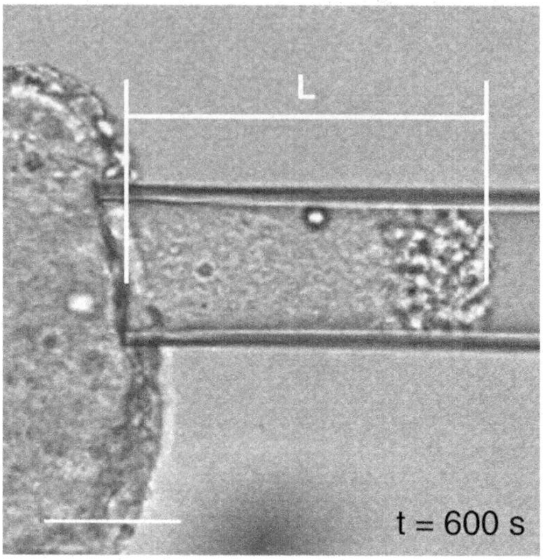

Fig. 4 Aspiration of medaka neural tube tissue. St. 22 neural tube was aspirated for 10 min with constant pressure. *L* length of measured tongue. Scale bar, 40 μm. Reproduced from Nature [1]

3.1.3 Micropipette Aspiration Assay for Medaka Neural Tube Tissue

The following steps should be done near or on the micropipette aspiration system:

1. Dissect out the whole neural tube from medaka embryos at stage 22, and cut it at the level of diencephalon-midbrain boundary using a tungsten needle under a stereoscope.
2. Transfer the dissected neural tube to the micropipette aspiration setup, and place the tip of the micropipette next to the open end of the neural tube.
3. Start recording a time-lapse movie, and then aspirate the neural tube tissue by the micropipette at a constant pressure ($\Delta P = 4.5$ mbar) for 10 min (Fig. 4).
4. Stop recording the movie, blow and remove the tissue from the micropipette by changing the pressure, and discard it.
5. Repeat **steps 1–4** using the same pipette.

3.1.4 Analyzing the Data

To measure the tongue length of the aspirated tissue, Fiji software is used.

1. Open a time-lapse movie on Fiji as a stack of images.
2. Measure the length between the tips of micropipette and tongue using a "Straight line" tool and a "Measure" function at each time point of the movie. Measured length will be shown in a "Results" window.
3. Once all the time points of a movie are analyzed, save the "Results" window as a CSV file for statistical analysis.

3.2 Analysis of Human 3D Spheroid

3.2.1 Cell Culture

1. hTERT-RPE1 cells are maintained with DMEM containing 10% FBS and antibiotics at 37 °C incubator with humidified atmosphere of 5% CO_2.
2. Passage the cells into an appropriate number of 10 cm plastic culture dish and use the cells before confluent.

3.2.2 Spheroid Formation with 12-Well Plates

When you want to make in large quantities of spheroid, we recommend this method.

1. Aspirate the media and wash with 5 ml of D-PBS.
2. Add 1 ml of 0.25% trypsin/1 mM EDTA to the dish and incubate for 2 min at 37 °C, and then detach the cells.
3. Add 5 ml of fresh media and suspend the cells and collect into 15 ml tube.
4. Centrifuge at 300 × *g* for 3 min.
5. Aspirate the supernatant and resuspend in fresh media.
6. Take a 20 μl sample of the cell suspension to determine the viable cell density.
7. Dilute the cell suspension to 2–4 × 10^5 cells/ml.
8. Add 0.5 ml of cell suspension to each well (1–2 × 10^5 cells/well) of low adherent cell culture plates with 12 wells (Hydro-Cell™ 12 well) (*see* **Notes 4** and **5**).
9. Incubate the cells at 37 °C incubator with humidified atmosphere of 5% CO_2, and spheroid will be formed within 12–24 h (Fig. 5a).

3.2.3 Spheroid Formation with Elplasia® Micro-Space Cell Culture Plate (24-Well Plate)

When you make a large number of spheroids uniform in size and shape, we recommend this method. Elplasia® Micro-Space Cell Culture Plate is a unique 3D cell culture tool, and each well has hundreds of dips, in which a spheroid is formed at the bottom (Fig. 5b).

1. Trypsinize the cells and prepare cell suspension as Subheading 3.2.2.
2. Count the cell density and dilute the cell suspension to appropriate cell number (e.g., 5.8 × 10^5 cells/ml for 500 cells spheroid).
3. Add 0.5 ml of cell suspension to each well (*see* **Notes 4** and **5**).
4. Incubate the cells at 37 °C incubator, and spheroid will be formed within 12–24 h (Fig. 5b) (*see* **Note 6**).

3.2.4 Spheroid Formation with Cellstar® Cell-Repellent (96-Well Plate)

When you make a small number of spheroids uniform in size and shape, we recommend this method.

1. Trypsinize the cells and prepare cell suspension as Subheading 3.2.2.

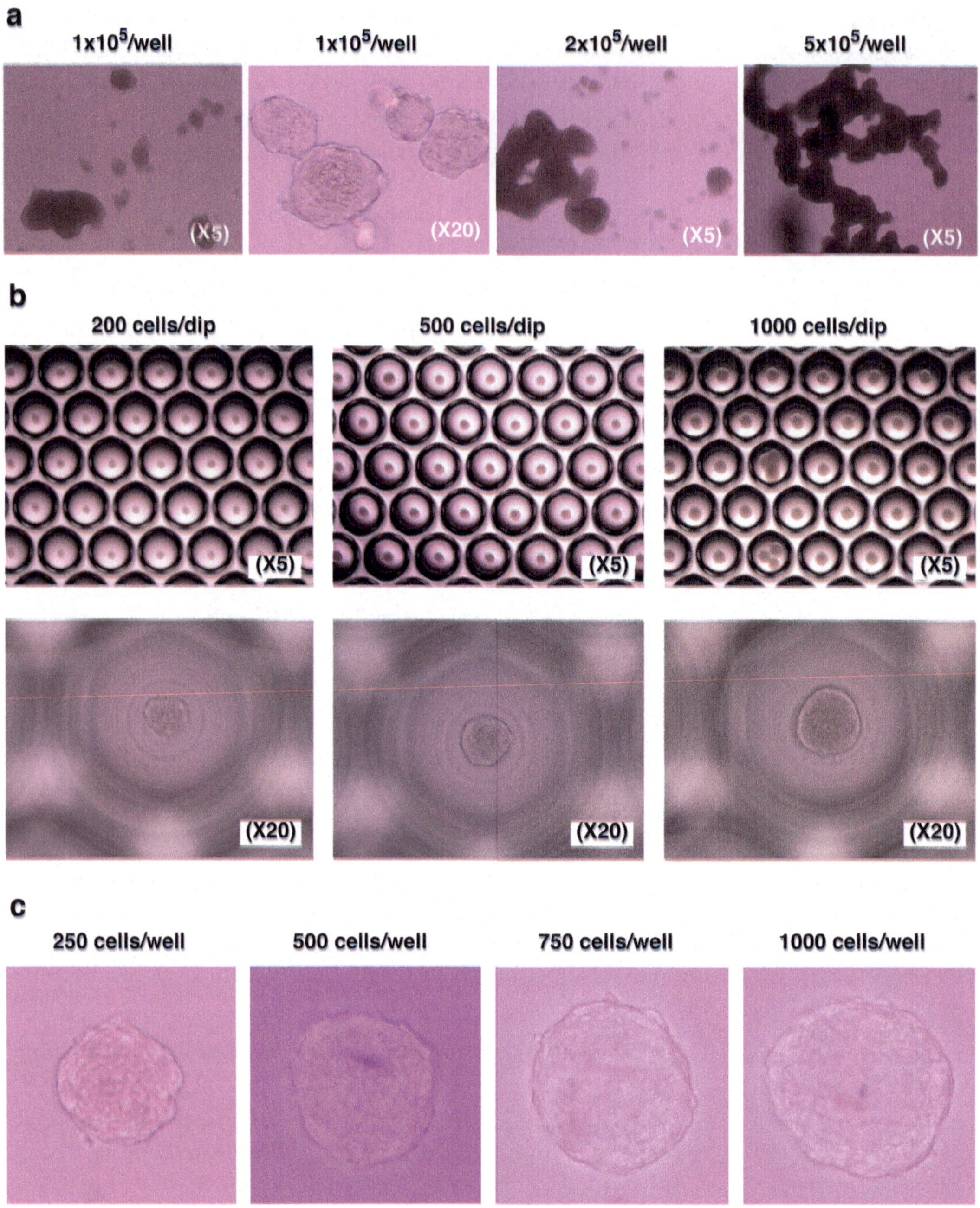

Fig. 5 3D spheroids developed from human RPE1 cell using HydroCell™ 12-well plate (**a**), Elplasia® 24-well plate (**b**), and Cellstar® Cell-Repellent 96-well plate (**c**)

2. Count the cell density and dilute the cell suspension to appropriate cell density ($0.25–2 \times 10^4$ cells/ml).
3. Add 100 μl of cell suspension to each well (*see* **Note 4**).
4. Incubate the cells at 37 °C incubator, and spheroid will be formed within 12–24 h (Fig. 5c).

3.2.5 Spheroid Gene Knockdown

1. Trypsinize the cells and prepare cell suspension as Subheading 3.2.2.
2. Count the cell density and dilute the cell suspension to 1×10^5 cells/ml.
3. Add 2 ml of cell suspension to each well in 6-well plates (2×10^5 cells per well).
4. Incubate the cells at 37 °C incubator.
5. Next day, transfection using Lipofectamine RNAiMax is performed.
6. Mix reagents in polystyrene tubes as below (for 6 wells):

a. Master Mix for siRNA	Opti-MEMI Stealth RNA (40 μM)	1500 μl 15 μl
b. Master Mix for RNAiMax	Opti-MEMI RNAiMax	1500 μl 36 μl

7. Mix by tapping and incubate for 5 min at room temperature.
8. Add **b** (Master Mix for RNAiMax) to **a** (Master Mix for siRNA) dropwise, and incubate for 20 min at room temperature.
9. Add 500 μl per well dropwise and swirl gently.
10. Incubate the cells at 37 °C incubator for 24 h.
11. Aspirate the media and wash with 2 ml of D-PBS.
12. Add 0.5 ml of 0.25% trypsin/1 mM EDTA per well and incubate for 2 min at 37 °C, and then detach the cells by smacking.
13. Add 5 ml of fresh media and suspend the cells and collect into 15 ml tube.
14. Centrifuge at 300 × *g* for 3 min.
15. Aspirate the supernatant and resuspend in 2 ml of fresh media.
16. Seed these resuspensions to 6 wells of a low adherent 12-well plate (Hydrocell™), and incubate for 24–48 h at 37 °C (*see* **Notes 4** and **5**).
17. Subject the spheroid with gene knockdown to immunostaining or other analysis.

3.2.6 Spheroid Immunostaining Using Anti-fibronectin and Rhodamine-Phalloidin

1. Prepare hTERT-RPE1 cell spheroids with Cellstar® Cell-Repellent, 96-well plate.
2. Aspirate the spheroid one by one by pipetting using wide bore tip to avoid mechanical damage, and collect them into 15 ml tube.
3. Leave the spheroids until the spheroids sink to the bottom of the tube.

4. Aspirate the medium and wash the spheroid 5 ml of D-PBS twice.
5. Add 1 ml of 4% paraformaldehyde solution and incubate for 6–24 h at 4 °C.
6. Centrifuge at 1000 × *g* for 3 min at 4 °C, and then aspirate the fix solution.
7. Add 5 ml of D-PBS and mix by inverting.
8. Centrifuge at 1000 × *g* for 3 min at 4 °C, and then aspirate the supernatant.
9. Repeat **steps** 7 and **8**.
10. Add 1 ml of 0.1% Triton X-100 in D-PBS, and transfer to a new 1.5 ml microcentrifuge tube.
11. Incubate for 15 min gently inverting.
12. Centrifuge at 1000 × *g* for 3 min, and then aspirate the supernatant.
13. Add 1 ml of D-PBS and mix by inverting.
14. Centrifuge at 1000 × *g* for 3 min, and then aspirate the supernatant.
15. Repeat **steps 13** and **14** twice.
16. Add 2 drops of Image-iT™ Fx Signal Enhancer and incubate for 30 min at RT.
17. Centrifuge at 1000 × *g* for 3 min, and then aspirate the supernatant.
18. Add 100 μl of Rhodamine-Phalloidin in PBS containing 1% BSA (1:200 dilution), and incubate for 30 min at 4 °C. Incubation overnight is also possible. From this step, incubation should be performed under the dark.
19. Wash the spheroid with 1 ml of D-PBS three times.
20. Add 100 μl of 1% normal goat serum in PBS containing 0.1% Tween-20, and incubate for 30 min at room temperature in the dark.
21. Centrifuge at 1000 × *g* for 3 min, and then aspirate the supernatant.
22. Add primary antibody solution prepared as below, and incubate for 2 hrs at room temperature. If you want to stop at this step, incubation overnight at 4 °C is also possible.

Anti-human fibronectin antibody (Rb)	0.5 μl
1% normal goat serum in PBS containing 0.1% Tween-20	100 μl

23. Add 0.1% Tween-20 PBS and incubate for 5 min with inverting by rocking mixer.

24. Centrifuge at 1000 × *g* for 3 min, and then aspirate the supernatant.
25. Repeat **steps 23** and **24** twice.
26. Add secondary antibody solution prepared as below, and incubate for 1 h at room temperature.

Alexa Fluor® 488 goat anti-rabbit IgG (H&L)	0.5 μl
1% normal goat serum in PBS containing 0.1% Tween-20	100 μl

27. For counterstaining, add 1 ml of 5 μg/ml DAPI solution, and incubate for 15 min at room temperature (*see* **Note 7**).
28. Centrifuge at 1000 × *g* for 3 min, and then aspirate the supernatant.
29. Add 0.1% Tween-20 PBS and incubate for 5 min with inverting by rocking mixer.
30. Centrifuge at 1000 × *g* for 3 min, and then aspirate the supernatant.
31. Repeat **steps 28** and **29** twice.
32. To prevent fading, mount the immunostained sample with anti-fade reagent (e.g., ProLong® Gold or SlowFade® Gold reagent) according to the manufacture's instruction.
33. View the localization of fibronectin and F-actin using confocal laser scanning microscopy (Fig. 6).

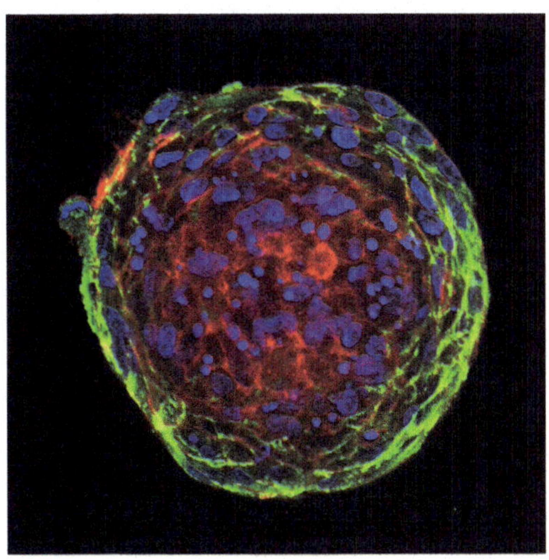

Fig. 6 Merged image of 3D human RPE1 spheroid stained for F-actin (red), fibronectin (green), and nuclei (blue)

4 Notes

1. The point of micropipette should be perpendicular to the flow path of the pipette. This is important for accurate measurement of aspirated tongue length.
2. We keep micropipettes hung from the inner surface of a 50 ml test tube filled with PBS using a sticky material such as clay. Be careful not to break the tip of the pipettes while storing them in the tube.
3. Make sure that no bubbles are inside of tubing of the system as they prevent precise control of the pressure for aspiration.
4. When you add cell suspension, do not touch the bottom of the well by the tip of pipet, since the well is fixed with cell repellent coating, such as poly-hydroxyethyl methacrylate.
5. Tap lightly a few times from the side of the plate to remove the air bubble.
6. After spheroid formation, handle the plate carefully. If you touch the plate roughly, the spheroids may transfer to the neighbor dips.
7. Other counterstaining reagents are available. (e.g., TO-PRO®-3).

Acknowledgments

We would like to thank Hiroshi Nishina (Tokyo Medical and Dental University, Japan), Hisato Kondoh (Kyoto Sangyo University, Japan), and other collaborators for the Kyoto medaka mutant screening. The research of gravity and YAP was partially supported by JSPS KAKENHI Grant Numbers 16H01643, 16H01644, and 17H05769. Yoichi Asaoka and Hitoshi Morita contributed equally to this chapter.

References

1. Porazinski S, Wang H, Asaoka Y, Behrndt M, Miyamoto T, Morita H et al (2015) YAP is essential for tissue tension to ensure vertebrate 3D body shape. Nature 521:217–221
2. Furutani-Seiki M, Sasado T, Morinaga C, Suwa H, Niwa K, Yoda H et al (2004) A systematic genome-wide screen for mutations affecting organogenesis in Medaka, *Oryzias latipes*. Mech Dev 121:647–658
3. Furutani-Seiki M, Wittbrodt J (2004) Medaka and zebrafish, an evolutionary twin study. Mech Dev 121:629–637
4. Watanabe T, Asaka S, Kitagawa D, Saito K, Kurashige R, Sasado T et al (2004) Mutations affecting liver development and function in Medaka, *Oryzias latipes*, screened by multiple criteria. Mech Dev 121:791–802
5. Nelson CM, Bissell MJ (2006) Of extracellular matrix, scaffolds, and signaling: tissue architecture regulates development, homeostasis, and cancer. Annu Rev Cell Dev Biol 22:287–309
6. Humphrey JD, Dufresne ER, Schwartz MA (2014) Mechanotransduction and extracellular matrix homeostasis. Nat Rev Mol Cell Biol 15:802–812

7. Halder G, Dupont S, Piccolo S (2012) Transduction of mechanical and cytoskeletal cues by YAP and TAZ. Nat Rev Mol Cell Biol 13:591–600
8. Calvo F, Ege N, Grande-Garcia A, Hooper S, Jenkins RP, Chaudhry SI et al (2013) Mechanotransduction and YAP-dependent matrix remodelling is required for the generation and maintenance of cancer-associated fibroblasts. Nat Cell Biol 15:637–646
9. Butcher DT, Alliston T, Weaver VM (2009) A tense situation: forcing tumour progression. Nat Rev Cancer 9:108–122
10. Antoni D, Burckel H, Josset E, Noel G (2015) Three-dimensional cell culture: a breakthrough in vivo. Int J Mol Sci 16:5517–5527
11. Heisenberg CP, Bellaïche Y (2013) Forces in tissue morphogenesis and patterning. Cell 153:948–962
12. Porazinski SR, Wang H, Furutani-Seiki M (2011) Essential techniques for introducing medaka to a zebrafish laboratory—towards the combined use of medaka and zebrafish for further genetic dissection of the function of the vertebrate genome. Methods Mol Biol 770:211–241
13. Maître JL, Berthoumieux H, Krens SF, Salbreux G, Jülicher F, Paluch E et al (2012) Adhesion functions in cell sorting by mechanically coupling the cortices of adhering cells. Science 338:253–256

Chapter 15

Regulation of YAP/TAZ Activity by Mechanical Cues: An Experimental Overview

Sirio Dupont

Abstract

YAP/TAZ activity is regulated by a complex network of signals that include the Hippo pathway, cell polarity complexes, and signaling receptors of the RTK, GPCR, and WNT pathways and by a seamlessly expanding number of intracellular cues including energy and mevalonate metabolism. Among these inputs, we here concentrate on mechanical cues embedded in the extracellular matrix (ECM) microenvironment, which are key regulators of YAP/TAZ activity. We review the techniques that have been used to study mechano-regulation of YAP/TAZ, including conceptual and practical considerations on how these experiments should be designed and controlled. Finally, we briefly review the most appropriate techniques to monitor YAP/TAZ activity in these experiments and their significance to study the mechanisms linking YAP/TAZ to mechanical cues.

Key words YAP/TAZ, Hippo, Mechanical cue, Mechano-regulation, Mechanotransduction

1 General Considerations on Mechanotransduction Experiments

The activity of YAP/TAZ transcriptional coactivators is regulated by multiple upstream inputs including the Hippo pathway, which remains by far the most studied and validated in vivo, and a number of different signals (GPCRs, cAMP/PKA, WNT, mevalonate, and energy metabolism) [1]. Among these, the discovery of YAP/TAZ regulation by extracellular matrix (ECM) mechanical cues was particularly significant, as it linked the activity of the Hippo pathway to a previously unrecognized regulator of cell and tissue homeostasis and because it provided the mechanobiology community with a missing biological readout to study mechanosensing at the cellular level and with a relevant transducer to explain the phenotypic effects of mechanical cues on cells and tissues [2, 3].

The study of the physical properties of cells and tissues in general, and of ECM adhesion-mediated mechanical cues in particular, requires an expertise at the border between cell biology, material sciences, and engineering. Indeed, while it is universally

Alexander Hergovich (ed.), *The Hippo Pathway: Methods and Protocols*, Methods in Molecular Biology, vol. 1893, https://doi.org/10.1007/978-1-4939-8910-2_15,

accepted that physical forces have profound effects on biological systems, designing experiments to faithfully manipulate the physical properties of the cell microenvironment without altering other variables remains a challenge [4]. In the case of ECM mechanical cues, for example, one important issue is the ability to dissociate the effects due to ECM "chemical" signaling (i.e., the different nature of distinct ECM adhesion molecules, the density of ligands, their spatial presentation, etc.) from the effects purely delivered by the physical properties of the ECM. Thus, in recent years several experimental systems have been developed and used to study the effects of ECM mechanical cues on cells and on YAP/TAZ activity that are briefly reviewed here.

2 ECM Mechanical Cues

In tissues, ECM elasticity is regulated by an array of factors, including ECM composition (e.g., what type of adhesive protein is produced), concentration, and posttranslational modifications (e.g., bundling and cross-linking of ECM protein monomers into complicate 3D networks, localized degradation by extracellular proteases). Further complicating the issue, ECM physicochemical properties are also regulated by other macromolecules such as glycosaminoglycans and proteoglycans, even if these are not always directly in contact with cells and their ECM receptors. Every and each of these factors act both as "chemical" signal and contribute to the overall physical properties of the ECM. Moreover, one variable that can be underestimated is the ability of cells themselves to produce ECM proteins and rearrange the pre-existing ECM, with the prototypical example of collagen gel contraction (and, thus, stiffening) by fibroblasts.

Several alternative approaches have been developed to study the effects of the physical properties of the ECM on cells; the degree to which each experimental system uncouples "chemical" from mechanical signals is usually directly proportional to the practical difficulties encountered to set up and subsequently utilize these systems.

2.1 Natural Extracellular Matrix Gels

Several alternatives are commercially available to assemble extracellular matrices composed of natural ECM proteins of varying stiffness. Usually, the easiest way to modulate ECM physical properties consists in assembling gels of different ECM concentrations. It is important to consider in this case that just coating the cell culture substrate (i.e., plastic or glass) with varying amount of an ECM protein is completely different from assembling a thick gel of ECM. When the layer of ECM is thin, what changes is only the "chemical" nature of the ECM: tissue-culture plastics or glass slides favor cell attachment mediated in general by integrins; coating it with fibronectin will favor adhesion mediated by integrin-β1 (and thus, in some cell types, a pro-migratory input), while coating with laminin or basal membrane (BM) equivalents will favor adhesion mediated by integrin-β4 (and thus, in some cell types, a cue instructing apicobasal polarity). However, the mechanical properties of the thin ECM layer reflect the properties of the underlying substratum, which are typically several orders of magnitude stiffer than any tissue in the human body. ECM-coated plastics can be thus conveniently used experimentally to study cells in a very stiff condition. When the layer of ECM becomes thick enough, cells will mainly sense the physical properties of the superficial portion of the ECM; in this case, variation of ECM concentration will produce hydrogels of varying elasticity or stiffness that can be then used to explore how cells respond to this.

Using natural extracellular matrices is thus limited to ECM proteins that can be obtained in a liquid form and then gelled by physical or chemical reactions. This includes high-concentration collagen-1 that is sold in its soluble form, and which can be precipitated, forming a hydrogel, by changing the pH of the solution, BM equivalents that are liquid below 4 °C and rapidly form hydrogels at room temperature, and hyaluronic acid. Any experiment aiming at studying the mechanical properties of the ECM with these reagents has two intrinsic limitations: (i) technically, unless one have easy access to an atomic force microscope with the appropriate cantilever tips and standard controls to calibrate the measure, it is difficult to provide an absolute quantification of the stiffness of the ECM hydrogels. This potentially disfavors the reproducibility of data in other labs, unless a positive control is provided to gauge the effects of the ECM; (ii) given that stiffness is regulated by the concentration of ECM molecules, the possibility always exists that cells read the number/concentration of ECM ligands, rather than their actual mechanical properties. This can only be sorted out by providing independent evidence based on alternative reagents and experimental setups (see below).

2.2 Synthetic Matrix Gels

The world of ECM engineering is seamlessly advancing, and we recommend reading more specialized reviews to explore the incredible possibilities put forward by synthetic matrices for cell and tissue biology. We will cover here only the most used systems, some of which are now becoming commercially available.

The basic idea of a synthetic matrix is to produce a hydrogel formed by a chemical to which cells cannot attach and that is very simple to polymerize and/or cross-link (by chemicals, heat, or even light) and then to coat the surface of this hydrogel with a natural ECM protein to enable cell attachment. In this way, the hydrogel sets the stiffness of the overlying ECM protein layer. These methods have the advantage of being more easily reproducible than natural ECM hydrogels, as it is easier to control the cross-linking reactions with few chemical components. The two most used synthetic hydrogel systems are those based on polyacrylamide (acrylamide and bis-acrylamide) and those based on polyethylene glycol (PEG), where the stiffness of the hydrogel is proportional to the concentration of the components/cross-linkers. The main difficulty in this case is to provide a stable attachment for cells by chemically cross-linking the ECM adhesive protein to the hydrogel: several reagents exist depending on the peculiar chemistry of the synthetic hydrogels, but it is still rather time-consuming to obtain a uniform coating within each hydrogel and to coat hydrogels of different stiffness with a similar amount of ECM protein (which is usually controlled by using fluorescent-labeled adhesive proteins). This is relevant because a stiff but insufficiently coated ECM hydrogel will not provide enough cell attachment for cells to spread, and will thus induce cell phenotypes typical of a soft ECM, even if it is stiff upon AFM measurements. So, the effects of any hydrogel should be measured at multiple levels (rheology, cell area/spreading, focal adhesion maturation) before measuring YAP/TAZ activity and interpreting the results. A good rule-of-the-thumb is to check cell adhesion and cell geometry 3–6 hours after plating cells on hydrogels: all hydrogels, irrespective of the stiffness, should promote cell adhesion (i.e., cells should not remain floating, such that cells move by just flowing the culture medium on them), and then, a stiff hydrogel must promote cell spreading compared to a softer one, where cells should progressively become smaller until they remain rounded up.

A possible caveat is that some cells produce high amounts of ECM proteins, which can be absorbed in the hydrogel even in absence of any chemical cross-linker. Thus, after prolonged incubation (e.g., overnight) cells will end up adhering to the secreted ECM and to the hydrogel and will likely take on a more spread or more rounded shape depending on its stiffness. However, care should be taken when interpreting results on this kind of hydrogels because one may confuse the differential effects of the physical properties of the ECM with the differential ability of a cell to produce ECM proteins. These are often two parts of the same process in real tissues but rely on different molecular mechanisms.

A second caveat when using synthetic hydrogels is that cells cannot remodel them. This is desirable when cells are plated on top of the hydrogel, because the mechanical properties of the ECM will be minimally affected by secretion of proteases or by cell traction forces. When cells are instead embedded into a synthetic hydrogel, this can lead to paradoxical results: if the cells are boxed by the hydrogel in a small geometry, and cells are not able to dig some space and spread, YAP/TAZ will remain inhibited even if the hydrogel is stiff [5, 6]. This can be overcome by including in the hydrogel network some peptides bearing the recognition motif for ECM proteases, such that cells can actively remodel the hydrogel.

The use of synthetic hydrogels also opened the way to more specific applications, including:

- Coupled to photolithography, the opportunity to pattern the adhesive ECM proteins in discrete geometries, including tracks for single vs. collective cell migration.
- The possibility of incorporating into the hydrogel some beads, and to functionalize only those for ECM attachment, so that it becomes possible to study at the nanometer scale how cells bind ECM adhesive proteins and develop cell-substratum adhesion complexes.
- Traction force microscopy, where fluorescent beads are incorporated into the hydrogel such that it is possible to visualize and measure the pulling forces that a cell exerts on its substratum.

An important variable in experiments where ECM elasticity is manipulated is cell crowding. This is particularly important in the case of YAP/TAZ, because cell crowding inhibits YAP/TAZ (through both mechanical and nonmechanical mechanisms – see below). We have previously shown that ECM elasticity and cell crowding can cooperate in regulating YAP/TAZ [7]. This on one side indicates that it is not absolutely required to perform these experiments by plating sparse single cells (even if it is desirable, as the effects will be maximal in these conditions), but on the other suggests that cell density (number of nuclei/area) should be carefully evaluated when comparing different experimental conditions. Moreover, the stiffening of the ECM does not only change cell proliferation rates but also their migratory behavior: for example, MCF10ATk1 cells embedded in Matrigel form filled spheres, while in Matrigel + collagen (which is stiffer) start migrating into the matrix due to the combined effect on proliferation and scattering. Similarly, if one compares an epithelial cell with its corresponding oncogene-transformed progeny that underwent epithelial-to-mesenchymal transition, these cells will form structures with intrinsic differences in cell architecture even if the matrix has the same mechanical properties.

Another variable to consider in these experiments is dimensionality, that is, the intrinsic differences between a 2D and a 3D microenvironment [8]. In 2D cells experience ECM rigidity directionally, through their basal surface, and assume a very flattened shape (on stiff ECM), while in 3D cells are completely embedded in ECM fibrils and likely contact them in all directions, such that the ECM is presented (and, likely, sensed) in a different manner. Moreover, other cell parameters including polarity, membrane curvature, and diffusion of cell-secreted factors might change between a between a 2D and a 3D ECM [9]. Finally, synthetic ECM gels exhibit a quasi-linear elastic behavior, while the natural ECMs are characterized by nonlinear elastic behavior with significant viscous components. These issues and their possible effects on YAP/TAZ have not been addressed yet in a formal experimental manner. Still, despite all possible complications, available evidence suggests that YAP/TAZ are readily regulated by ECM stiffness, independently of ECM presentation, 2D vs. 3D, and whether the ECM is natural or synthetic.

3 ECM Micropatterns

Initial studies on the physical properties of the ECM were based on the use adhesive substrata (glass or culture plastics) made partially inaccessible to cell-adhesion receptors, for example, by using poly-HEMA or integrin-blocking reagents [10–13]. It is possible in this way to modulate cell adhesion in a subtler way than just comparing the two extremes, adhesion vs. suspension. It soon became apparent that decreasing cell adhesion proportionally decreases cell spreading, raising the issue of what was the underlying key parameter: the geometry of the ECM, or the amount of ECM? This was elegantly solved by developing ECM micropatterning techniques, which showed unequivocally that cell geometry regulates cell traction forces (or actomyosin contractility) and cellular responses accordingly [14–16].

Parallel studies performed on ECM-coated hydrogels [17] suggested a similarity between ECM stiffness and cell shape: decreasing ECM elasticity proportionally decreases cell spreading. Strikingly, the phenotypes induced by a soft ECM and by a small (yet, extremely stiff) micropattern are very similar, indicating that cells "read" these two different properties at the same manner, and through common mechanisms: in both conditions cells are unable to mature focal adhesions, to develop stress fibers, to build high levels of actomyosin contractility, and to activate YAP/TAZ [18]. This provides a strong rationale for the use of ECM micropatterns as a convenient tool to study ECM mechanical properties; at the same time, it indicates that cell geometry can be dominant over ECM stiffness in controlling cell phenotypes.

Photolithography and microcontact printing techniques enable the very precise deposition of ECM proteins onto glass slides or other materials such as silicone (PDMS) or hydrogels, with basically any desired bidimensional geometry at micrometer scale. The surface is coated with a photo-inducible chemical cross-linker, which is then reacted with the desired ECM protein in solution; empty spaces are instead protected by coating with chemicals preventing non-specific sticking of ECM proteins and/or cell adhesion receptors. ECM micropatterns can be obtained in collaboration with material scientists and engineers, commercially, or through recent systems that have been developed to make photolithography more accessible in a cell-biology laboratory, based on the use of a standard optical microscope and of commercial reagents.

ECM micropatterning can be used to study cell behavior both on single cells and on cell clusters. For example, it is possible to pattern cell-cell contacts in cell doublets and study the relative importance of cell-substratum vs. cell-cell adhesion structures (and forces) for any given phenotype. Alternatively, this has been used in the past to study how spatial constraints can influence cell behavior in a monolayer: multicellular micropatterns of varying shape can induce local stretching or compressions in cells, which correspond to patterned responses [7, 19]. These data should however be reconsidered also in light of the growing appreciation that cells in monolayers can actively migrate [20], raising the possibility of differential migration between the external and internal regions of those multicellular micropatterns.

4 Cell Crowding

When cells are grown in vitro for a long time without passaging, they occupy the whole available surface and then increase their density until they undergo growth arrest, in a process generally known as contact inhibition of growth. This process is observed both in epithelial and in mesenchymal cells, and escape from this behavior is considered a sign of oncogene transformation. The notion that YAP/TAZ are negatively regulated when cells become confluent is well known and precedes the discovery of mechanical signals as upstream YAP/TAZ inputs [21]. The formation of cell monolayers entails the remodeling of many cell structures: cells form cell-cell adhesions (adherens junctions mediated by cadherins, tight junctions in the case of some epithelial cells, plus other types of junctions), change their polarity, and remodel their cortical cytoskeleton. Moreover, formation of cell-cell junctions is accompanied by an overall reduction of cell-substrate adhesions [22]. This can be due both to an inhibitory cross talk between cell-cell and cell-substrate adhesions and also to the simple reduction of cell area available for cell-substrate adhesion in very dense monolayers, which is visible by the disappearance of basal stress fibers. The analogy between reduction of cell area in dense monolayers and the effects of a small area attained by micropatterning suggested the involvement of similar regulatory mechanisms based on F-actin remodeling [15, 23] and finally led to the formal demonstration that cell crowding entails a mechanical checkpoint [7, 24]. Indeed the sole stretching of a dense monolayer is sufficient to reactivate YAP/TAZ, and depletion of F-actin capping/severing proteins can boost YAP/TAZ both in dense monolayers and on a soft ECM, indicating a common mechanism based on actin [7]. Thus, cell crowding can be also used as a mean to change the cells' mechanical properties and to modulate YAP/TAZ activity accordingly.

Clearly, however, formation of a dense monolayer entails also nonmechanical regulations of YAP/TAZ. Cell-cell adhesion is generally thought to be an important subcellular compartment for Hippo/LATS signaling to take place, even if MST1/MST2 DKO MEFs still undergo density-dependent YAP regulation [25], and LATS1/LATS2 KO MEFs remain surprisingly untested in the context of contact inhibition. Formation of adherens junctions recruits alpha-catenin to the membrane, where it anchors cadherins to the actin cytoskeleton and where it also plays direct or indirect YAP-/TAZ-inhibitory functions [26, 27]. Finally, formation of cell-cell adhesions is also coupled, in many cell systems, to establishment or reinforcing of apicobasal cell polarity cues that contribute at many levels in the regulation of YAP/TAZ: negative regulation by the Scribble and Crumbs complexes has been reported by experiments in *Drosophila* and mammals [28, 29]; but also positive regulation by the PAR/MARK polarity regulators was observed in the early mouse embryo, where the most polarized tissue is the one displaying the strongest activation of YAP/TAZ, in *Drosophila* and in some mammalian cellular systems [30–33]. Thus, the use of cell crowding as "mechanical cue" requires a series of controls to first establish whether or not F-actin remodeling is experimentally relevant in the cells and conditions used, as opposed to many other inputs that contribute to inhibit YAP/TAZ in conditions of high density (e.g., YAP/TAZ activity should be rescued by depletion of F-actin capping/severing proteins or by stretching).

Another important technical issue, often poorly considered when performing cell crowding experiments, is how uniform (internal to each sample) and how consistent (across different experimental conditions) is crowding. Contact inhibition of YAP requires a much higher density of cells compared to what usually considered as a "100% confluent" culture and time to get established (in our hands, with MCF10A cells, at least 48 hours – see below). Indeed contact inhibition of YAP/TAZ is exquisitely dependent on cell density, and it is common observation from many labs that even slight differences in local density can lead to significant differences in YAP localization (and likely activity). Thus, care should be taken when performing this type of experiments:

1. Titrate the number of cells plated, and establish the time needed to attain a complete and uniform YAP exclusion from nuclei by immunofluorescence.
2. If using YAP immunofluorescence localization as readout, several pictures should be taken at random in different areas of each sample and quantified for nuclear/cytoplasmic ratio.

3. If comparing different cultures (e.g., cells transfected with different siRNAs), the density of nuclei/area should be quantified to be sure this is not changing in the first place; this should also apply when YAP/TAZ activity is measured by qPCR, as in this case differences in cell density might not be apparent because cells are lysed in bulk.

A last consideration on the use of cell crowding as "mechanical cue" relates to what type of mechanical cue is being considered. Most data on regulation of YAP/TAZ were obtained on sparse cell cultures or isolated single cells and thus with ECM- and integrin-borne mechanical cues. Other types of mechanical forces however exist, which may still regulate YAP/TAZ, also by different mechanisms. For example, shear stresses induce YAP/TAZ nuclear localization and activity in endothelial cell monolayers, likely owing to remodeling of the actin cytoskeleton to better resist to the forces applied by a flowing liquid. Another example is the myosin-dependent tension observed in some epithelia across cell–cell junctions: in this case it remains unclear whether tension is instrumental to establish the correct epithelial architecture, and thus to inhibit YAP/TAZ, or whether tension sustains YAP/TAZ activity, compensating in part the reduction of forces exerted through cell-substrate adhesions. In general, it is thus essential to use crowding as a "mechanical cue" in combination with another experimental technique (ECM hydrogels or micropatterning) based on single-cell cultures, to ensure that the observed phenotypes and/or mechanisms are widespread and common, as opposed to being specific for crowding.

5 Cell Detachment

Another manner to study ECM adhesion-mediated mechanical cues is to completely detach cells from the substratum, by culturing cells in nonadhesive vessels. This condition entails the complete inhibition of integrin-mediated cell adhesion and signaling. Upon detachment, the F-actin cytoskeleton is dramatically remodeled, and YAP is rapidly excluded from the nucleus, such that any round of trypsinization/reseeding of an adherent cell line can be considered as a round of YAP/TAZ inactivation/reactivation. Of note, this YAP-centered mechanism plays a fundamental role to induce anoikis, i.e., cell death that ensues cell detachment in most cells normally growing in 2D [23, 34]. It remains unknown whether it is possible to rescue YAP/TAZ activity in detached cells by knockdown of F-actin capping/severing proteins, or by physically changing cell shape (e.g., by compression between non-adherent plates), which would formally indicate a mechanism related to F-actin contractility and/or to nuclear mechanics [35] also in this experimental setup.

Cell detachment likely activates YAP-/TAZ-independent signaling events, as demonstrated by the observation that MCF10A mammary epithelial cells embedded in a 3D very soft ECM (Matrigel) grow, but cells losing the contact with the ECM and growing within the other cells undergo apoptosis [36]. This might still be due to quantitative differences in YAP/TAZ activity (i.e., on a soft ECM, a residual amount of YAP/TAZ remains active, while in detached cells, all YAP/TAZ are inactivated), but also to the activation/inactivation of parallel pathways specifically sensitive to the presence/absence of adhesions.

When cells are kept in suspension and in isolation, this will rapidly lead to cell death. If instead cells can form multicellular clusters, this enables to some extent cell survival and growth. This assay, in the presence of very rich and strong growth-promoting media supplements, is commonly used in cancer biology to assay for the presence of "cancer stem cells," because cancer cell populations often contain subpopulations selectively able to resist growth in suspension, forming the so-called cancer stem cell spheroids. It remains unknown what is the role of the ECM produced by the cells themselves in these clusters, and thus possibly a role for ECM-mediated (or cell-cell mediated?) mechanical signals, to promote survival.

6 Flow/Shear/Pressure Forces

It is also possible to directly apply forces on cells without changing the properties of the ECM to which cells are attached. This is physiologically relevant for tissues such as endothelia or kidney cells that are naturally exposed to flowing liquids and thus to shear stresses (i.e., forces that are applied tangentially to the cell surface). Nuclear translocation of YAP under flow has been observed by many groups in different experimental systems, indicating it is a general response. Data remain less clear on whether YAP/TAZ are activated by any kind of flow, or preferentially by turbulent vs. laminar flows, which elicit different degrees of cytoskeletal rearrangements and different phenotypic responses. Moreover, these experiments did not address the effects of pulsatile flow (s), which is the most significant from the point of view of blood vessel biology, and the possible differences between physiological venous and arteriosus shear stress levels.

The experiments on shear stresses are usually performed in vitro after endothelial cells establish a monolayer, as this is the normal arrangement of cells in the corresponding epithelium. It is however possible to apply forces directly to cells, even in isolation, also by stretching the substratum [7] or by applying forces at the single-cell level by micromanipulation systems. These systems have so far been used more in the context of mechanobiology studies than in studies aimed at understanding the effects on YAP/TAZ activity specifically. For example, it is possible to culture cells in the presence of adhesive microscopic beads and to move these beads relative to the cell body by an optical trap or a magnetic tweezer. Another possibility is to apply compressive forces on cells by using a non-adherent surface pressed on cells or by using an atomic force microscope tip to induce local deformations, for example, in the region of the nucleus [35].

7 Intracellular Actomyosin Contractility

A wealth of information is available on what are the main ECM receptors (integrins) and the cellular structures by which cells sense and measure ECM mechanical properties (focal adhesions). This provides a series of experimental tools by which it is possible to hinder the machinery underlying cells' mechanosensing ability, and thus indirectly study the effects of mechanical signals on cells, and/or to study the mechanism(s) by which mechanical cues regulate YAP/TAZ. These treatments have effects on cells that may go beyond the ones strictly related to ECM mechanosensing (e.g., a cell on soft ECM has still plenty of F-actin, while inhibition of F-actin polymerization affects any pool of F-actin). Moreover, almost all the treatments described below, with the exception of capping/severing protein knockdown, have the same predicted effect, precipitating cells in a state of "low tension" equivalent to a very soft ECM. Importantly, while any of these systems can be used as an entry point to study ECM mechanotransduction, any result should then be repeated with as many of these independent reagents as possible to differentiate between a bona fide mechanical response and a function of F-actin unrelated to ECM forces.

7.1 Integrins

Integrins are key for cells to measure extracellular forces [37], and can be inhibited by incubating cells with RGD peptides or blocking antibodies that compete for the ECM ligand, thus reducing cell adhesion strength. Integrin inhibition can be used only if cells have no alternative manner to attach to the substratum and spread. For example, attachment to polylysine-coated glass is generally considered an integrin-independent process (as it does not lead to maturation of focal adhesions) but sustains F-actin bundling, contractility, and thus YAP/TAZ nuclear localization [34].

7.2 Talins

Force-dependent maturation of focal adhesions can be prevented by deletion/knockdown of talins. Talin is considered one key mechanosensing mechanism at focal adhesions, and it has been shown that deletion of talins compromises the cells ability to develop mature focal adhesions and to develop internal contractility forces proportional to ECM elasticity [38, 39]. Deletion of talins is incompatible with cell proliferation in vitro (like deletion of YAP/-TAZ), requiring the use of talin1 knockout cells with inducible talin2 shRNA, and is thus difficult to use in human cell lines.

7.3 Focal Adhesion-Associated Kinases

Inhibition of focal adhesion-associated signaling kinases such as FAK or SRC has been shown, in some systems, to cause a YAP/-TAZ inhibition [40–42]. This is overall coherent with regulation of the core substratum mechanosensing structure in cells, i.e., focal adhesions. Inhibition of SRC family kinases is usually carried out by treating cells with dasatinib, but analysis of the results is complicated by the wide spectrum of activity of this compound toward non-SRC tyrosine kinases, and experiments are still lacking in triple knockout cells for Src/Fyn/Yes sibling kinases [43]. Inhibition of FAK would be, in principle, a good manner to disable mechanosensing, but data in the field are still contradictory as Chen and colleagues showed that FAK deletion rescued proliferation and stress fiber formation on a small micropattern, thus suggesting FAK as a negative (rather than positive) regulator of focal adhesion mechanosensing [44].

7.4 Actomyosin Contractility

Actomyosin contractility, and thus the ability of cells to sense extracellular mechanical cues, can be prevented by treating cells with small molecules targeting non-muscle myosin (NMII) isoforms (blebbistatin). Alternatively, similar effects can be reached by inhibiting the ROCK and MLCK kinases, which phosphorylate and activate NMII complexes. In this case, as with any kinase, data should be based on the use of different small-molecule inhibitors to avoid confounding off-target effects. Moreover, it might be sometimes recommended to inhibit both ROCK and MLCK to obtain a more quantitative effect, depending on what is the prevalent endogenous NMII-activating input.

7.5 The RHO Small GTPase

The RHO GTPase is generally considered as a master regulator of cell mechanosensing. RHO is required and to some extent sufficient to induce formation of stress fibers in cells, and it controls the activity of ROCK, such that inhibition of RHO (by the use of the C3 toxin, by expressing dominant-negative cDNAs, or by RNAi) can recapitulate the phenotypes induced by low ECM stiffness/small cell geometry. However, the relationship between RHO and ECM mechanosensing is not a linear one [45]: expression of an activated RHO (or of an activated ROCK) in cells plated on a soft ECM is not sufficient to induce phenotypes observed in a stiff ECM. Likely, this depends on the fact that RHO activity is very dynamic and undergoes constant cycles of activation/deactivation. Moreover, RHO activity is usually confined to very specific subcellular compartments, such that its constitutive activation might precipitate cells in a nonphysiological condition. Perhaps even more relevantly for this discussion, RHO activity is mutually dependent with that of F-actin and on F-actin contractility, because F-actin inhibition also inhibits RHO: in other words, RHO is required to maintain contractile F-actin structures when the microenvironment mechanical properties sustain actomyosin contractility but are unable to do so by itself. This is very different from a classical signaling pathway, where, for example, the absence of the extracellular ligand can be compensated by providing a constitutively activated receptor (e.g., FGF, TGF-β, WNT, Notch). Finally, while the effects of RHO on YAP/TAZ were discovered in the context of ECM mechanical cues [18, 34], later studies indicated that RHO regulates YAP/TAZ downstream of other signaling events and through mechanisms that are partially different from ECM stiffness and F-actin inhibition: phosphorylation-mutant YAP-5SA is still sensitive to ECM stiffness and F-actin inhibition, while it is completely resistant to inhibition of RHO by the C3 toxin [46].

7.6 F-Actin

Direct inhibition of F-actin by latrunculin A or cytochalasin B induces by far the most efficient YAP/TAZ inhibition in cell culture. Of note, treatment of cells with latrunculin A to perform epistasis studies (i.e., to establish whether or not regulation of F-actin is a relevant mechanism to regulate YAP/TAZ downstream of other signals) is poorly informative, because F-actin is so strongly required for basal YAP/TAZ activity that any upstream input can be shown to "depend on F-actin," even if it does not really modulate F-actin dynamics.

Clearly, F-actin has also multiple effects on several other cellular processes, ranging from signaling to endo- and exocytosis and from autophagosome formation to mitochondria fission. For example, F-actin directly regulates the activity of another transcriptional complex composed of SRF and the MAL/MRTF; in this case, however, at difference with YAP/TAZ, the effects of latrunculin A and cytochalasin B are opposite: latA dissolves F-actin and liberates G-actin monomers that sequester MAL/MRTF and preclude their binding to SRF, while cytochalasin B dissolves F-actin but remains bound to G-actin, masking its interaction with MAL/MRTF [47], such that cytochalasin B is among most potent activators of SRF activity in vitro. This can be relevant because SRF regulates similar target genes to YAP/TAZ, including *CTGF*, *CYR61*, and *ANKRD1* [48].

7.7 F-Actin Capping/Severing Proteins

F-actin capping and severing proteins such as CapZ, Cofilins, and Gelsolin play an important role in setting the cell's responsiveness to ECM mechanical cues and YAP/TAZ activity. Indeed these were isolated as YAP/TAZ inhibitors by unbiased loss-of-function screens in *Drosophila* and mammalian cells [7, 49]. Knockdown of these F-actin-binding and remodeling proteins is sufficient rescue, at least in human MCF10A mammary epithelial cells, YAP/TAZ activity that has been inhibited by a soft ECM or by cell crowding and has a net effect on cell proliferation in these systems. Moreover, knockdown or mutation of CapZ cause *Drosophila* imaginal disc cells to overgrow, a typical phenotype caused by Yorkie activation. This indicates that knockdown of F-actin capping and severing proteins can be used as a tool to restore YAP/TAZ activity in conditions of "low tension," which is conceptually different from all treatments described in this section. Indeed, a similar rescue has not been obtained so far by expression of activated ROCK or of any other player of the cell mechanosensitive structure(s).

It remains unknown whether capping and severing proteins control YAP/TAZ activity in mammalian tissues and how general is the role of these proteins. It is possible that distinct capping/severing proteins dominate the responsiveness of specific tissues to mechanical cues, depending on the relative levels of expression. Finally, it is possible that other factors with similar function, including destrin/ADF (actin-depolymerizing factor), CapG, and others may regulate YAP/TAZ activity in different cell types.

8 Phenotypes to Study YAP/TAZ Regulation by Mechanical Forces

The readouts used experimentally to follow YAP/TAZ activity merit a final discussion, as different readouts might provide different information. One of the most convenient techniques to monitor YAP/TAZ activity is to follow by immunofluorescence their nucleo/cytoplasmic localization. This is based on the common observation that when YAP/TAZ are inhibited, they are almost invariably relocated preferentially toward the cytoplasm. Early studies proposed that this was due to retention of phosphorylated YAP and TAZ in the cytoplasm, for example, by interaction with 14-3-3 proteins mediated by Ser127 phosphorylation; however, several data indicate that if YAP/TAZ are found in the cytoplasm in steady-state conditions, this actually depends on increased export from the nucleus, rather than on decreased nuclear import. This explains why Ser127-phosphorylated YAP can be easily observed in the nucleus by immunofluorescence [23] and why treatment with the nuclear export inhibitor leptomycin B rapidly induces nuclear localization of YAP in cells where it is mainly cytoplasmic [18, 50]; moreover, this has been directly measured for YAP and TAZ by photobleaching experiments [51, 52]. Thus, even when YAP is cytoplasmic, it keeps passing through the nucleus but is not retained there by a nuclear anchor, for example, by TEAD transcription factors.

Immunofluorescence for YAP localization (by confocal microscopy) has several advantages:

- It reads YAP localization in a shorter time scale (as little as 20 minutes in the case of latrunculin A treatment) compared to luciferase assays or qPCR for endogenous target genes.
- It reads YAP localization on a single-cell basis and has the advantage to inform the user on how uniform vs. localized is the effect throughout the culture (see the warnings above on contact inhibition).
- It is quantitatively and proportionately modulated by YAP-/TAZ-regulatory inputs, including by ECM mechanical cues such as stiffness and micropatterning.
- Commercial antibodies are available to monitor the relative nucleo/cytoplasmic distribution of endogenous YAP, which have been independently validated by several laboratories, such as the sc101199 and the CST #4912. Of note, the quantifications of YAP localization with the two antibodies can differ slightly when performed back-to-back on the same samples, likely to different background staining, and some problems have been reported for the CST antibody as it can display non-specific nucleolar staining [53].

YAP localization can be usually quantitated by counting the percentage of cells displaying nuclear-accumulated YAP (when the nucleus can be recognized by only looking at the YAP IF channel), evenly distributed YAP (when the nucleus cannot be recognized because the intensity is even throughout the cell), or cytoplasmic YAP (when the nucleus can be recognized by only looking at the YAP IF channel, because it is darker than the cytoplasm) [18, 54]. This system might not be useful in conditions in which all cells display nuclear YAP and more/less cytoplasmic staining [55], on when cells display cytoplasmic YAP and minor variations in the nuclear pool. In these cases, it might be useful to use automated imaging systems able to detect the nucleus and to extract the ratio of the YAP signal intensity between the nucleus and the cytoplasm, either by considering as cytoplasm a ring surrounding the nucleus or by taking into consideration the whole cytoplasm area and by normalizing each signal to the relative areas considered. Measuring a single ROI in the nucleus and a single ROI in the cytoplasm is not advisable, because YAP levels are not uniform in the cytoplasm, likely due to different heights of the cell body, organelles, etc. Moreover, since these automated systems strongly rely on the setting of thresholds and parameters to detect nuclei and cytoplasms faithfully, these data should be always presented with a "standard" nucleo/cytoplasmic distribution quantification in parallel, at least on selected positive and negative controls, to compare the two methods and, most importantly, to faithfully indicate the magnitude of the observed phenotypes.

YAP localization as a method to assess YAP/TAZ activity has some intrinsic limitations:

1. It is generally more time-consuming than a general readout of YAP/TAZ transcriptional activity (qPCR or luciferase).
2. It is mainly limited to YAP, because antibodies detecting endogenous TAZ by immunofluorescence can give inconsistent results. This might be due to poor antibodies but also likely depends on the fact that TAZ is mainly regulated at the protein level by proteasomal degradation. This is apparent in some cells and with some antibodies, where TAZ is always found in the nucleus, but with variable nuclear intensity (e.g., with small micropatterns or latrunculin A treatment). On the other hand, this gives the opportunity to monitor TAZ levels/activity by Western blotting [18, 28, 56, 57].
3. Perhaps most importantly, while it is true that cytoplasmic accumulation is consistently associated with conditions leading to functional inactivation of YAP/TAZ, *nuclear localization of YAP/TAZ is not sufficient to claim they are active or reactivated.* Indeed, YAP/TAZ are not able to directly bind DNA, and need assistance from TEAD factors to do so and regulate gene transcription, which is the ultimate and (so far) prevalent function of YAP/TAZ. Thus, conditions in which TEAD binding is compromised might lead to inhibition of YAP/TAZ activity, without their nuclear exclusion. Moreover, YAP/TAZ activity can be regulated at the promoter level by the recruitment of coactivators and corepressors [58]. Finally, it is possible to rescue YAP nuclear localization by treating cells with latrunculin A together with leptomycin B (an inhibitor of nuclear export), but this is not associated with a rescue of YAP/TAZ transcriptional activity (personal unpublished observations). This has important implications for the interpretation of mechanisms, because YAP might be rescued as far as localization is considered, but still inactive in the nucleus. It follows that, in practical terms, it would be preferable to accompany any nucleo/cytoplasmic distribution result with a readout of YAP/TAZ transcriptional activity, especially in the cases where localization and activity can be uncoupled. For example, while some data based on immunofluorescence suggest that inactivation of LATS1/LATS2 kinases can be sufficient to rescue nuclear localization of YAP in conditions of low actomyosin tension, whether this is also sufficient to rescue YAP/TAZ activity has not been addressed, leaving open a significant question in the field.

Apart from nucleo/cytoplasmic localization, multiple functional outputs of YAP/TAZ activity can and should be used to study their regulation by mechanical cues, depending on the cellular context and experimental setup. This includes luciferase assays, qPCR for bona fide direct YAP/TAZ target genes (better if multiple targets in parallel), microarray, or RNA sequencing approaches, but also more indirect readouts of YAP/TAZ activity such as proliferation, apoptosis, or differentiation. This is described in details in other sections of this book. These assays are more indirect than YAP immunofluorescence localization or TAZ protein stability and are thus, potentially, less specific. However, they remain a fundamental piece of information in building any experimental evidence, as they inform on what is the overall strength and relevance of the observed regulation of YAP/TAZ for the buildup of a mechano-regulated cell phenotype. These assays and phenotypes have been used to build the initial experimental evidence that mechanical cues regulate cell biology, and should be incorporated in any study dealing with mechanical regulation of YAP/TAZ to gauge what is the general relevance of the new findings for mechanotransduction, other than YAP/TAZ themselves. Collectively, it is thus important to provide evidence for a coherent regulation of YAP/TAZ both at the protein level and by monitoring their downstream phenotypic readouts.

References

1. Piccolo S, Dupont S, Cordenonsi M (2014) The biology of YAP/TAZ: hippo signaling and beyond. Physiol Rev 94:1287–1312. https://doi.org/10.1152/physrev.00005.2014
2. Halder G, Dupont S, Piccolo S (2012) Transduction of mechanical and cytoskeletal cues by YAP and TAZ. Nat Rev Mol Cell Biol 13:591–600. https://doi.org/10.1038/nrm3416
3. Dupont S (2016) Role of YAP/TAZ in cell-matrix adhesion-mediated signalling and mechanotransduction. Exp Cell Res 343:42–53. https://doi.org/10.1016/j.yexcr.2015.10.034
4. Eyckmans J, Boudou T, Yu X, Chen CS (2011) A Hitchhiker's guide to mechanobiology. Dev Cell 21:35–47. https://doi.org/10.1016/j.devcel.2011.06.015
5. Tang Y, Rowe RG, Botvinick EL, Kurup A, Putnam AJ, Seiki M et al (2013) MT1-MMP-dependent control of skeletal stem cell commitment via a β1-integrin/YAP/TAZ signaling axis. Dev Cell 25:402–416. https://doi.org/10.1016/j.devcel.2013.04.011
6. Gjorevski N, Sachs N, Manfrin A, Giger S, Bragina ME, Ordóñez-Morán P et al (2016) Designer matrices for intestinal stem cell and organoid culture. Nature 539:560. https://doi.org/10.1038/nature20168
7. Aragona M, Panciera T, Manfrin A, Giulitti S, Michielin F, Elvassore N et al (2013) A mechanical checkpoint controls multicellular growth through YAP/TAZ regulation by actin-processing factors. Cell 154:1047–1059. https://doi.org/10.1016/j.cell.2013.07.042
8. Schwartz MA, Chen CS (2013) Cell biology. Deconstructing dimensionality, Science (New York, NY) 339:402–404. https://doi.org/10.1126/science.1233814.
9. Nelson CM, VanDuijn MM, Inman JL, Fletcher DA, Bissell MJ (2006) Tissue geometry determines sites of mammary branching morphogenesis in Organotypic cultures. Science (New York, NY) 314:298–300. https://doi.org/10.1126/science.1131000.
10. Folkman J, Moscona A (1978) Role of cell shape in growth control. Nature 273:345–349
11. Mooney D, Hansen L, Vacanti J, Langer R, Farmer S, Ingber D (1992) Switching from differentiation to growth in hepatocytes:

control by extracellular matrix. J Cell Physiol 151:497–505. https://doi.org/10.1002/jcp.1041510308
12. Ingber DE (1990) Fibronectin controls capillary endothelial cell growth by modulating cell shape. Proc Natl Acad Sci U S A 87:3579–3583
13. Roskelley CD, Desprez PY, Bissell MJ (1994) Extracellular matrix-dependent tissue-specific gene expression in mammary epithelial cells requires both physical and biochemical signal transduction. Proc Natl Acad Sci U S A 91:12378–12382
14. Chen CS, Mrksich M, Huang S, Whitesides GM, Ingber DE (1997) Geometric control of cell life and death. Science (New York, NY) 276:1425–1428
15. McBeath R, Pirone DM, Nelson CM, Bhadriraju K, Chen CS (2004) Cell shape, cytoskeletal tension, and RhoA regulate stem cell lineage commitment. Dev Cell 6:483–495. https://doi.org/10.1016/S1534-5807(04)00075-9
16. Watt FM, Jordan PW, O'Neill CH (1988) Cell shape controls terminal differentiation of human epidermal keratinocytes. Proc Natl Acad Sci U S A 85:5576–5580
17. Engler AJ, Sen S, Sweeney HL, Discher DE (2006) Matrix elasticity directs stem cell lineage specification. Cell 126:677–689. https://doi.org/10.1016/j.cell.2006.06.044
18. Dupont S, Morsut L, Aragona M, Enzo E, Giulitti S, Cordenonsi M et al (2011) Role of YAP/TAZ in mechanotransduction. Nature 474:179–183. https://doi.org/10.1038/nature10137
19. Nelson CM, Jean RP, Tan JL, Liu WF, Sniadecki NJ, Spector AA et al (2005) Emergent patterns of growth controlled by multicellular form and mechanics. Proc Natl Acad Sci U S A 102:11594–11599. https://doi.org/10.1073/pnas.0502575102
20. Malinverno C, Corallino S, Giavazzi F, Bergert M, Li Q, Leoni M et al (2017) Endocytic reawakening of motility in jammed epithelia. Nat Mater 16:587–596. https://doi.org/10.1038/nmat4848
21. Zhao B, Wei X, Li W, Udan RS, Yang Q, Kim J et al (2007) Inactivation of YAP oncoprotein by the hippo pathway is involved in cell contact inhibition and tissue growth control. Genes Dev 21:2747–2761. https://doi.org/10.1101/gad.1602907
22. Puliafito A, Hufnagel L, Neveu P, Streichan S, Sigal A, Fygenson DK et al (2012) Collective and single cell behavior in epithelial contact inhibition. Proc Natl Acad Sci 109:739–744. https://doi.org/10.1073/pnas.1007809109
23. Wada K-I, Itoga K, Okano T, Yonemura S, Sasaki H (2011) Hippo pathway regulation by cell morphology and stress fibers. Development 138:3907–3914. https://doi.org/10.1242/dev.070987
24. Benham-Pyle BW, Pruitt BL, Nelson WJ (2015) Mechanical strain induces E-cadherin-dependent Yap1 and -catenin activation to drive cell cycle entry. Science (New York, NY) 348:1024–1027. https://doi.org/10.1126/science.aaa4559.
25. Zhou D, Conrad C, Xia F, Park J-S, Payer B, Yin Y et al (2009) Mst1 and Mst2 maintain hepatocyte quiescence and suppress hepatocellular carcinoma development through inactivation of the Yap1 oncogene. Cancer Cell 16:425–438. https://doi.org/10.1016/j.ccr.2009.09.026
26. Silvis MR, Kreger BT, Lien W-H, Klezovitch O, Rudakova GM, Camargo FD et al (2011) α-Catenin is a tumor suppressor that controls cell accumulation by regulating the localization and activity of the transcriptional coactivator Yap1. Sci Signal 4:ra33. https://doi.org/10.1126/scisignal.2001823
27. Schlegelmilch K, Mohseni M, Kirak O, Pruszak J, Rodriguez JR, Zhou D et al (2011) Yap1 acts downstream of α-catenin to control epidermal proliferation. Cell 144:782–795. https://doi.org/10.1016/j.cell.2011.02.031
28. Cordenonsi M, Zanconato F, Azzolin L, Forcato M, Rosato A, Frasson C et al (2011) The hippo transducer TAZ confers cancer stem cell-related traits on breast cancer cells. Cell 147:759–772. https://doi.org/10.1016/j.cell.2011.09.048
29. Szymaniak AD, Mahoney JE, Cardoso WV, Varelas X (2015) Crumbs3-mediated polarity directs airway epithelial cell fate through the Hippo pathway effector Yap. Dev Cell 34:283–296. https://doi.org/10.1016/j.devcel.2015.06.020
30. Lv X-B, Liu C-Y, Wang Z, Sun Y-P, Xiong Y, Lei Q-Y et al (2015) PARD3 induces TAZ activation and cell growth by promoting LATS1 and PP1 interaction. EMBO Rep 16:975. https://doi.org/10.15252/embr.201439951
31. Huang H-L, Wang S, Yin M-X, Dong L, Wang C, Wu W et al (2013) Par-1 regulates tissue growth by influencing hippo phosphorylation status and hippo-Salvador association. PLoS Biol 11:e1001620. https://doi.org/10.1371/journal.pbio.1001620

32. Heidary Arash E, Shiban A, Song S, Attisano L (2017) MARK4 inhibits hippo signaling to promote proliferation and migration of breast cancer cells. EMBO Rep 18:420–436. https://doi.org/10.15252/embr.201642455
33. Hirate Y, Hirahara S, Inoue K-I, Kiyonari H, Niwa H, Sasaki H (2015) Par-aPKC-dependent and -independent mechanisms cooperatively control cell polarity, hippo signaling, and cell positioning in 16-cell stage mouse embryos. Develop Growth Differ 57:544. https://doi.org/10.1111/dgd.12235
34. Zhao B, Li L, Wang L, Wang C-Y, Yu J, Guan K-L (2012) Cell detachment activates the hippo pathway via cytoskeleton reorganization to induce anoikis. Genes Dev 26:54–68. https://doi.org/10.1101/gad.173435.111
35. Elosegui-Artola A, Andreu I, Beedle AEM, Lezamiz A, Uroz M, Kosmalska AJ et al (2017) Force triggers YAP nuclear entry by regulating transport across nuclear pores. Cell 171:1397–1410.e14. https://doi.org/10.1016/j.cell.2017.10.008
36. Debnath J, Mills KR, Collins NL, Reginato MJ, Muthuswamy SK, Brugge JS (2002) The role of apoptosis in creating and maintaining luminal space within normal and oncogene-expressing mammary acini. Cell 111:29–40
37. Schiller HB, Hermann M-R, Polleux J, Vignaud T, Zanivan S, Friedel CC et al (2013) β1- and αv-class integrins cooperate to regulate myosin II during rigidity sensing of fibronectin-based microenvironments. Nat Cell Biol 15:625–636. https://doi.org/10.1038/ncb2747
38. Elosegui-Artola A, Oria R, Chen Y, Kosmalska A, Pérez-González C, Castro N et al (2016) Mechanical regulation of a molecular clutch defines force transmission and transduction in response to matrix rigidity. Nat Cell Biol 18:540–548. https://doi.org/10.1038/ncb3336
39. Roca-Cusachs P, Gauthier NC, del Rio A, Sheetz MP (2009) Clustering of alpha(5)beta(1) integrins determines adhesion strength whereas alpha(v)beta(3) and Talin enable mechanotransduction. Proc Natl Acad Sci 106:16245–16250. https://doi.org/10.1073/pnas.0902818106.
40. Taccioli C, Sorrentino G, Zannini A, Caroli J, Beneventano D, Anderlucci L et al (2015) MDP, a database linking drug response data to genomic information, identifies dasatinib and statins as a combinatorial strategy to inhibit YAP/TAZ in cancer cells. Oncotarget 6:38854–38865. https://doi.org/10.18632/oncotarget.5749
41. Oku Y, Nishiya N, Shito T, Yamamoto R, Yamamoto Y, Oyama C et al (2015) Small molecules inhibiting the nuclear localization of YAP/TAZ for chemotherapeutics and chemosensitizers against breast cancers. FEBS Open Bio 5:542–549. https://doi.org/10.1016/j.fob.2015.06.007
42. Kim N-G, Gumbiner BM (2015) Adhesion to fibronectin regulates hippo signaling via the FAK-Src-PI3K pathway. J Cell Biol 210:503–515. https://doi.org/10.1083/jcb.201501025
43. Stein PL, Vogel H, Soriano P (1994) Combined deficiencies of Src, Fyn, and yes tyrosine kinases in mutant mice. Genes Dev 8:1999–2007. https://doi.org/10.1101/gad.8.17.1999
44. Pirone DM, Liu WF, Ruiz SA, Gao L, Raghavan S, Lemmon CA et al (2006) An inhibitory role for FAK in regulating proliferation: a link between limited adhesion and RhoA-ROCK signaling. J Cell Biol 174:277–288. https://doi.org/10.1083/jcb.200510062
45. Geiger B, Bershadsky A (2001) Assembly and mechanosensory function of focal contacts. Curr Opin Cell Biol 13:584–592. https://doi.org/10.1016/S0955-0674(00)00255-6
46. Sorrentino G, Ruggeri N, Specchia V, Cordenonsi M, Mano M, Dupont S et al (2014) Metabolic control of YAP and TAZ by the mevalonate pathway. Nat Cell Biol 16:357–366. https://doi.org/10.1038/ncb2936
47. Miralles F, Posern G, Zaromytidou A-I, Treisman R (2003) Actin dynamics control SRF activity by regulation of its coactivator MAL. Cell 113:329–342
48. Esnault C, Stewart A, Gualdrini F, East P, Horswell S, Matthews N et al (2014) Rho-actin signaling to the MRTF coactivators dominates the immediate transcriptional response to serum in fibroblasts. Genes Dev 28:943–958. https://doi.org/10.1101/gad.239327.114
49. Sansores-Garcia L, Bossuyt W, Wada K-I, Yonemura S, Tao C, Sasaki H et al (2011) Modulating F-actin organization induces organ growth by affecting the hippo pathway. EMBO J 30:2325–2335. https://doi.org/10.1038/emboj.2011.157
50. Ren F, Zhang L, Jiang J (2010) Hippo signaling regulates Yorkie nuclear localization and activity through 14-3-3 dependent and independent mechanisms. Dev Biol 337:303–312. https://doi.org/10.1016/j.ydbio.2009.10.046

51. Chan SW, Lim CJ, Loo LS, Chong YF, Huang C, Hong W (2009) TEADs mediate nuclear retention of TAZ to promote oncogenic transformation. J Biol Chem 284:14347–14358. https://doi.org/10.1074/jbc.M901568200
52. Nil Ege, Anna M Dowbaj, Ming Jiang, Michael Howell, Robert P Jenkins, Erik Sahai. Actin and Src-family kinases regulate nuclear YAP1 and its export. https://doi.org/10.1101/201004
53. Finch ML, Passman AM, Strauss RP, Yeoh GC, Callus BA (2015) Sub-cellular localisation studies may spuriously detect the yes-associated protein, YAP, in nucleoli leading to potentially invalid conclusions of its function. PLoS One 10:e0114813. https://doi.org/10.1371/journal.pone.0114813
54. Enzo E, Santinon G, Pocaterra A, Aragona M, Bresolin S, Forcato M et al (2015) Aerobic glycolysis tunes YAP/TAZ transcriptional activity. EMBO J 34:1349–1370. https://doi.org/10.15252/embj.201490379
55. Sorrentino G, Ruggeri N, Zannini A, Ingallina E, Bertolio R, Marotta C et al (2017) Glucocorticoid receptor signalling activates YAP in breast cancer. Nat Comms 8:14073. https://doi.org/10.1038/ncomms14073
56. Azzolin L, Zanconato F, Bresolin S, Forcato M, Basso G, Bicciato S et al (2012) Role of TAZ as mediator of Wnt signaling. Cell 151:1443–1456. https://doi.org/10.1016/j.cell.2012.11.027
57. Liu C-Y, Zha Z-Y, Zhou X, Zhang H, Huang W, Zhao D et al (2010) The hippo tumor pathway promotes TAZ degradation by phosphorylating a phosphodegron and recruiting the SCF{beta}-TrCP E3 ligase. J Biol Chem 285:37159–37169. https://doi.org/10.1074/jbc.M110.152942
58. Kim M, Kim T, Johnson RL, Lim D-S (2015) Transcriptional co-repressor function of the hippo pathway transducers YAP and TAZ. Cell Rep 11:270. https://doi.org/10.1016/j.celrep.2015.03.015

Chapter 16

CRISPR-Mediated Approaches to Regulate YAP/TAZ Levels

Ryan J. Quinton and Neil J. Ganem

Abstract

The advent of CRISPR has revolutionized genomic engineering, and harnessing its power to regulate levels of the transcriptional co-activators YAP and TAZ represents an exciting new opportunity in the field of Hippo signaling. Initially repurposed from the microbial immune system to perform highly specific gene knockouts, CRISPR technology has now been expanded to modulate the transcriptional activity of any gene of interest in mammalian systems. Here, we describe strategies to employ CRISPR to genetically knock out the genes encoding for YAP (*YAP1*) or TAZ (*WWTR1*) in mammalian cell lines, as well as briefly outline an approach for utilizing CRISPR to transcriptionally modulate YAP/TAZ levels.

Key words Cas9, sgRNA, Hippo, LATS, CRISPRa, CRISPRi, KRAB, VP64

1 Introduction

The mammalian Hippo tumor suppressor pathway comprises a cascade of kinases that negatively regulate the transcriptional co-activator paralogs, Yes-associated protein (YAP) and transcriptional co-activator with PDZ-binding motif (TAZ) [1, 2]. Upon Hippo pathway activation, large tumor suppressors 1 and 2 (LATS1/2) phosphorylate YAP/TAZ to induce their cytoplasmic sequestration and degradation. By contrast, when the Hippo pathway is inactive, YAP and TAZ remain unphosphorylated, allowing them to translocate to the nucleus, bind to the TEAD family of transcription factors, and induce the expression of an array of genes that promote cellular proliferation, migration, and survival [3]. The Hippo pathway was first identified as a highly conserved canonical regulator of organ size; however, recent discoveries situate this pathway as a highly integrated signaling hub, responding to complex inputs that monitor cell-cell adhesion, cell-matrix adhesion, GPCR signaling, contractile tension from the actin cytoskeleton, cell polarity, and changes in the cell cycle [3–11]. These cues converge on LATS/YAP/TAZ signaling to regulate embryological and organ development as well as adult tissue maintenance [3, 12].

Alexander Hergovich (ed.), *The Hippo Pathway: Methods and Protocols*, Methods in Molecular Biology, vol. 1893, https://doi.org/10.1007/978-1-4939-8910-2_16, © Springer Science+Business Media, LLC, part of Springer Nature 2019

With their capacity to drive proliferation, invasion, dedifferentiation, and survival, YAP/TAZ can be exploited to favor neoplastic development [13]. It is therefore unsurprising that functional impairment of the Hippo tumor suppressor pathway, with subsequent hyperactivation and/or amplification of YAP/TAZ, has been observed in many cancers [14–16]. Indeed, aberrant YAP/TAZ activity has been observed in glioblastoma, lung, colorectal, ovarian, breast, liver, and prostate cancers and is often correlated with poor prognosis [17–19].

The importance of Hippo signaling in development and tumorigenesis places it at the center of investigative efforts across many fields. Within this context, modulating YAP/TAZ activity is vital to elucidating their many roles and regulations. A recent innovation in gene editing, termed clustered regularly interspaced short palindromic repeats (CRISPR), has enabled novel methods of modulating gene activity with radically improved efficacy and efficiency, thus opening many new and exciting avenues of investigation.

CRISPR was discovered as a mechanism of microbial adaptive immunity wherein Cas9, a RNA-guided bacterial endonuclease, cleaves and destroys foreign DNA from invading viruses based on 20-nucleotide targeting sequences [20–26]. This biological tool has been adapted to allow for the specific targeting and cleavage of any region of the genome in eukaryotic cells, enabling the introduction of double strand breaks (DSBs) at a given locus of interest. These DSBs are most commonly repaired using non-homologous end joining (NHEJ), which is error-prone and frequently results in insertion/deletion (indel) mutations that can introduce frameshifts or premature stop codons, ultimately rendering a gene of interest inactive (Fig. 1a) [27]. More recently, CRISPR technology been developed to go beyond gene knockouts to enable the recruitment of transcriptional activators or repressors to the promoter region of any gene (Fig. 1b, c) [27–34]. Herein, we outline the methods for using CRISPR to knock out the genes encoding for YAP (*YAP1*) and TAZ (*WWTR1)* in mammalian cell lines as well as briefly describe the use of CRISPR activator and inhibitor systems to tightly regulate endogenous YAP/TAZ expression.

2 Materials

We use non-transformed telomerase-immortalized retinal pigmented epithelial cells (hRPE-1) for the method described below. However, this protocol can be adapted to any adherent cell line so long as cell culture conditions are adjusted as necessary.

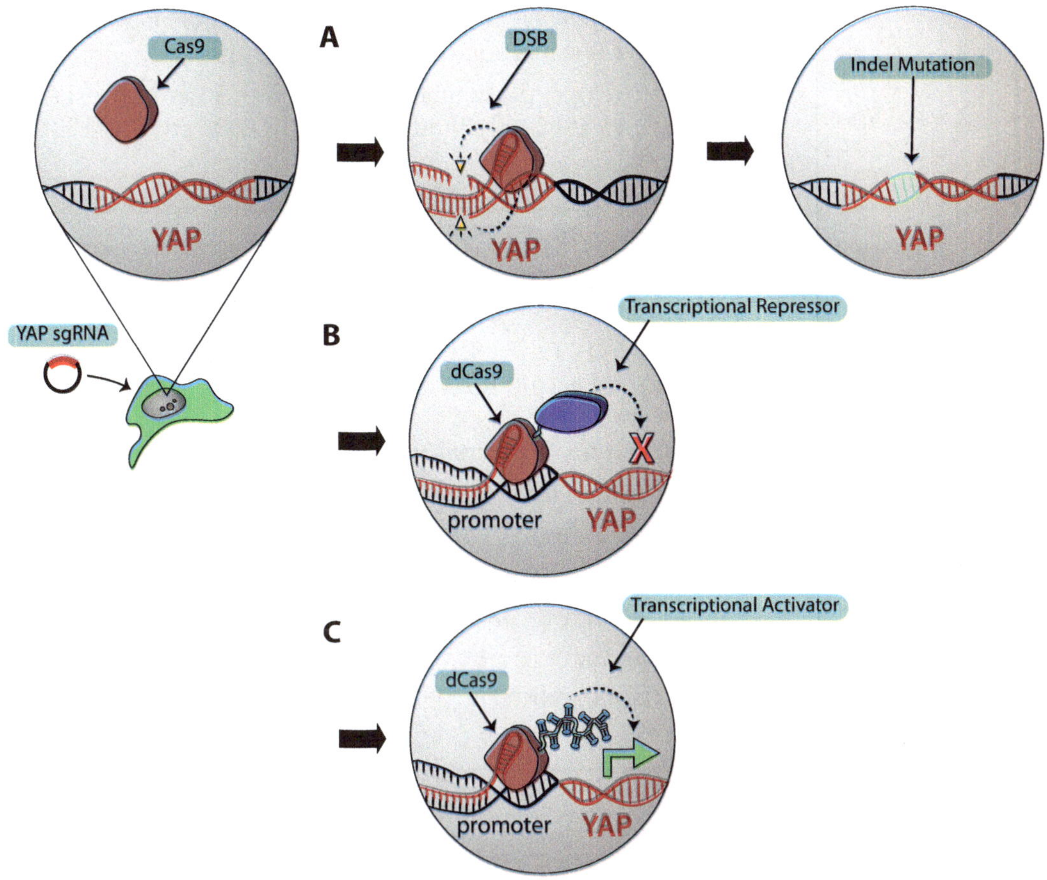

Fig. 1 CRISPR technology can be used to modulate gene expression. (**a**) Cas9 can be specifically targeted to a region of the gene encoding YAP by a 20 nucleotide sgRNA, where it subsequently introduces DSBs that are repaired by error-prone NHEJ leading to indel mutations and gene inactivation. (**b**) Catalytically inactive Cas9 (dCas9) fused to a transcriptional repressor (e.g. KRAB) can be specifically targeted to the promoter region of *YAP1* by a sgRNA, thus repressing YAP transcription. (**c**) Catalytically inactive Cas9 (dCas9) fused to a transcriptional activator (e.g. VP64*)* can be specifically targeted to the promoter region of *YAP1* by a sgRNA, thus promoting YAP transcription

2.1 Cell Culture

1. 10 cm polystyrene tissue culture plates.
2. 15 cm polystyrene tissue culture plates.
3. 6-well polystyrene tissue culture plates.
4. Sterile 1.5 mL microcentrifuge tubes.
5. Complete DMEM/F12: DME/F12 media supplemented with 10% fetal bovine serum (FBS), 100 IU/mL penicillin, and 100 μg/mL streptomycin. Store at 4 °C in the dark.
6. 0.25% Trypsin/EDTA (0.53 mM). Store at 4 °C in the dark.
7. Sterile phosphate-buffered saline (PBS: 1.06 mM KH_2PO_4, 155.17 mM NaCl, 2.97 mM $Na_2PO_4 \cdot 7H_2O$). Store at 4 °C.

8. Transfection reagent: Polyethylenimine (PEI) in water at a working concentration of 1 μg/mL. Store at 4 °C.
9. Viral envelope/packaging plasmids (for example: psPAX2 (Addgene, #12260) and pMD2.G (Addgene, #12259) plasmids).
10. CRISPR plasmid: lentiCRISPR v2 (Addgene, #52961) (or any other applicable CRISPR vector).
11. Transduction reagent: Polybrene (10 mg/mL stock concentration). Store at −20 °C.
12. RPE-1 (ATCC, #CRL-4000) and HEK293FT cell lines (thermo fisher, #R7007).

2.2 Equipment

1. Tissue culture incubator set at 37 °C with 5% humidified CO_2.
2. Tissue culture hood.
3. Hemocytometer.
4. High-speed centrifuge.
5. Phase contrast microscope equipped with a ×10 objective for cell counting.
6. Sterile cloning discs.

3 Methods

3.1 Design sgRNA Sequences Optimized to Target YAP or TAZ for Knockout

1. In order to use CRISPR to successfully knock out a gene, cells must express both Cas9 endonuclease and a single guide RNA (sgRNA) that targets Cas9 to the gene of interest. The design of a specific, functional sgRNA sequence is critical to the success of this approach and must adhere to certain parameters (*see* **Note 1**).
2. Custom synthesized sgRNAs should be purified using standard desalting procedures. It is also important to include additional nucleotides flanking the sgRNA sequence to facilitate proper ligation and cloning into the expression vector of choice (*see* **Note 2**).
3. We designed and/or purchased several sgRNAs to target *YAP1* or *WWTR1* and found these sequences to be the most efficient for gene knockout in RPE-1 cells:

   YAP1: 5′-TGCCCCAGACCGTGCCCAT-3′

   WWTR1: 5′-ATCCGAAGCCTAGCTCGTGG-3′
4. Numerous lentiviral expression plasmids are available to integrate both sgRNA and Cas9 expression cassettes into the genomes of infected hosts. Some all-in-one plasmids express both the sgRNA and Cas9 from the same vector [35]. Alternatively,

Cas9 and sgRNAs can be expressed from different vectors [35]. These CRISPR elements can also be transiently expressed using standard expression plasmids. Several factors impact which expression system is optimal (*see* **Note 3**). Here, we describe methods to express Cas9 and a sgRNA from an all-in-one lentiviral vector (*see* Subheading 3.2 below).

3.2 Production of Virus Containing Cas9 and sgRNA Expression Vectors

Ensure that all institutional biosafety rules and regulations related to working with virus are strictly followed.

1. Seed 293FT cells in complete DMEM/F12 media into a 6-well plate. Cells are ready for transfection when they are attached and roughly 80% confluent.
2. When cells are ready to transfect, change the media on 293FT cells to serum supplemented media without penicillin/streptomycin added. Gently add 2 mL per well.
3. Prepare transfection mixture as follows:

   (Volumes provided are for 1 well within a 6-well plate; scale volumes as necessary).

   (a) Mix 200 μL of serum free media with 16 μL of 1 μg/mL PEI (*see* **Note 4**) in a 1.5 mL microcentrifuge tube. Mix gently. Wait 5 min.

   (b) Add 2 μg of plasmid DNA (e.g., the LentiCRISPR-v2 all-in-one plasmid containing a YAP targeting sgRNA).

   (c) Add 1.5 μg of psPAX or equivalent lentiviral packaging plasmid.

   (d) Add 500 ng of pMD2.G or equivalent lentiviral envelope plasmid.

   (e) Mix transfection mixture gently; avoid pipetting up and down. Wait 15 min.
4. Add 200 μL of solution gently in dropwise manner to 293FT cells in penicillin/streptomycin free media.
5. Place in 37 °C incubator.
6. After 4 h, gently replace media with 2 mL of complete DMEM/F12 media.
7. After 48 h, collect supernatant containing virus and filter through a 0.45 μm filter. Collect virus in a 15 mL conical vial and use for infection immediately or store at −80 °C.

3.3 Infection

1. On the day before the infection, plate RPE-1 cells in complete DMEM/F12 media in a 6-well plate so that they will be roughly 40–50% confluent the next day.
2. Dilute viral supernatant in fresh complete DMEM/F12 media at a ratio of 1 mL of viral supernatant: 1 mL of media and add polybrene at a final concentration of 10 μg/mL. Aim for a

multiplicity of infection (MOI) of ~1, which must be empirically determined (*see* **Note 5**).

3. Aspirate media from RPE-1 cells and replace with 2 mL of prepared viral supernatant/media mixture containing polybrene.
4. Optionally, spin the 6-well plate at 1200 × *g* for 60 min at 32 °C to improve infection efficiency.
5. Place in 37 °C incubator.
6. Add antibiotic selection to infected cells and non-infected controls after 24 h.
7. Once resistant cells have grown sufficiently to passage, and uninfected control cells have died in antibiotic, passage cells and verify knockdown of YAP (or TAZ) via western blot. If YAP (or TAZ) levels appear significantly decreased by western blot, proceed to isolating single cell knockout clones.

3.4 Selecting Knockout Clones

1. Wash the population of cells infected with YAP (or TAZ) CRISPR reagents with sterile PBS, then trypsinize cells with 0.25% trypsin/EDTA for 5 min (or until all cells are detached from the plate).
2. Collect cells in complete medium containing 10% FBS and pellet cells at 200 × *g* for 5 min. Resuspend cells in complete growth medium, and use a hemocytometer (or similar counting device) to count cells.
3. Plate cells sparsely on a 15 cm tissue culture plate so that they will grow up as individual colonies (*see* **Note 6**). Alternatively, use FACS to sort single cells into a 96-well tissue culture plate. Place plate in 37 °C incubator.
4. Change media every 3–4 days, and wait 1–2 weeks for distinct colonies of cells to become visible.
7. Once clonal populations are visible and sufficiently grown, mark where they are on the bottom of the tissue culture dish with a marker (*see* **Note** 7).
8. Aspirate media and wash the tissue culture plate with sterile, room temperature PBS.
9. Immerse sterile cloning discs in 0.25% trypsin/EDTA that has been pre-warmed to 37 °C.
10. Aspirate the PBS from the tissue culture plate containing the clones, and then use forceps to place trypsin saturated discs on top of each marked colony. Move quickly to avoid cells drying out (*see* **Note** 7).
11. Wait 5 min for cells to detach, and then use forceps to transfer individual cloning discs into separate wells of 6-well tissue culture plates containing 2 mL of pre-warmed complete

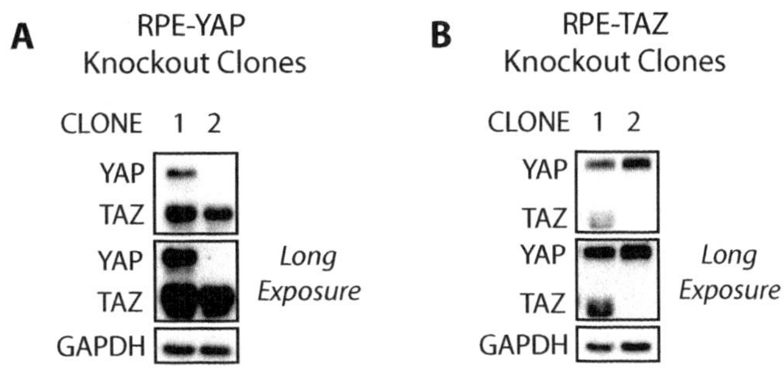

Fig. 2 Genetic knockout of YAP/TAZ using CRISPR. (**a**) Two clones of RPE-1 cells stably expressing Cas9 and a sgRNA targeting the *YAP1* gene were isolated and total YAP and TAZ levels were assessed by western blotting. Clone 2 shows complete loss of YAP1 protein. (**b**) Two clones of RPE-1 cells stably expressing Cas9 and a sgRNA targeting the *WWTR1* gene were isolated and total YAP and TAZ levels were assessed by western blotting. Clone 2 shows complete loss of TAZ protein

media. Do not agitate the plate while waiting for cells to detach.

12. Allow clonal populations to grow in 6-well dishes, passage, and verify YAP knockout in clones via western blot (Fig. 2) (*see* **Note 8**).

3.5 CRISPR-Mediated Transcriptional Regulation

Recent developments have allowed for the use of CRISPR technology in modulating the transcriptional activity of a gene of interest. CRISPR-mediated transcriptional regulation is typically accomplished by fusing a catalytically inactive Cas9 (dCas9) that is devoid of endonuclease activity to transcriptional activator or repressor proteins, such as VP64 or KRAB, respectively (Fig. 1). This allows for sgRNA-mediated targeting of transcriptional activators/repressors to the promoter region of any gene of interest and its resultant activation or repression [29–34]. Excitingly, this technology confers significant investigative advantage over current methods in performing genome wide gene activation/repression screens, allowing for their simple and cost effective execution [33, 37, 38].

While many of the considerations when using CRISPR as a means of transcriptional regulation are similar to performing CRISPR mediated gene knockout, the sgRNA targeting region shifts from the coding region of a gene to the upstream native promoter region. As such, alternate parameters regarding off-target effects and optimal sgRNA design must be accounted for. Publically available tools, such as this website furnished by the Broad institute, https://portals.broadinstitute.org/gpp/public/analysis-tools/sgrna-design-crisprai, incorporate the necessary considerations regarding transcriptional start sites and DNase 1 hypersensitive sites into their sgRNA design algorithms.

CRISPR activator/inhibitor systems can also be utilized to induce YAP/TAZ expression/inhibition in a way that allows for temporal control. This is accomplished by placing dCas9, fused to a transcriptional activator or inhibitor, under the control of an inducible promoter, allowing for the tunable activation of CRISPR transcriptional modulation over time. Several plasmids encoding for dCas9 fused to transcriptional activators or repressors are publically available, as are plasmids containing CRISPR transcriptional modulation elements under the inducible control of both light and chemical activation [39, 40].

4 Notes

1. When using *Streptococcus pyogenes* Cas9, the most commonly used Cas9 ortholog, the 20-nt DNA sequence to be targeted by the sgRNA must be followed at the 3′ end by the protospacer adjacent motif (PAM) sequence NGG in order for the Cas9 nuclease to perform its cleavage function [41]. In addition, it is advised to select sgRNA sequences that target early exons that are common to all transcript variants and that occur upstream of functional protein domains [42, 43]. There are multiple online tools to assist with the design of sgRNAs targeting specific genes of interest, such as http://tools.genome-engineering.org. It is recommended to test 1–3 different sgRNAs for each gene of interest.
2. Custom sgRNA sequences can be cloned into numerous available vectors following cloning strategies described for each vector. A protocol for cloning sgRNA sequences into commonly used lentiviral CRISPR constructs is available here: https://media.addgene.org/data/plasmids/52/52961/52961-attachment_B3xTwla0bkYD.pdf. Alternatively, pre-designed sgRNAs cloned into commercially available vectors are available for purchase from several companies.
3. In general, stable integration of Cas9 and sgRNA using lentiviral vectors, leading to prolonged expression of CRISPR elements, may increase the likelihood of successful on-target modifications [38]. Alternatively, prolonged expression of sgRNA may also lead to increased off-target effects, and since the location of genome integration is random, there is also the potential for phenotypic artifacts to arise [44]. The delivery of recombinant Cas9 protein, sgRNA delivery using non-integrating adeno-associated viral (AAV)-based vectors, delivery via transient transfection of sgRNA plasmids, or the delivery of sgRNAs as synthetic oligonucleotides are viable alternatives to that may limit off-target effects while still maintaining a high degree of efficacy [44].

4. It is important to optimize the ratio of DNA:PEI to be used for each batch of PEI. Commercially available DNA transfection reagents can also be used in place of PEI.
5. Polybrene (hexadimethrine bromide) can be used to enhance transduction efficiency, however, this chemical is not required for infection and has been shown to be toxic for some cell types. It is ideal to aim for a multiplicity of infection (MOI) of ~1. This can be achieved by testing multiple dilutions of viral supernatant: media added to cells. Typically, dilutions from 1:1 to 1:8 are performed.
6. The number of cells plated can vary based on how well they adhere after suspension and how fast they proliferate. For RPE-1 cells, plate roughly 500 cells on a 15 cm tissue culture plate. This generates dozens of clonal populations after 1–2 weeks. This method is not ideal for highly motile cells, and sorting cells via FACS into individual wells of a 96-well plate may be preferred.
7. Clonal outgrowths are typically ready for transfer when they have reached numbers in the hundreds. As a general rule for RPE cells, trypsinize colonies when the outer boundaries of the clonal population nearly fill the field of view of a ×10 objective. Keep in mind that knockdown of YAP or TAZ impairs cell proliferation rates; therefore, it is important to also collect slow-growing clones that may take longer to develop (e.g. 2–4 weeks). It is recommended to collect at least 12 clones.
8. When imaging a blot to verify gene knockout, overexpose the blot to ensure knockout is complete. There are additional methods to verify gene knockout, such as sequencing, mismatch cleavage assays, and others [36].

Acknowledgements

R.J.Q. is funded by the Canadian Institutes of Health Research Doctoral Foreign Study Award. N.J.G. is a member of the Shamim and Ashraf Dahod Breast Cancer Research Laboratories and is supported by NIH grants GM117150 and CA-154531, the Karin Grunebaum Foundation, the Smith Family Awards Program, the Searle Scholars Program, and the Melanoma Research Alliance.

References

1. Huang J, Wu S, Barrera J, Matthews K, Pan D (2005) The Hippo signaling pathway coordinately regulates cell proliferation and apoptosis by inactivating Yorkie, the Drosophila Homolog of YAP. Cell 122(3):421–434. https://doi.org/10.1016/j.cell.2005.06.007
2. Lei QY, Zhang H, Zhao B, Zha ZY, Bai F, Pei XH, Zhao S, Xiong Y, Guan KL (2008) TAZ

promotes cell proliferation and epithelial-mesenchymal transition and is inhibited by the hippo pathway. Mol Cell Biol 28 (7):2426–2436. https://doi.org/10.1128/MCB.01874-07

3. Meng Z, Moroishi T, Guan K-L (2016) Mechanisms of Hippo pathway regulation. Genes Dev 30(1):1–17. https://doi.org/10.1101/gad.274027.115
4. Yu F-X, Guan K-L (2013) The Hippo pathway: regulators and regulations. Genes Dev 27 (4):355–371. https://doi.org/10.1101/gad.210773.112
5. Ganem Neil J, Cornils H, Chiu S-Y, O'Rourke Kevin P, Arnaud J, Yimlamai D, Théry M, Camargo Fernando D, Pellman D (2014) Cytokinesis failure triggers hippo tumor suppressor pathway activation. Cell 158 (4):833–848. https://doi.org/10.1016/j.cell.2014.06.029
6. Aragona M, Panciera T, Manfrin A, Giulitti S, Michielin F, Elvassore N, Dupont S, Piccolo S (2013) A mechanical checkpoint controls multicellular growth through YAP/TAZ regulation by actin-processing factors. Cell 154 (5):1047–1059. https://doi.org/10.1016/j.cell.2013.07.042
7. Halder G, Dupont S, Piccolo S (2012) Transduction of mechanical and cytoskeletal cues by YAP and TAZ. Nat Rev Mol Cell Biol 13:591. https://doi.org/10.1038/nrm3416
8. Paramasivam M, Sarkeshik A, Yates JR, Fernandes MJG, McCollum D (2011) Angiomotin family proteins are novel activators of the LATS2 kinase tumor suppressor. Mol Biol Cell 22(19):3725–3733. https://doi.org/10.1091/mbc.E11-04-0300
9. Wada K-I, Itoga K, Okano T, Yonemura S, Sasaki H (2011) Hippo pathway regulation by cell morphology and stress fibers. Development 138(18):3907–3914. https://doi.org/10.1242/dev.070987
10. Zhao B, Li L, Wang L, Wang C-Y, Yu J, Guan K-L (2012) Cell detachment activates the Hippo pathway via cytoskeleton reorganization to induce anoikis. Genes Dev 26(1):54–68. https://doi.org/10.1101/gad.173435.111
11. Zhao B, Wei X, Li W, Udan RS, Yang Q, Kim J, Xie J, Ikenoue T, Yu J, Li L, Zheng P, Ye K, Chinnaiyan A, Halder G, Lai Z-C, Guan K-L (2007) Inactivation of YAP oncoprotein by the Hippo pathway is involved in cell contact inhibition and tissue growth control. Genes Dev 21(21):2747–2761. https://doi.org/10.1101/gad.1602907
12. Nishioka N, Inoue K, Adachi K, Kiyonari H, Ota M, Ralston A, Yabuta N, Hirahara S, Stephenson RO, Ogonuki N, Makita R, Kurihara H, Morin-Kensicki EM, Nojima H, Rossant J, Nakao K, Niwa H, Sasaki H (2009) The Hippo signaling pathway components Lats and Yap pattern Tead4 activity to distinguish mouse trophectoderm from inner cell mass. Dev Cell 16(3):398–410. https://doi.org/10.1016/j.devcel.2009.02.003
13. Moroishi T, Hansen CG, Guan KL (2015) The emerging roles of YAP and TAZ in cancer. Nat Rev Cancer 15(2):73–79. https://doi.org/10.1038/nrc3876
14. Johnson R, Halder G (2014) The two faces of Hippo: targeting the Hippo pathway for regenerative medicine and cancer treatment. Nat Rev Drug Discov 13(1):63–79. https://doi.org/10.1038/nrd4161
15. Harvey KF, Zhang X, Thomas DM (2013) The Hippo pathway and human cancer. Nat Rev Cancer 13(4):246–257. https://doi.org/10.1038/nrc3458
16. Zender L, Spector MS, Xue W, Flemming P, Cordon-Cardo C, Silke J, Fan S-T, Luk JM, Wigler M, Hannon GJ, Mu D, Lucito R, Powers S, Lowe SW (2006) Identification and validation of oncogenes in liver cancer using an integrative oncogenomic approach. Cell 125 (7):1253–1267. https://doi.org/10.1016/j.cell.2006.05.030
17. Steinhardt AA, Gayyed MF, Klein AP, Dong J, Maitra A, Pan D, Montgomery EA, Anders RA (2008) Expression of Yes-associated protein in common solid tumors. Hum Pathol 39 (11):1582–1589. https://doi.org/10.1016/j.humpath.2008.04.012
18. Bhat KP, Salazar KL, Balasubramaniyan V, Wani K, Heathcock L, Hollingsworth F, James JD, Gumin J, Diefes KL, Kim SH, Turski A, Azodi Y, Yang Y, Doucette T, Colman H, Sulman EP, Lang FF, Rao G, Copray S, Vaillant BD, Aldape KD (2011) The transcriptional coactivator TAZ regulates mesenchymal differentiation in malignant glioma. Genes Dev 25(24):2594–2609. https://doi.org/10.1101/gad.176800.111
19. Cordenonsi M, Zanconato F, Azzolin L, Forcato M, Rosato A, Frasson C, Inui M, Montagner M, Parenti AR, Poletti A, Daidone MG, Dupont S, Basso G, Bicciato S, Piccolo S (2011) The Hippo transducer TAZ confers cancer stem cell-related traits on breast cancer cells. Cell 147(4):759–772. https://doi.org/10.1016/j.cell.2011.09.048
20. Horvath P, Barrangou R (2010) CRISPR/Cas, the immune system of bacteria and archaea. Science 327(5962):167–170. https://doi.org/10.1126/science.1179555

21. Makarova KS, Haft DH, Barrangou R, Brouns SJ, Charpentier E, Horvath P, Moineau S, Mojica FJ, Wolf YI, Yakunin AF, van der Oost J, Koonin EV (2011) Evolution and classification of the CRISPR-Cas systems. Nat Rev Microbiol 9(6):467–477. https://doi.org/10.1038/nrmicro2577
22. Bhaya D, Davison M, Barrangou R (2011) CRISPR-Cas systems in bacteria and archaea: versatile small RNAs for adaptive defense and regulation. Annu Rev Genet 45:273–297. https://doi.org/10.1146/annurev-genet-110410-132430
23. Cong L, Ran FA, Cox D, Lin S, Barretto R, Habib N, Hsu PD, Wu X, Jiang W, Marraffini LA, Zhang F (2013) Multiplex genome engineering using CRISPR/Cas systems. Science 339(6121):819–823. https://doi.org/10.1126/science.1231143
24. Mali P, Yang L, Esvelt KM, Aach J, Guell M, DiCarlo JE, Norville JE, Church GM (2013) RNA-guided human genome engineering via Cas9. Science 339(6121):823–826. https://doi.org/10.1126/science.1232033
25. Jinek M, East A, Cheng A, Lin S, Ma E, Doudna J (2013) RNA-programmed genome editing in human cells. elife 2:e00471. https://doi.org/10.7554/eLife.00471
26. Cho SW, Kim S, Kim JM, Kim JS (2013) Targeted genome engineering in human cells with the Cas9 RNA-guided endonuclease. Nat Biotechnol 31(3):230–232. https://doi.org/10.1038/nbt.2507
27. Ran FA, Hsu PD, Wright J, Agarwala V, Scott DA, Zhang F (2013) Genome engineering using the CRISPR-Cas9 system. Nat Protoc 8:2281. https://doi.org/10.1038/nprot.2013.143
28. Hsu PD, Lander ES, Zhang F (2014) Development and applications of CRISPR-Cas9 for genome engineering. Cell 157 (6):1262–1278. https://doi.org/10.1016/j.cell.2014.05.010
29. Qi LS, Larson MH, Gilbert LA, Doudna JA, Weissman JS, Arkin AP, Lim WA (2013) Repurposing CRISPR as an RNA-guided platform for sequence-specific control of gene expression. Cell 152(5):1173–1183. https://doi.org/10.1016/j.cell.2013.02.022
30. Tanenbaum ME, Gilbert LA, Qi LS, Weissman JS, Vale RD (2014) A protein-tagging system for signal amplification in gene expression and fluorescence imaging. Cell 159(3):635–646. https://doi.org/10.1016/j.cell.2014.09.039
31. Gilbert LA, Larson MH, Morsut L, Liu Z, Brar GA, Torres SE, Stern-Ginossar N, Brandman O, Whitehead EH, Doudna JA, Lim WA, Weissman JS, Qi LS (2013) CRISPR-mediated modular RNA-guided regulation of transcription in eukaryotes. Cell 154 (2):442–451. https://doi.org/10.1016/j.cell.2013.06.044
32. Perez-Pinera P, Kocak DD, Vockley CM, Adler AF, Kabadi AM, Polstein LR, Thakore PI, Glass KA, Ousterout DG, Leong KW, Guilak F, Crawford GE, Reddy TE, Gersbach CA (2013) RNA-guided gene activation by CRISPR-Cas9-based transcription factors. Nat Methods 10:973. https://doi.org/10.1038/nmeth.2600
33. Gilbert LA, Horlbeck MA, Adamson B, Villalta JE, Chen Y, Whitehead EH, Guimaraes C, Panning B, Ploegh HL, Bassik MC, Qi LS, Kampmann M, Weissman JS (2014) Genome-scale CRISPR-mediated control of gene repression and activation. Cell 159(3):647–661. https://doi.org/10.1016/j.cell.2014.09.029
34. Maeder ML, Linder SJ, Cascio VM, Fu Y, Ho QH, Joung JK (2013) CRISPR RNA-guided activation of endogenous human genes. Nat Methods 10:977. https://doi.org/10.1038/nmeth.2598
35. Sanjana NE, Shalem O, Zhang F (2014) Improved vectors and genome-wide libraries for CRISPR screening. Nat Methods 11:783. https://doi.org/10.1038/nmeth.3047
36. Zischewski J, Fischer R, Bortesi L (2017) Detection of on-target and off-target mutations generated by CRISPR/Cas9 and other sequence-specific nucleases. Biotechnol Adv 35(1):95–104. https://doi.org/10.1016/j.biotechadv.2016.12.003
37. Konermann S, Brigham MD, Trevino AE, Joung J, Abudayyeh OO, Barcena C, Hsu PD, Habib N, Gootenberg JS, Nishimasu H, Nureki O, Zhang F (2014) Genome-scale transcriptional activation by an engineered CRISPR-Cas9 complex. Nature 517:583. https://doi.org/10.1038/nature14136
38. Shalem O, Sanjana NE, Hartenian E, Shi X, Scott DA, Mikkelsen TS, Heckl D, Ebert BL, Root DE, Doench JG, Zhang F (2014) Genome-scale CRISPR-Cas9 knockout screening in human cells. Science 343(6166):84–87. https://doi.org/10.1126/science.1247005
39. Nihongaki Y, Yamamoto S, Kawano F, Suzuki H, Sato M (2015) CRISPR-Cas9-based photoactivatable transcription system. Chem Biol 22(2):169–174. https://doi.org/10.1016/j.chembiol.2014.12.011
40. Lu J, Zhao C, Zhao Y, Zhang J, Zhang Y, Chen L, Han Q, Ying Y, Peng S, Ai R, Wang Y (2018) Multimode drug inducible CRISPR/Cas9 devices for transcriptional activation and

genome editing. Nucleic Acids Res 46(5):e25. https://doi.org/10.1093/nar/gkx1222

41. Mojica FJ, Diez-Villasenor C, Garcia-Martinez J, Almendros C (2009) Short motif sequences determine the targets of the prokaryotic CRISPR defence system. Microbiology 155. (Pt 3:733–740. https://doi.org/10.1099/mic.0.023960-0

42. Doench JG, Hartenian E, Graham DB, Tothova Z, Hegde M, Smith I, Sullender M, Ebert BL, Xavier RJ, Root DE (2014) Rational design of highly active sgRNAs for CRISPR-Cas9-mediated gene inactivation. Nat Biotechnol 32(12):1262–1267. https://doi.org/10.1038/nbt.3026

43. Doench JG, Fusi N, Sullender M, Hegde M, Vaimberg EW, Donovan KF, Smith I, Tothova Z, Wilen C, Orchard R, Virgin HW, Listgarten J, Root DE (2016) Optimized sgRNA design to maximize activity and minimize off-target effects of CRISPR-Cas9. Nat Biotechnol 34(2):184–191. https://doi.org/10.1038/nbt.3437

44. Agrotis A, Ketteler R (2015) A new age in functional genomics using CRISPR/Cas9 in arrayed library screening. Front Genet 6:300. https://doi.org/10.3389/fgene.2015.00300

Chapter 17

Hippo Pathway Regulation by Tyrosine Kinases

Nina Reuven, Matan Shanzer, and Yosef Shaul

Abstract

The Hippo pathway utilizes a well-characterized Ser/Thr kinase cascade to control the downstream effectors, Yap and Taz. In addition, Yap/Taz and other Hippo pathway components are directly regulated by tyrosine kinases (TKs). The methodological strategies described here use the example of the c-Abl non-receptor TK and the Yap substrate to outline the steps used to identify and to validate tyrosine phosphorylation sites, including bioinformatic approaches, ectopic expression of proteins in transfected tissue culture cells, and mutagenesis of endogenous proteins by CRISPR-Cas9. These general strategies can be applied to investigate regulation of protein signaling moieties by tyrosine phosphorylation in the context of distinct TKs.

Key words Hippo pathway, Cell fate determination, Tyrosine phosphorylation, Non-receptor tyrosine kinases, Abl, Src, Yap, Yes-associated protein, Taz, Transcriptional co-activator with PDZ-binding motif, WWTR1, CRISPR-mediated gene editing

1 Introduction

The Hippo pathway is an evolutionarily conserved pathway from flies to humans regulating development, organ size and cell fate. The pathway is activated upon high cell density, energy stress, serum deprivation, and F-actin disassembly and triggers a kinase cascade with the serine/threonine (S/T) kinases Mst1/2 and Lats1/2 at its core (reviewed in [1]). Lats phosphorylation of the downstream effectors Yap and Taz serves to prevent their nuclear accumulation by promoting cytoplasmic sequestration and proteasomal degradation. Yap and Taz are transcriptional coactivators, and interact with TEAD and several other transcription factors (TFs). Interaction with the TEAD TF family is mediated through the Yap and Taz TEAD binding domain (Fig. 1), while interaction with other TFs such as Runx and the tumor suppressor p73 (a paralog of p53) is mediated through the Yap and Taz WW domain, and the TFs PPxY motif.

While the Hippo pathway induces Yap and Taz S/T phosphorylation by Lats, Yap and Taz are also independently regulated by

Alexander Hergovich (ed.), *The Hippo Pathway: Methods and Protocols*, Methods in Molecular Biology, vol. 1893, https://doi.org/10.1007/978-1-4939-8910-2_17, © Springer Science+Business Media, LLC, part of Springer Nature 2019

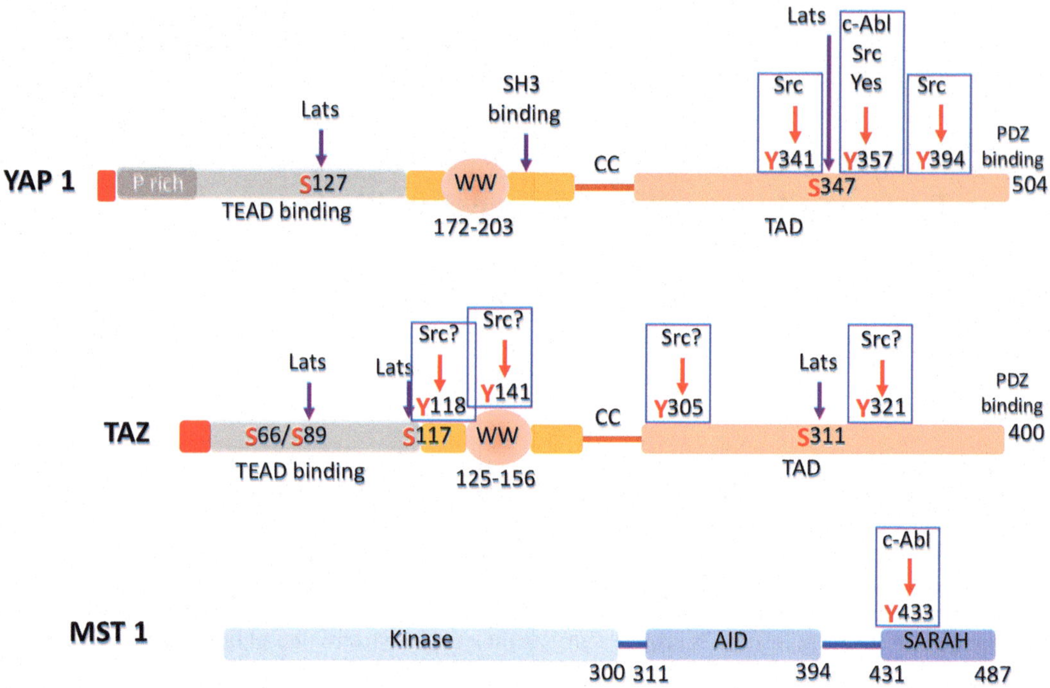

Fig. 1 Schematic representation of human Yap1, Taz, and MST1. Indicated are functional domains and selected serine and tyrosine phosphorylation sites. The kinases known or suspected to phosphorylate the sites are indicated. P rich (proline-rich), TEAD-binding, WW, TAD (trans-activation domain), kinase, AID, and SARAH domains are indicated

TK. Yap and Taz are phosphorylated by non-receptor tyrosine kinases (nRTK), mainly by the members of the Src kinase family and c-Abl, under physiological conditions whereby these nRTK are activated. In contrast to the Hippo pathway regulation, the tyrosine phosphorylated Yap and Taz are nuclear-retained and functional in supporting transcription. In this chapter, we will cover the role of nRTKs in regulating Yap and Taz (reviewed in [2]).

The nRTKs are made up of sub-families, including the Abl, Src, Ack, Csk, Fak, Fes, Frk, Jak, Tec and Syk. The Src family kinases (SFK) include Src, Fgr, Fyn, Yes, Blk, Hck, Lck, and Lyn kinases. nRTKs, in addition to the kinase domain (SH1) often bear the Src homology 3 (SH3) domain. TKs recognize substrates via the SH3 domain that physically binds a short proline rich motif. In general nRTKs are under inhibition and are activated in response to specific signaling cues or stresses. The N-terminal residues of c-Abl inhibit the kinase domain and therefore mutant c-Abl Δ1–81 is used to obtain constitutively active c-Abl [3]. Src is inhibited by another kinase named CSK that phosphorylates Src at Y527, and is activated by dephosphorylation of this residue [4]. Src Y527F is therefore constitutively active.

Yap was first identified by its ability to bind to the nRTK Yes, and other adaptor proteins and TKs having an SH3-domain, including Nck, Crk, Src, and Abl [5]. Subsequently, several nRTKs were shown to phosphorylate and regulate Yap and Taz activity. The role of nRTKs on Yap/Taz was recognized while investigating the role of Runx2 in bone development. Yap phosphorylation by Src/Yes promotes Yap/Runx2 interaction that inhibits expression of osteocalcin in bone [6]; whereas c-Abl promotes the assembly and activation of Taz/Runx2 that is required for osteoblastogenesis [7]. These nRTKs target the same substrate, namely Runx2, yet the outcome is disparate. Such divergent outcomes in response to a biochemically similar cue, i.e. tyrosine phosphorylation, exemplify the importance of identifying the involved nRTKs in the process under study.

Yap and Taz play pivotal roles in cell fate determination in an nRTKs-dependent manner. Active nRTKs have been found to modify Yap/Taz-mediated proliferation at times overriding canonical Hippo suppressive function. Thus, it is important to investigate not only the Lats-mediated serine phosphorylation status of Yap/-Taz, but also their tyrosine phosphorylation state. For example, Yap phosphorylation by Yes facilitates the association of a Yap/beta-catenin/TBX5 complex to the promoters of anti-apoptotic genes, leading to proliferation and survival of beta-catenin driven cancers [8]. In sporadic human malignancies alternate RASSF1 transcripts promote Src/Yes localization to epithelial cell-cell junctions, where it phosphorylates Yap and beta-catenin, promoting tumorigenesis [9]. Furthermore, inflammation in the gut triggers Yap phosphorylation by Src, which stabilizes Yap and promotes its nuclear localization to activate tissue regeneration genes [10]. In colorectal cancer, APC loss triggers SFK (Src family kinase) activation and Yap activation, and sustained Yap activity is dependent on its tyrosine phosphorylation [11]. Src phosphorylation of Taz has similar effects on intestinal regeneration and in colorectal cancer [12]. In contrast to the pro-oncogenic function of Yap and Taz in these examples Yap phosphorylation by c-Abl promotes cell death, in response to DNA damage. Under DNA damage conditions c-Abl is activated and phosphorylates Yap and p73 and increases their association and function in inducing the expression of pro-apoptotic genes [13–15]. In general, SFKs appear to support a role for Yap/Taz in proliferation whereas c-Abl is more active in differentiation and cell growth-arrest and death programs. Consistent with this emerging picture is the observation that Src and c-Abl are each activated under distinct signaling cues.

Y357 is the major phosphorylation residue for the Yap1-1 isoform, possessing one WW domain, or Y407, for the Yap1-2-isoform, having two WW domains (Yap isoforms are described in [16] (Fig. 1). The corresponding phosphorylation site in human Taz is Y321 (Y316 in mouse, the system used in the reference cited above [12]). If SFKs and c-Abl both phosphorylate Yap Y357 the question is then how distinctive cell fate determination is obtained. One possibility is that the physiological context determines the function of the phosphorylation, such as determining a specific subcellular localization or generating unique protein–protein interaction and functional complexes. Consistent with this possibility is the fact that the nRTKs not only phosphorylate Yap/Taz but also their target TFs, namely Runx and p73 [17, 18]. In addition, lately it has been reported that Src phosphorylates three other Yap tyrosine residues in the transactivation domain that are important for Src-Yap-mediated transformation [19].

Yap and Taz are not the only members of Hippo pathway regulated by non-receptor tyrosine kinases. For example, it has been reported that Abl phosphorylates MST1 [20] and MST2 [21] in the process of neuronal cell death under oxidative stress. Of much interest is the finding that Lats phosphorylates and inactivates c-Abl [22]. This crosstalk between the two regulators of Yap and Taz is important in reaching homeostasis. An example of violation of the described coherent picture of SFKs promoting oncogenesis and Abl promoting cell death was observed with the polyomavirus middle T-antigen (PyMT) oncogene. During cellular transformation PyMT counter-intuitively activates the core Hippo pathway tumor suppressor Lats, in a Src-dependent manner, causing Yap/Taz cytoplasmic retention, thereby inhibiting their reported pro-proliferative functions [23]. In addition, Src activation by PyMT inhibits Taz degradation by the SCF-beta-TrCP E3-ubiquitin ligase [24].

In this chapter, we outline a basic strategy for identifying and confirming tyrosine phosphorylation of proteins of interest by non-receptor tyrosine kinases. Due to space constraints, we can describe in detail only a portion of the methodology involved in this task. Therefore, aspects of the methodology are more briefly described, such as database mining, and the use of genetic and pharmacological inhibitors. We will outline in detail the identification of phosphorylation by using overexpression of the target proteins and TKs of interest, using the Yap/c-Abl system as a model. We also provide a protocol for endogenous genome editing using CRISPR-Cas9 to produce the Yap Y357F mutation in MCF10A cells.

2 Materials

2.1 Tissue Culture

For this protocol basic tissue culture equipment and supplies are required, including incubators, laminar flow hoods, centrifuges, inverted microscopes, and other standard equipment and plasticware.

1. Plasmids expressing Yap and active c-Abl (for example, pCDNA Flag-Yap1 (Addgene 18,881) [25], pCDNA Flag-Yap1 Y357F (Addgene 18,882) [25], pCDNA c-Abl Δ1–81 [22]).
2. HEK293 cells.
3. MCF10A cells.
4. Growth medium for HEK293 cells: DMEM (high glucose, 4500 mg/l), 8% (v/v) Fetal Bovine Serum, 100 U/ml penicillin, 0.1 mg/ml streptomycin.
5. Growth medium for MCF10A: DMEM-F12 (1:1), 5% (v/v) Horse serum, 100 U/ml penicillin, 0.1 mg/ml streptomycin, 2 mM glutamine, 20 ng/ml epidermal growth factor (EGF), 10 μg/ml insulin, 0.5 μg/ml hydrocortisone, 100 ng/ml cholera toxin.
6. Trypsin solution for passaging HEK293 (0.25% (w/v) Trypsin, (0.05% w/v) EDTA in solution B (*see* **Note 1**)).
7. Trypsin solution for passaging MCF10A (0.05% (w/v) Trypsin, 0.02% (w/v) in solution C (*see* **Note 1**)).
8. Polyethylenimine (PEI), 1 mg/ml (*see* **Note 2**).
9. 150 mM NaCl solution, 0.22 μ filter-sterilized.

2.2 SDS-PAGE and Western Blotting

1. Gel-running and transfer apparatus such as Bio-Rad Mini-Protean III system with power supply.
2. PBS: 137 mM NaCl,, 2.7 mM KCl, 10 mM Phosphate, pH 7.4.
3. RIPA buffer: 150 mM NaCl, 50 mM Tris–Cl pH 7.5, 1% (v/v) NP-40, 0.5% (w/v) deoxycholate, 0.1% (w/v) SDS.
4. Protease Inhibitor cocktail (for example Sigma P8340 or equivalent (*see* **Note 3**)).
5. Tyrosine Phosphatase Inhibitor cocktail (for example Sigma P5726 or equivalent (*see* **Note 4**)).
6. SDS-PAGE gels (10%).
7. Prestained molecular weight standards (Bio-Rad or equivalent).
8. 3× Laemmli protein gel sample buffer: 6% (w/v) SDS, 30% (v/v) glycerol, 0.1875 M Tris–Cl pH 6.8, 15% (v/v) β-mercaptoethanol, 0.03% (w/v) bromphenol blue.
9. 10× TG (Tris–Glycine): 0.25 M Tris, 1.9 M glycine, pH approximately at 8.3.
10. Gel running buffer: 1× TG, 0.1% (w/v) SDS.
11. Gel transfer buffer: 1× TG, 20% (v/v) methanol.

12. Ponceau Red (Ponceau S) solution: 0.1% Ponceau S (w/v) in 5% (v/v) acetic acid.
13. Nitrocellulose membrane 0.45 μm.
14. 3MM chromatography (blotting) paper.
15. PBST: PBS with 0.1% (v/v) Tween-20.
16. TBST: 10 mM Tris–Cl pH 7.5, 100 mM NaCl, 0.1% (v/v) Tween-20.
17. Non-fat dry milk powder.
18. Primary and secondary antibodies (*see* **Note 5**).
19. Anti-FlagM2 agarose (Sigma).
20. Reagents for enhanced chemiluminescence such as EZ-ECL (Biological Industries).
21. X-ray film and developer, or digital camera system such as ImageQuant LAS4000.
22. 50 mM citric acid solution.

2.3 Molecular Biology Equipment and Supplies

1. Thermocycler.
2. TAE (Tris–acetate–EDTA): 40 mM Tris–acetate (pH 8.3), 1 mM M EDTA.
3. Ethidium bromide (0.625 mg/ml stock solution).
4. Agarose, molecular biology grade.
5. Agarose gel apparatus.
6. Restriction endonucleases.
7. Purification kits for DNA fragments from agarose gels.
8. DNA polymerases for PCR (*see* **Note 6**).
9. Oligonucleotide primers.

3 Methods

3.1 Using Web-Based Tools to Find Evidence of Phosphorylation

It is advised to collect as much information as possible before designing the wet lab experiments. Today there are freely available datasets showing phosphorylation sites of proteins. The Phosphosite Plus website [26] (https://www.phosphosite.org/homeAction.action, Accessed 9 Jan. 2018) [27] curates data from unbiased Mass-Spectrometry screens for protein modifications, as well as from the published literature. This tool provides information on type of phosphorylation and specific residue that was experimentally detected on a protein of interest. Detection of a putative phosphorylation site in multiple datasets is a tell-tale sign of a biologically important site, although absence of a record does not exclude the possibility of phosphorylation in vivo. A site may be particularly labile, rare, difficult to detect by Mass Spectrometry or not induced by the conditions employed in the reported experimental systems. Additional tools are available for prediction of tyrosine phosphorylation and employ a multitude of bioinformatics methodologies ranging from search of short linear motifs (SLiMs) that serve as TK phosphorylation consensus sites in the substrate primary sequence to neural network based predictions. Some examples of useful tools include: NetPhos3.1 [28] (http://www.cbs.dtu.dk/services/NetPhos/) [29], Phospho ELM [30] (http://phospho.elm.eu.org/links.html) [31], and GPS [32] (http://gps.biocuckoo.org/) [33]. Another approach is to search for functional protein modules such as protein–protein interaction modules. Our experience suggests that clustering of functional protein modules (such as overlapping phosphorylation consensus sites, SH2/SH3-domain proline rich binding motifs, phospho-degron motifs, etc.) may be indicative of a regulatory hotspot. Such juxtaposed elements increase the likelihood of a putative phosphorylation site to be functionally valid. Scansite [34] (http://scansite.mit.edu/) [35] is an especially useful tool for searching short linear motifs against the proteome. Motifs are also easily detected using the Eukaryotic Linear Motif resource [36] (ELM, http://elm.eu.org/) [37]. Using these tools will help to focus the study on the phosphorylation sites most likely to be biologically relevant.

The tools above might also help in identifying the TKs that are likely to phosphorylate the identified tyrosine residues, based on the consensus phosphorylation motifs. It is important to remember, however, that there are always exceptions to this rule, and there are specific sites known to be phosphorylated by kinases that do not match the defined consensus. While the database mining may offer a number of potential kinases of interest in an unbiased manner, your system will dictate which kinases are most likely to be relevant, and should be examined first. In the example of Yap phosphorylation by c-Abl, our system was the DNA damage response, where we had prior evidence of c-Abl activation, and phosphorylation by c-Abl of the Yap binding partner p73 [18]. This allowed the investigation of physiological consequences of inhibition of phosphorylation, using genetic knockdown or knockout tools to target the kinase. As part of validating the phosphorylation and the kinase responsible, specific kinase inhibitors can be used. Since many of the compounds target more than one kinase (albeit with different IC50 values) or may have other off-target effects, it is a good idea to try several different inhibitors and also to complement this approach with genetic knockdown or knockout strategies. For c-Abl, Imatinib (Gleevec) [38] has been a useful tool, although it is not very specific and also inhibits PDGFR and c-Kit, among others. More specific and potent inhibitors also targeting the Abl ATP-binding site have been developed, such as nilotinib [39], and more recently allosteric inhibitors, such as GNF-5 [40] have become available. Dasatinib inhibits both Src and Abl kinases

[41], while PP2 [42] is an inhibitor of SFKs. These kinase inhibitors are useful for determining whether tyrosine phosphorylation is involved in mediating a biological effect in the cells. Having established through the database searches that a phosphorylation event is likely, and through inhibitors/knockdown/knockout strategies have shown that phosphorylation influences a phenotype of interest in the cells, we can use the methods outlined below to focus on specific phosphorylation sites. In addition, Mass Spectrometry can be used to identify phosphorylation sites in your protein of interest [43].

3.2 Validating Phosphorylation and the Involved Tyrosine Residues

To verify protein phosphorylation by a suspected kinase, transient transfection is used to express an active form of the kinase together with the protein of interest. Transfected cell extracts are then subjected to immunoprecipitation for the isolation of the protein of interest. The immunoprecipitates are analyzed by SDS-PAGE and immunoblotting, using an anti-phospho-tyrosine antibody to detect the putative phosphorylation. To verify a specific suspected site of phosphorylation, mutagenesis to create a Y-to-F mutation is used, which eliminates the possibility of phosphorylation at the mutated site (phospho-dead mutant), but should minimally affect the structure of the protein. In this chapter we describe the methods of detecting phosphorylation of Yap1 at Y357 by c-Abl. In the following protocol, we will use the Flag-tagged version of Yap1, and the corresponding Y357F mutant, together with the constitutively active c-Abl construct c-Abl Δ1-81, which is missing its autoinhibitory N-terminus [3] (*see* **Note 7**). This protocol does not address cloning strategies for preparing the plasmid constructs for expression.

1. Seed HEK293 cells in 6-well dishes (800,000 cells/well) 16 h prior to transfection, so that the cells are 50–60% confluent at the time of transfection (*see* **Note 8**).
2. Dilute the DNA for each transfection into sterile 150 mM NaCl to give a final volume of 100 μl. A total of 4 μg of DNA will be used for each well. To verify transfection efficiency, include 0.05 μg of a plasmid expressing a fluorescent protein, such as pEGFP. Use 0.5 μg of pCDNA3 c-Abl Δ1–81, and 1 μg of the pCDNA3 Flag-Yap or Flag-Yap Y357F. In control wells, use the corresponding empty vector (pCDNA3). Use a plasmid without a mammalian promoter on it, such as pBluescript, to bring the total amount of DNA to 4 μg per tube. Vortex to mix.
3. Prepare a mix of diluted PEI sufficient for all of the tubes. Use 16 μl PEI into 100 μl of 150 mM NaCl per tube. Vortex.
4. Add 100 μl of the diluted PEI solution all at once to 100 μl of the DNA solution (do not reverse this order). Vortex, and let tube stand at room temperature 15–30 min.

5. Add the DNA/PEI mix dropwise to the medium of the cells (*see* **Note 9**). There is usually no need to change the HEK293 cells' medium with this transfection procedure.
6. Verify transfection by checking for expression of the control fluorescent protein using a fluorescent microscope. This should be clearly visible 16–24 h post-transfection.
7. Harvest cells 24–48 h post-transfection by scraping cells into PBS. Centrifuge at low speed (3000 × *g*), and wash with PBS (*see* **Note 10**).
8. Resuspend cells in 300 μl RIPA buffer containing protease inhibitors and tyrosine phosphatase inhibitors. We usually use commercially available 100× cocktails of the inhibitors (*see* **Notes 3** and **4**).
9. Incubate on ice 10 min, then centrifuge at maximum speed (16,000 × *g*) in a microcentrifuge.
10. During the incubations of **step 9** above, prepare the anti-Flag-agarose beads that you will need for the experiment. Typically, 15 μl of slurry per assay tube is sufficient. Wash the beads for the experiment twice with an excess (600 μl) of RIPA buffer. If a large number of assay tubes are involved, you may wish to wash the beads for the experiment in a larger volume. To pellet beads, centrifuge at low speed (3000 × *g*). Use a syringe with a 21G needle to suck off the supernatant. Resuspend beads in RIPA buffer with inhibitors, in a volume that will provide 50 μl per assay tube. For all pipetting of the beads, use a wide-bore tip, to avoid shearing. Aliquot 50 μl washed bead slurry per tube into fresh 1.5 ml tubes.
11. Remove 50 μl of the supernatant from **step 9** into a fresh 1.5 ml tube. Add 25 μl 3× gel loading buffer. This whole cell extract (WCE) sample is needed as a control to demonstrate expression of the constructs in the cells. This amount is sufficient for 4–8 gel loadings, depending on the size of the wells.
12. Take the remainder of the supernatant from **step 9** above and add it to the anti-Flag-agarose beads for immunoprecipitation (IP). Incubate these tubes at 4 °C for 2–4 h, with slow rotation, 10–15 rpm.
13. Centrifuge the IP samples for 2 min at 3000 × *g*, and save the supernatant for further analysis, if desired (*see* **Note 11**). Wash beads four times with 400 μl RIPA. Invert tubes at each wash to completely disrupt and wash the pellet, and pellet beads by centrifuging at 3000 × *g* for 1 min. Do not vortex.
14. After the last wash, leave approximately 30 μl of buffer on the beads. Add 15 μl 3× sample buffer.
15. Boil WCE samples for 5 min, and IP samples for 2 min (*see* **Note 12**). Centrifuge samples for 1 min at 16,000 × *g*, and

load supernatant on a 10% SDS-PAGE gel. Load a pre-stained protein ladder marker in one of the lanes.

16. Electrophorese at 80 V (for example Bio-Rad Mini protean III system), until the samples enter the resolving gel, then raise voltage to 175 V. Run the gel until the bromphenol blue reaches the bottom of the gel, or until you have achieved satisfactory separation, as seen by the migration of the pre-stained markers.
17. Transfer the proteins to nitrocellulose membrane (for example Bio-Rad system) using gel transfer buffer, at 300 mA for 1 h.
18. Stain the membrane with Ponceau Red (this takes less than 1 min) followed by washing with water. This enables the visualization of the antibody chains in the IP samples, which will guide you in cutting off this portion of the blot. The Ponceau staining of the WCE samples can be used to verify equal loading of the samples, and can also help guide in cutting the blot in order to allow probing with different antibodies–anti-Flag to visualize Yap expression, and anti-Abl.
19. Remove the Ponceau staining by washing the blots in TBST.
20. Block the membrane by incubating in TBST with 2% (w/v) milk powder 30–60 min at room temperature (*see* **Note 13**).
21. Incubate the membrane with the IP samples overnight at 4 °C in a 1:1000 dilution of anti-phosphotyrosine PY20 in blocking solution. Probe the relevant pieces of the WCE membrane separately with anti-Flag and anti-Abl.
22. Remove antibody solution, and wash blot with three changes of TBST at room temperature (15 min, then 2× 5 min), with gentle shaking.
23. Incubate the membranes with HRP-conjugated secondary antibody in blocking solution (1:10–20,000 dilution) for 1 h, followed by washing, as in **step 22** above. For best results, incubate each membrane in its own separate antibody solution.
24. Use commercially available reagents for enhanced chemiluminescence (ECL) to detect the proteins on the blot. The blot can be exposed to X-Ray film for detection, or by using a digital camera system designed for this purpose.
25. To verify the amount of immunoprecipitated protein, the IP blot should be stripped and re-probed. Strip by incubation in 50 mM citric acid for 30 min-1 h at room temperature, followed by two 5 min washes with PBST. The blot can then be re-probed with the anti-FlagM2 antibody, to detect the total immunoprecipitated protein.

Figure 2 shows the results for Flag-Yap1 and Flag-Yap1 Y357 phosphorylation by c-Abl described in this section. Tyrosine

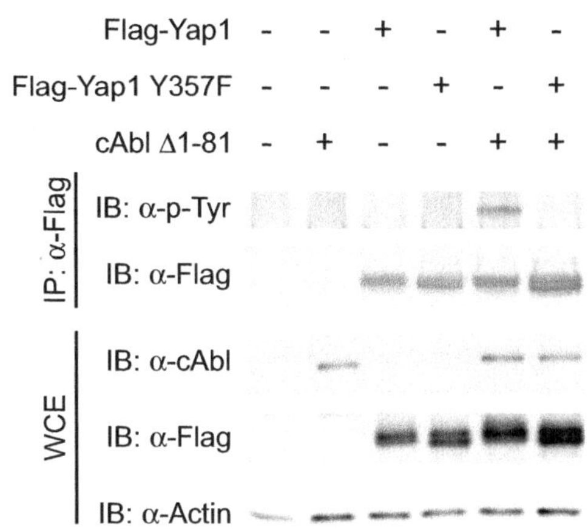

Fig. 2 Yap1 phosphorylation by c-Abl at Y357: Results of the immunoblot described in Subheading 3. HEK293 cells were transfected with plasmids expressing Flag-Yap1, Flag-Yap1 Y357F, and c-Abl Δ1–81, as indicated. As described in Subheading 3, phosphorylation of the immunoprecipitated Flag-Yap1 was detected by the anti-phosphotyrosine antibody PY20. The whole cell extracts (WCE) were probed with anti-FlagM2, anti-cAbl 8E9 and anti-actin, as a loading control. Following probing with PY20, the IP blot was stripped and re-probed with anti-FlagM2

phosphorylation of Flag-Yap1 is detected with co-expression of active c-Abl, but it is not detected with the Flag-Yap1 Y357F mutant. Figure 3 shows a similar experiment, showing tyrosine phosphorylation of HA-tagged p73α (control), and the Hippo pathway proteins Lats2, Mst1, and WW45 by c-Abl. The phosphorylation of Mst1 (band marked by an asterisk) by c-Abl was much weaker than the other proteins, but was clearly visible upon overexposure (not shown).

At this stage, anti-phospho antibodies recognizing the specific sites can be raised, which will be useful in further analysis of the physiological effects of the phosphorylation. There are many commercial services available for this (*see* **Note 14**).

3.3 Creating an Endogenous Phospho-Dead Mutant Using CRISPR

Once a phosphorylation site has been identified and confirmed, its biological activity in the context of the endogenous protein can be examined using genome editing. CRISPR-Cas9 technology enables the creation of endogenous protein mutations. CRISPR editing relies on the CRISPR-Cas9 system to make a site-specific double-strand break in the genome, which is directed by a 20 bp guide RNA sequence (reviewed in [44]). The cellular DNA repair enzymes are then recruited to repair the break. In order to engineer a specific change in the genome, we provide a repair template with the desired sequence, so that with recombinational repair, the new sequence can be incorporated. The success rate depends upon the expression of the Cas9 and guide RNA, the efficiency of the guide

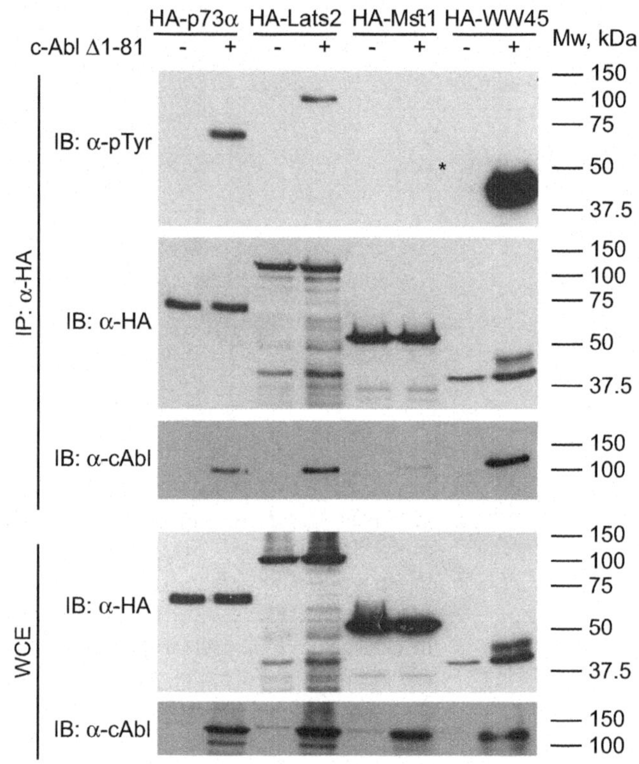

Fig. 3 Active c-Abl phosphorylates Hippo pathway components. HEK293 cells were transfected with plasmids expressing HA-tagged constructs of p73α (control), Lats2, Mst1, and WW45, and with cAbl Δ1–81, as indicated. Cells were harvested 48 h following the transfection, and anti-HA agarose was used for the IP. IP and WCE samples were analyzed by SDS-PAGE and immunoblot with the indicated antibodies. The asterisk marks a faint band of phosphorylated Mst1, which was more clearly observed upon overexposure (not shown)

RNA, the propensity to repair the damage by homologous recombination, and the cell line used. There are many useful tools and protocols available today for CRISPR. Our strategy is based primarily on protocols from the Zhang laboratory at MIT (http://www.genome-engineering.org/crispr/). The following protocol is for engineering the Y357F mutation in MCF10A cells, giving the specific reagents and protocols that we used for this modification, although the strategy can be applied to other sites.

1. Before choosing a guide RNA sequence, it is a good idea to sequence the genomic DNA of the cell line in the region that will be targeted. The reason for this is that the 20 base guide must match the targeted sequence exactly to achieve efficient cleavage. Therefore, choose PCR primers flanking the mutation site that will result in a 500–600 bp amplicon. Be sure to use the genomic (and not cDNA) sequence of the gene. Choose primers that have similar Tm and that are specific to

your gene, when analyzed by NCBI BLAST (https://blast.ncbi.nlm.nih.gov/Blast.cgi).

2. For flanking Y357 in Yap, to produce a 523 bp fragment, we used:

   forward primer: CCTTAGGAGAAATCTTGGTTCCTTAGAGCTG

   reverse primer: GTTACAAATCACCGTATAGAGACAAAGCAGG

3. Amplify the 523 bp fragment as follows: 1 μg genomic DNA, 12.5 pmol each primer, in a 50 μl reaction using iPfu PCR mix. PCR program: 94 °C 5 min, 36 cycles of (94 °C 30 s, 60 °C 30 s, 72 °C 1 min). PCR conditions should be adjusted as needed to suit your specific primers and template. The genomic DNA was made by using a kit (we used Sigma GenElute™ Mammalian Genomic DNA Miniprep Kit). It is also possible to use the simple genomic DNA preparation outlined below (**step 11**).
4. Separate the PCR fragment on a 1.5% (w/v) agarose gel in TAE buffer with 0.5 μg/ml ethidium bromide, and excise the band. Purify the DNA from the gel using a commercial gel purification kit, such as Qiagen Qiaquick, and sequence the DNA by Sanger sequencing (institutional or commercial service) using one of the PCR primers. Confirm that the cell line sequence matches the database sequence.
5. Using a web-based tool such as DeskGen [45] (https://www.deskgen.com/landing/cloud.html), choose a guide RNA for SpCas9 that will generate a double-strand break close to the site to be edited (in our example, it is 13 bases away). If there are several options for guide RNAs, choose the guide that has the highest score for cleavage efficiency, and that is predicted to cut a minimum of off-target sites. The following guide RNA sequence was used: GGAAGTCATCTGGGGTTCGA.
6. To introduce the Y357F mutation, a long single-stranded oligonucleotide donor (ssODN) was used. Aside from the mutation changing the tyrosine TAC codon to TTC to encode phenylalanine, we included three other silent point mutations that would prevent further targeting of the site by the guide RNA, and that would introduce a new restriction site (StyI) that could be used later to screen for the mutation. The mutations introduced are indicated by lowercase letters. See also Fig. 4. Sequence of ssODN: CCTATCACTCTCGAGATGAGAGTACAGACAGTGGACTAAGCATGAGCAGCTtCAGTGTgCCTaGgACCCCAGATGACTTCCTGAACAGTGTGGATGAGATGGA.
7. For expression of the SpCas9 nuclease and guide RNA by transient transfection, we used the pX330 plasmid from the

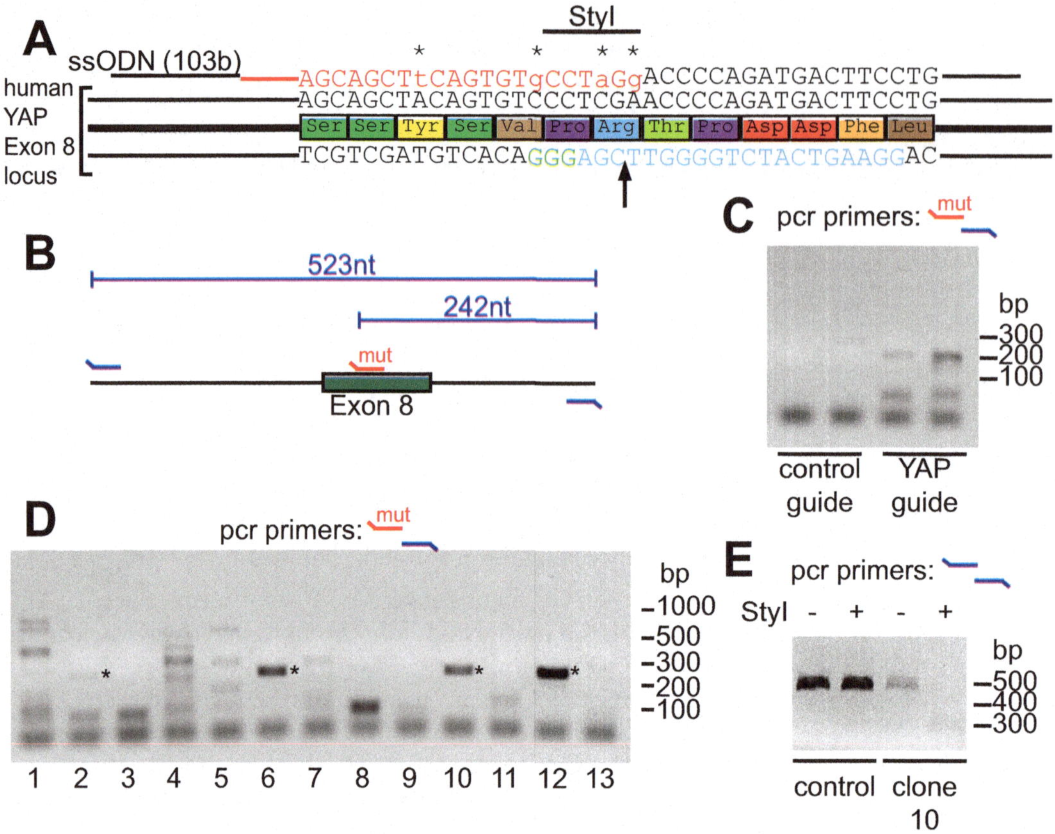

Fig. 4 CRISPR-mediated mutagenesis of Yap Y357. (**a**) Sequence of the targeted Yap locus. The guide RNA sequence is marked in blue, and the PAM sequence in green. The CRISPR-Cas9 cleavage site is marked by an arrow. The donor oligonucleotide (ssODN) is shown above, and the mutated sites are written in lowercase and indicated by asterisks. The first mutation (TAC to TTC) changes Tyr to Phe, while the remaining mutations are silent. The StyI site introduced by the mutations is indicated. The sequence of the mutated primer used for PCR analysis is marked in red. (**b**) Schematic of the targeted Yap locus, with the pcr primers indicated. The primer matching the mutated sequence is marked in red. (**c**) Results of the PCR on genomic DNA from four independent pools of transfected cells, two that were transfected with plasmid encoding a control guide sequence, and two with the Yap-specific guide. The desired 242 bp product indicating successful incorporation of the mutations appears only in the samples that were transfected with the Yap guide. (**d**) PCR analysis of genomic DNA from clones. Asterisks mark the desired 242 bp product. (**e**) StyI digests of the 523 bp pcr product amplified from genomic DNA from control cells, and from clone 10

Zhang lab (Addgene 42230) [46]. The guide RNA sequence was cloned into pX330 as described on the reagent page [47] (http://www.genome-engineering.org/crispr/wp-content/uploads/2014/05/CRISPR-Reagent-Description-Rev20140509.pdf).

8. Plate 10^5 MCF10A cells per well in a 12-well dish, to give 50–60% confluency the following day.
9. Transfect with the following: 0.05 μg pEGFP, 0.5 μg pX330 expressing the guide RNA, 1 μl of 10 μM ssODN, and 0.4 μg of

pEFIRES-p [48] (to provide resistance to puromycin—*see* **Note 15**) diluted into 150 mM NaCl to a final volume of 50 μl. Dilute 3 μl of 1 mg/ml PEI solution into 50 μl of 150 mM NaCl. Add the PEI solution to the DNA solution at once, and vortex. Let stand 15–30 min, and add dropwise to the medium of the cells.

10. Change the medium on the cells 30 h post-transfection to medium with 5 μg/ml puromycin.
11. Change medium again to medium without puromycin after 24 h. Allow cells to recover, and replate as necessary.
12. The efficiency of endogenous editing varies greatly, depending on the site edited and other conditions described above. We used PCR with a mutation-specific primer (described in **step 12** below) to detect evidence of proper integration of the mutation. Once the proper PCR conditions have been established, enabling detection of the mutation-specific band, this assay will be used as a first step in screening single-cell clones. A simple and cost-effective protocol for preparing genomic DNA for this purpose is as follows: Harvest 5 × 10^4–10^5 cells, and resuspend the cell pellet in 18 μl of 50 mM NaOH. Incubate at 95 °C for 10 min. Add 2 μl of 1 M Tris–Cl pH 8, and 140 μl of water. This crude genomic DNA preparation can be used in PCR reactions, although a range of dilutions may be required in order to find the conditions producing a clear band.
13. PCR of the genomic DNA is used to verify incorporation of the mutation. A primer spanning the mutation site is used (mutated sites indicated by lowercase): forward mutant primer: GGACTAAGCATGAGCAGCTtCAGTGTgCCTaGg.
14. This primer is used together with the reverse primer described above to amplify a 242 bp fragment. Conditions for pcr (each tube): 2 μl of 1:10 dilution of genomic DNA preparation, 5 pmol of each primer, 2 μl of 5× PCR mix (Larova Red Load Taq Master mix), and water to 10 μl. PCR program: 95 °C 5 min, 35 cycles of (95 °C 30 s, 65 °C 30 s, 72 °C 30 s). The pcr reactions are loaded directly on a 2% (w/v) agarose gel in TAE buffer with 0.5 μg/ml ethidium bromide. Figure 4c shows the proper 242 bp band was produced in samples from two independent wells of the transfected cells described above, whereas in two control wells, using a control guide RNA in pX330, this band was not produced.
15. To isolate clones from single cells, plate the cells in 96 well dishes at a very low density-calculated at 0.5 cells/well. Re-feed the growing colonies with 50 μl growth medium as needed.
16. Analyze DNA from colonies of cells that grew from single cells using the PCR assay as above (*see* **step 12**). Our results showed that 8 colonies out of 24 analyzed produced a positive PCR result. Figure 4d shows some of the clones analyzed.

17. To further verify the mutation, we checked for the StyI restriction site. For this we amplified the 525 bp genomic fragment, and cleaved with StyI. Complete cleavage suggests that both alleles have incorporated the mutation. For this amplification, we used Terra polymerase (Clontech), because the pcr reaction mix could be added directly to the restriction digest without inhibiting the restriction enzyme. The procedure is as follows: The genomic DNA is amplified in a 20 μl final volume, using 5 pmol of each primer. PCR program: 98 °C 5 min, 35 cycles of (98 °C 10 s, 60 °C 15 s, 68 °C 1 min). Following the PCR, 5 μl of the PCR reaction mix is incubated in a 20 μl restriction digest, using NEB buffer 3.1, with or without 1 μl of StyI, for 1 h at 37 °C. Figure 4e shows the digest of DNA from a control (wild-type) sample, and from clone 10 (Fig. 4d). Clone 10 shows complete digest.
18. Based on the results of the restriction digest in **step 15**, purify the 524 bp PCR fragment from the remainder of the PCR reaction, and sequence it by Sanger sequencing. Figure 5 shows sequencing of wild type cells, and two clones—one that is heterozygous for the mutation, and one that appears to be homozygous. By these criteria, one of the eight clones from **step 14** appeared to be homozygous for the desired mutations, with no additional extraneous mutations. *See* **Note 16**.

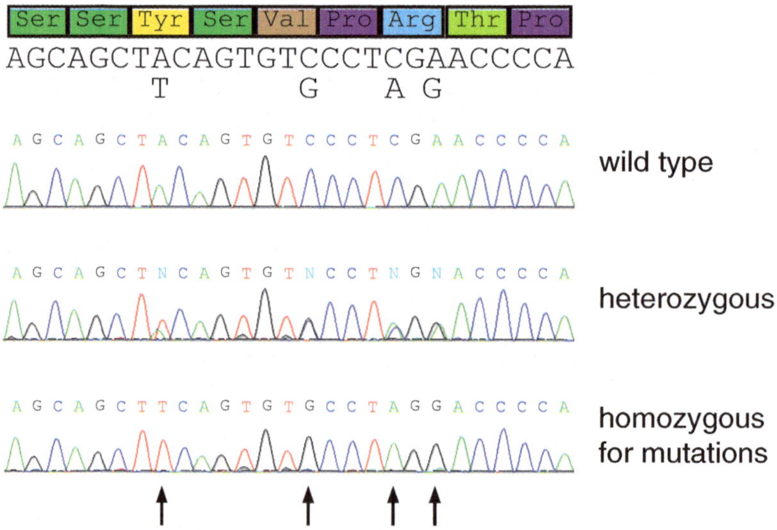

Fig. 5 Sequencing results from CRISPR mutagenesis. The wild type amino acid and DNA sequence of the targeted locus is presented above, with the mutated nucleotides indicated below, and with arrows. The Sanger sequencing results of amplified genomic DNA from wild-type cells, and from two clones, one heterozygous and one homozygous for the desired mutations, are shown

4 Notes

1. We typically purchase Trypsin solutions B and C from Biological Industries. Their formulations are as follows: Trypsin B: 2.5 mg/l trypsin, 0.5 mg/l EDTA $Na_2{\cdot}2H_2O$, 8 mg/l NaCl, 0.4 mg/l KCl, 1 mg/l D-glucose, 0.01 mg/l Phenol Red, 0.35 $NaHCO_3$; Trypsin C: 0.5 mg/l trypsin, 0.2 mg/l EDTA Na_2 $2H_2O$, 8 mg/l NaCl, 0.2 mg/l KCl, 0.01 mg/l Phenol Red, 1.15 mg/l Na_2HPO_4, 0.2 mg/l KH_2PO_4.
2. Transient transfection using polyethylenimine (PEI) can be achieved using commercially prepared solutions. However, a cost-effective option is to prepare the PEI solution as follows (according to link from Polysciences http://www.polysciences.com/skin/frontend/default/polysciences/pdf/23966_procl.pdf): Dissolve 1 g of PEI 25 K (Polysciences cat. no. 23966) in 900 ml biological-grade water. With stirring, bring the pH to <2 with dropwise addition of HCl. Stir for up to 3 h until the solid is fully dissolved. Add NaOH dropwise until the pH is 6.9–7.1. Bring the volume to 1 l, and filter-sterilize. Freeze aliquots at −80 °C. Although the manufacturer does not recommend repeated freeze-thaws, we have found that it is possible to freeze-thaw at least five times without a loss of effectiveness. Paradoxically, although the commercially prepared PEI reagents should perform similarly to this reagent, we have found that transfection of certain cell lines is more efficient with the "homemade" PEI over commercial, and with other cell lines the opposite is true. MCF10A cells are difficult to transfect, and we found that this protocol for transfection worked better than commercial PEI reagent, in terms of transfection efficiency and toxicity.
3. Protease inhibitor cocktail components: AEBSF (Aminoethyl benzenesulfonyl fluoride hydrochloride), Aprotinin, Bestatin hydrochloride, E64 [N-(trans-Epoxysuccinyl)-L-leucine 4-guanidinobutylamide], Leupeptin hemisulfate salt, Pepstatin A.
4. Phosphatase inhibitor cocktail components: sodium orthovanadate, sodium molybdate, sodium tartrate, imidazole.
5. The primary antibodies used in the protocol: Anti-phosphotyrosine PY20 (Santa Cruz), Anti-FlagM2 (Sigma), Anti-c-Abl 8E9 and K12 (Santa Cruz). Secondary antibodies: HRP-linked Goat anti-Mouse and anti-Rabbit antibodies (Jackson Laboratories).
6. We used several different polymerases for PCR. For amplifying genomic DNA for sequencing, we used high-fidelity polymerases such as 2× iPfu mix (iNtRON Biotechnology). We have found that certain amplicons, particularly when they must

be amplified from genomic DNA, give a poor result with PCR, either in terms of specificity or quantity. In addition to troubleshooting the conditions of the PCR program (annealing temperature, extension time, etc.), occasionally the problem can be solved by trying a different polymerase. Terra polymerase (Clontech) is good for direct PCR from cells or from partially purified DNA, or for two-step analyses such as the one described in **step 15**. For screening colonies we usually use less costly Taq polymerase mixes, such as Larova Red Load Taq Master mix (5×).

7. Other constructs of constitutively active Abl, such as the Abl-PP (P242E/P249E) mutant [49] are also in widespread use, and may be used instead of the construct described here.
8. HEK293 cells are used here because of the ease of transfection. Most transfection reagents will work well with HEK293 cells, and can be used instead of the polyethylenimine (PEI) method provided here.
9. Alternatively, cells can be plated and transfected at the same time. After aliquoting the trypsinized and resuspended cells to the wells, the transfection mixes can be added to the medium. This procedure works, but more of the cells die from the transfection than if they had been plated the day before.
10. At this point, cell pellets can be frozen quickly in liquid nitrogen and stored at −80 °C for later use.
11. This "flow through" fraction can be analyzed by western blot to determine how efficient the incubation with the antibody was in depleting the protein of interest.
12. The suggestion here is to boil and load all IP samples on the gel is simple, but may not be appropriate for all applications. When the IP samples are boiled in sample buffer, the S-S bonds connecting the antibody chains are reduced, and the antibody chains are released into the sample. Since the heavy chain migrates at 50 kDa, this can interfere with detection of target proteins in this region. If the protein of interest migrates far enough away from the 50 kDa region, the blot can be cut to avoid this region. In addition, if the antibody used for the IP is of mouse origin, and the antibody used to probe the blot is from rabbit or another species, then the secondary antibody used to probe the blot should not detect the mouse heavy chain. If same-species antibodies are used for IP and probing, and the protein of interest is too close to the heavy or light chain so that these cannot be avoided by cutting the blot, specific anti-light chain or anti-heavy chain secondary antibodies are available. Alternatively, the proteins can be eluted from the IP rather than boiling. For example, Flag-tagged proteins can be eluted from the beads by incubating with 0.1 mg/ml

Flag peptide in TBS or wash buffer for several hours at 4 °C or 10 min at 30 °C (or as empirically determined to be best for your system). The bead-free eluted protein is loaded on the gel, which produces a cleaner blot, and avoids the problems with the antibody chains. The disadvantage with this is that the elution is less efficient than boiling the sample in releasing the protein from the beads.

13. Different blocking agents, at different concentrations, can be used for this step. The best conditions should be determined empirically for the antibody to be used later. However, in our experience, optimization of the blocking solution is more critical for the antibody solution, while the prior blocking step is more forgiving. For the anti-phosphotyrosine antibody PY20, we have found that 2% (w/v) milk powder in TBST works well. For the anti-phospho antibodies we use TBST rather than PBST for incubation and washes, although this might not be critical. Membranes can also be blocked overnight at 4 °C.
14. There are many commercial services for preparing custom-made anti-phosphotyrosine antibodies for your specific phosphorylation site. Some examples include: Creative Biolabs, PrecisionAntibody, Biomatik, ThermoFisher, and others.
15. MCF10A cells are difficult to transfect. We have found that these conditions provide a low (estimated at 10%) but reasonable level of transfection with lower toxicity than other methods. In order to select for the transfected cells, we include a vector that has an antibiotic resistance gene on it - here we used pEFIRES-p, which provides puromycin resistance. Alternatively, we could have used an updated version of the pX330 plasmid (pX459, Addgene 62988) [50] that provides puromycin resistance. Another option is to use fluorescent protein-expressing vectors and FACS sorting to isolate the transfected cells.
16. The Sanger sequencing result suggests that the cells are homozygous for the mutation, since only one sequence is seen—with no overlapping sequence peaks. However, if the second allele had suffered a large deletion that would prevent the amplification of the 523 bp fragment by PCR, then such an event would be invisible to us by these assays. To further show that the point mutations were incorporated as expected, with no other gross rearrangements, a Southern blot should be performed.

References

1. Meng Z, Moroishi T, Guan KL (2016) Mechanisms of Hippo pathway regulation. Genes Dev 30(1):1–17. https://doi.org/10.1101/gad.274027.115
2. Reuven N, Shanzer M, Shaul Y (2015) Tyrosine phosphorylation of WW proteins. Exp Biol Med (Maywood) 240(3):375–382. https://doi.org/10.1177/1535370214565991

3. Pluk H, Dorey K, Superti-Furga G (2002) Autoinhibition of c-Abl. Cell 108(2):247–259
4. Brown MT, Cooper JA (1996) Regulation, substrates and functions of src. Biochim Biophys Acta 1287(2–3):121–149
5. Sudol M (1994) Yes-associated protein (YAP65) is a proline-rich phosphoprotein that binds to the SH3 domain of the Yes proto-oncogene product. Oncogene 9 (8):2145–2152
6. Zaidi SK, Sullivan AJ, Medina R, Ito Y, van Wijnen AJ, Stein JL, Lian JB, Stein GS (2004) Tyrosine phosphorylation controls Runx2-mediated subnuclear targeting of YAP to repress transcription. EMBO J 23(4):790–799. https://doi.org/10.1038/sj.emboj.7600073
7. Matsumoto Y, La Rose J, Kent OA, Wagner MJ, Narimatsu M, Levy AD, Omar MH, Tong J, Krieger JR, Riggs E, Storozhuk Y, Pasquale J, Ventura M, Yeganeh B, Post M, Moran MF, Grynpas MD, Wrana JL, Superti-Furga G, Koleske AJ, Pendergast AM, Rottapel R (2016) Reciprocal stabilization of ABL and TAZ regulates osteoblastogenesis through transcription factor RUNX2. J Clin Invest 126(12):4482–4496. https://doi.org/10.1172/JCI87802
8. Rosenbluh J, Nijhawan D, Cox AG, Li X, Neal JT, Schafer EJ, Zack TI, Wang X, Tsherniak A, Schinzel AC, Shao DD, Schumacher SE, Weir BA, Vazquez F, Cowley GS, Root DE, Mesirov JP, Beroukhim R, Kuo CJ, Goessling W, Hahn WC (2012) Beta-catenin-driven cancers require a YAP1 transcriptional complex for survival and tumorigenesis. Cell 151 (7):1457–1473. https://doi.org/10.1016/j.cell.2012.11.026
9. Vlahov N, Scrace S, Soto MS, Grawenda AM, Bradley L, Pankova D, Papaspyropoulos A, Yee KS, Buffa F, Goding CR, Timpson P, Sibson N, O'Neill E (2015) Alternate RASSF1 transcripts control SRC activity, E-cadherin contacts, and YAP-mediated invasion. Curr Biol 25 (23):3019–3034. https://doi.org/10.1016/j.cub.2015.09.072
10. Taniguchi K, Wu LW, Grivennikov SI, de Jong PR, Lian I, Yu FX, Wang K, Ho SB, Boland BS, Chang JT, Sandborn WJ, Hardiman G, Raz E, Maehara Y, Yoshimura A, Zucman-Rossi J, Guan KL, Karin M (2015) A gp130-Src-YAP module links inflammation to epithelial regeneration. Nature 519(7541):57–62. https://doi.org/10.1038/nature14228
11. Taniguchi K, Moroishi T, de Jong PR, Krawczyk M, Grebbin BM, Luo H, Xu RH, Golob-Schwarzl N, Schweiger C, Wang K, Di Caro G, Feng Y, Fearon ER, Raz E, Kenner L, Farin HF, Guan KL, Haybaeck J, Datz C, Zhang K, Karin M (2017) YAP-IL-6ST autoregulatory loop activated on APC loss controls colonic tumorigenesis. Proc Natl Acad Sci U S A 114(7):1643–1648. https://doi.org/10.1073/pnas.1620290114
12. Byun MR, Hwang JH, Kim AR, Kim KM, Park JI, Oh HT, Hwang ES, Hong JH (2017) SRC activates TAZ for intestinal tumorigenesis and regeneration. Cancer Lett 410:32–40. https://doi.org/10.1016/j.canlet.2017.09.003
13. Levy D, Reuven N, Shaul Y (2008) A regulatory circuit controlling Itch-mediated p73 degradation by Runx. J Biol Chem 283 (41):27462–27468. https://doi.org/10.1074/jbc.M803941200
14. Keshet R, Adler J, Ricardo Lax I, Shanzer M, Porat Z, Reuven N, Shaul Y (2015) C-Abl antagonizes the YAP oncogenic function. Cell Death Differ 22(6):935–945. https://doi.org/10.1038/cdd.2014.182
15. Keshet R, Reuven N, Shaul Y (2015) C-Abl forces YAP to switch sides. Mol Cell Oncol 2 (3):e995006. https://doi.org/10.4161/23723556.2014.995006
16. Sudol M (2013) YAP1 oncogene and its eight isoforms. Oncogene 32(33):3922. https://doi.org/10.1038/onc.2012.520
17. Huang H, Woo AJ, Waldon Z, Schindler Y, Moran TB, Zhu HH, Feng GS, Steen H, Cantor AB (2012) A Src family kinase-Shp2 axis controls RUNX1 activity in megakaryocyte and T-lymphocyte differentiation. Genes Dev 26 (14):1587–1601. https://doi.org/10.1101/gad.192054.112
18. Agami R, Blandino G, Oren M, Shaul Y (1999) Interaction of c-Abl and p73alpha and their collaboration to induce apoptosis. Nature 399 (6738):809–813. https://doi.org/10.1038/21697
19. Li P, Silvis MR, Honaker Y, Lien WH, Arron ST, Vasioukhin V (2016) alphaE-catenin inhibits a Src-YAP1 oncogenic module that couples tyrosine kinases and the effector of Hippo signaling pathway. Genes Dev 30(7):798–811. https://doi.org/10.1101/gad.274951.115
20. Xiao L, Chen D, Hu P, Wu J, Liu W, Zhao Y, Cao M, Fang Y, Bi W, Zheng Z, Ren J, Ji G, Wang Y, Yuan Z (2011) The c-Abl-MST1 signaling pathway mediates oxidative stress-induced neuronal cell death. J Neurosci 31 (26):9611–9619. https://doi.org/10.1523/JNEUROSCI.0035-11.2011
21. Liu W, Wu J, Xiao L, Bai Y, Qu A, Zheng Z, Yuan Z (2012) Regulation of neuronal cell death by c-Abl-Hippo/MST2 signaling pathway. PLoS One 7(5):e36562. https://doi.org/10.1371/journal.pone.0036562
22. Reuven N, Adler J, Meltser V, Shaul Y (2013) The Hippo pathway kinase Lats2 prevents

DNA damage-induced apoptosis through inhibition of the tyrosine kinase c-Abl. Cell Death Differ 20(10):1330–1340. https://doi.org/10.1038/cdd.2013.83
23. Shanzer M, Ricardo-Lax I, Keshet R, Reuven N, Shaul Y (2015) The polyomavirus middle T-antigen oncogene activates the Hippo pathway tumor suppressor Lats in a Src-dependent manner. Oncogene 34 (32):4190–4198. https://doi.org/10.1038/onc.2014.347
24. Shanzer M, Adler J, Ricardo-Lax I, Reuven N, Shaul Y (2017) The nonreceptor tyrosine kinase c-Src attenuates SCF(beta-TrCP) E3-ligase activity abrogating Taz proteasomal degradation. Proc Natl Acad Sci U S A 114 (7):1678–1683. https://doi.org/10.1073/pnas.1610223114
25. Levy D, Adamovich Y, Reuven N, Shaul Y (2008) Yap1 phosphorylation by c-Abl is a critical step in selective activation of proapoptotic genes in response to DNA damage. Mol Cell 29(3):350–361. https://doi.org/10.1016/j.molcel.2007.12.022
26. PhosphositePlus. http://www.phosphosite.org/homeAction.action. Accessed 9 Jan 2018
27. Hornbeck PV, Zhang B, Murray B, Kornhauser JM, Latham V, Skrzypek E (2015) PhosphoSitePlus, 2014: mutations, PTMs and recalibrations. Nucleic Acids Res 43(Database issue): D512–D520. https://doi.org/10.1093/nar/gku1267
28. Server N. http://www.cbs.dtu.dk/services/NetPhos/
29. Blom N, Sicheritz-Ponten T, Gupta R, Gammeltoft S, Brunak S (2004) Prediction of post-translational glycosylation and phosphorylation of proteins from the amino acid sequence. Proteomics 4(6):1633–1649. https://doi.org/10.1002/pmic.200300771
30. PhosphoELM. http://phospho.elm.eu.org/links.html
31. Dinkel H, Chica C, Via A, Gould CM, Jensen LJ, Gibson TJ, Diella F (2011) Phospho.ELM: a database of phosphorylation sites—update 2011. Nucleic Acids Res 39(Database issue): D261–D267. https://doi.org/10.1093/nar/gkq1104
32. GPS. http://gps.biocuckoo.org/
33. Xue Y, Liu Z, Cao J, Ma Q, Gao X, Wang Q, Jin C, Zhou Y, Wen L, Ren J (2011) GPS 2.1: enhanced prediction of kinase-specific phosphorylation sites with an algorithm of motif length selection. Protein Eng Des Sel 24 (3):255–260. https://doi.org/10.1093/protein/gzq094
34. Scansite. http://scansite.mit.edu/
35. Obenauer JC, Cantley LC, Yaffe MB (2003) Scansite 2.0: proteome-wide prediction of cell signaling interactions using short sequence motifs. Nucleic Acids Res 31(13):3635–3641
36. ELM. http://elm.eu.org/
37. Dinkel H, Van Roey K, Michael S, Kumar M, Uyar B, Altenberg B, Milchevskaya V, Schneider M, Kuhn H, Behrendt A, Dahl SL, Damerell V, Diebel S, Kalman S, Klein S, Knudsen AC, Mader C, Merrill S, Staudt A, Thiel V, Welti L, Davey NE, Diella F, Gibson TJ (2016) ELM 2016—data update and new functionality of the eukaryotic linear motif resource. Nucleic Acids Res 44(D1): D294–D300. https://doi.org/10.1093/nar/gkv1291
38. Druker BJ (2002) Perspectives on the development of a molecularly targeted agent. Cancer Cell 1(1):31–36
39. Weisberg E, Manley P, Mestan J, Cowan-Jacob S, Ray A, Griffin JD (2006) AMN107 (nilotinib): a novel and selective inhibitor of BCR-ABL. Br J Cancer 94(12):1765–1769. https://doi.org/10.1038/sj.bjc.6603170
40. Gray NS, Fabbro D (2014) Discovery of allosteric BCR-ABL inhibitors from phenotypic screen to clinical candidate. Methods Enzymol 548:173–188. https://doi.org/10.1016/B978-0-12-397918-6.00007-0
41. Lombardo LJ, Lee FY, Chen P, Norris D, Barrish JC, Behnia K, Castaneda S, Cornelius LA, Das J, Doweyko AM, Fairchild C, Hunt JT, Inigo I, Johnston K, Kamath A, Kan D, Klei H, Marathe P, Pang S, Peterson R, Pitt S, Schieven GL, Schmidt RJ, Tokarski J, Wen ML, Wityak J, Borzilleri RM (2004) Discovery of N-(2-chloro-6-methyl- phenyl)-2-(6-(4-(2-hydroxyethyl)- piperazin-1-yl)-2-methylpyrimidin-4- ylamino)thiazole-5-carboxamide (BMS-354825), a dual Src/Abl kinase inhibitor with potent antitumor activity in preclinical assays. J Med Chem 47 (27):6658–6661. https://doi.org/10.1021/jm049486a
42. Hanke JH, Gardner JP, Dow RL, Changelian PS, Brissette WH, Weringer EJ, Pollok BA, Connelly PA (1996) Discovery of a novel, potent, and Src family-selective tyrosine kinase inhibitor. Study of Lck- and FynT-dependent T cell activation. J Biol Chem 271(2):695–701
43. Dephoure N, Gould KL, Gygi SP, Kellogg DR (2013) Mapping and analysis of phosphorylation sites: a quick guide for cell biologists. Mol Biol Cell 24(5):535–542. https://doi.org/10.1091/mbc.E12-09-0677
44. Hsu PD, Lander ES, Zhang F (2014) Development and applications of CRISPR-Cas9 for

genome engineering. Cell 157(6):1262–1278. https://doi.org/10.1016/j.cell.2014.05.010

45. Deskgen. http://www.deskgen.com/landing/cloud.html

46. Cong L, Ran FA, Cox D, Lin S, Barretto R, Habib N, Hsu PD, Wu X, Jiang W, Marraffini LA, Zhang F (2013) Multiplex genome engineering using CRISPR/Cas systems. Science 339(6121):819–823. https://doi.org/10.1126/science.1231143

47. page Cr. http://www.genome-engineering.org/crispr/wp-content/uploads/2014/05/CRISPR-Reagent-Description-Rev20140509.pdf

48. Hobbs S, Jitrapakdee S, Wallace JC (1998) Development of a bicistronic vector driven by the human polypeptide chain elongation factor 1alpha promoter for creation of stable mammalian cell lines that express very high levels of recombinant proteins. Biochem Biophys Res Commun 252(2):368–372. https://doi.org/10.1006/bbrc.1998.9646

49. Barila D, Superti-Furga G (1998) An intramolecular SH3-domain interaction regulates c-Abl activity. Nat Genet 18(3):280–282. https://doi.org/10.1038/ng0398-280

50. Ran FA, Hsu PD, Wright J, Agarwala V, Scott DA, Zhang F (2013) Genome engineering using the CRISPR-Cas9 system. Nat Protoc 8 (11):2281–2308. https://doi.org/10.1038/nprot.2013.143

Part III

Structure Biology and Biochemistry to Study the Hippo Pathway

Chapter 18

Structural and Biochemical Analyses of the Core Components of the Hippo Pathway

Lisheng Ni and Xuelian Luo

Abstract

The Hippo pathway controls organ size and maintains tissue homeostasis through a central MST–LATS kinase cascade. When Hippo signaling is on, activated MST1/2 partner with SAV1 to phosphorylate and activate the LATS1/2–MOB1 complexes, which in turn phosphorylate and inactivate YAP/TAZ transcription co-activators. This process halts the expression of Hippo-responsive genes, thereby inhibiting cell proliferation and promoting apoptosis. Our studies have shown that two core adaptor proteins MOB1 and SAV1 use distinctive mechanisms to enhance Hippo signaling. MOB1 promotes MST-dependent LATS activation through dynamic scaffolding and allosteric regulation. SAV1 promotes MST activation by antagonizing the PP2A phosphatase activity. Here we describe the detailed methods for the purification and crystallization of the MST2–SAV1 and pMOB1–LATS1 complexes, for assaying the SAV1-dependent inhibition of PP2A, and for analyzing LATS1 kinase activation using in vitro reconstitution.

Key words Hippo pathway, Kinase cascade, Protein purification, X-ray crystallography, Kinase assay, In vitro reconstitution, Autophosphorylation, PP2A

1 Introduction

In the past two decades, genetic studies in *Drosophila* and mammals have established the critical functions of the Hippo signaling pathway in tissue homeostasis [1–7]. It does so by coordinately regulating cell growth, proliferation and apoptosis. Dysregulation of the Hippo pathway leads to tumorigenesis [3, 8, 9]. Central to the Hippo pathway is an evolutionarily conserved kinase cascade. In mammals, this cascade includes four core components: STE20-like MST1/2 kinases, Large Tumor Suppressor LATS1/2 kinases, the WW-domain containing scaffold protein Salvador (SAV1), and the adaptor protein MOB1 [2, 10, 11]. Upon activation by extracellular stimuli, such as cell-cell contact, MST1/2 become activated and act together with SAV1 to phosphorylate and activate the LATS1/2-MOB1 complexes [12, 13]. Activated LATS1/2-MOB1 then phosphorylate and inactivate YAP/TAZ transcription

Alexander Hergovich (ed.), *The Hippo Pathway: Methods and Protocols*, Methods in Molecular Biology, vol. 1893,
https://doi.org/10.1007/978-1-4939-8910-2_18,

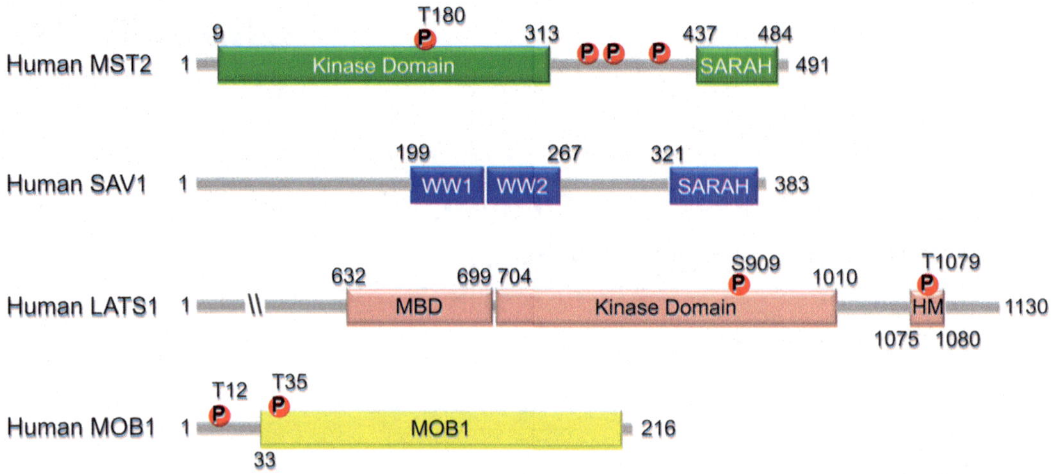

Fig. 1 Domain organization of human MST2, SAV1, LATS1, and MOB1. Residue numbers for the boundaries of corresponding domains are labeled. The regulatory phosphorylation sites of MST2, LATS1, and MOB1 are shown in red circles and labeled, respectively

co-activators and block YAP/TAZ–TEAD-mediated transcription of genes whose products promote cell proliferation and inhibit apoptosis [14–18].

Recent biochemical and structural studies have provided new insights into the functional relationships among the four core components of the central MST-LATS kinase cascade [19–29]. MST1/2 contain an N-terminal Ser/Thr kinase domain, a C-terminal SARAH dimerization domain, and a long flexible linker connecting these two domains (Fig. 1). We have shown that MST2 forms a constitutive homodimer through the SARAH-SARAH interaction, which is essential for its autophosphorylation and activation [21]. MST2 undergoes trans-autophosphorylation at T180 in the activation loop (T-loop) to turn on the kinase activity. Activated MST2 then phosphorylates and activates LATS1. LATS1 kinase activation requires the sequential phosphorylation of two conserved sites (Fig. 1) [30, 31]. First, MST2 phosphorylates the C-terminal hydrophobic motif (HM) at T1079. Second, phosphorylated HM (pHM) triggers a conformational change in the kinase domain, which strongly stimulates autophosphorylation at S909 in the T-loop and LATS1 kinase activation [29].

Activated MST2 also auto-phosphorylates several di-peptide TM motifs in its linker to create the phospho-TM (pTM) docking sites for MOB1 (Fig. 1) [29]. Binding to the pTM sites of MST2 causes a conformational change of MOB1 from the auto-inhibited state to the open state [32]. Open MOB1 binds to LATS1 and forms an MST2-MOB1-LATS1 ternary complex. Within this complex, MST2 efficiently phosphorylates both MOB1 at residues T12

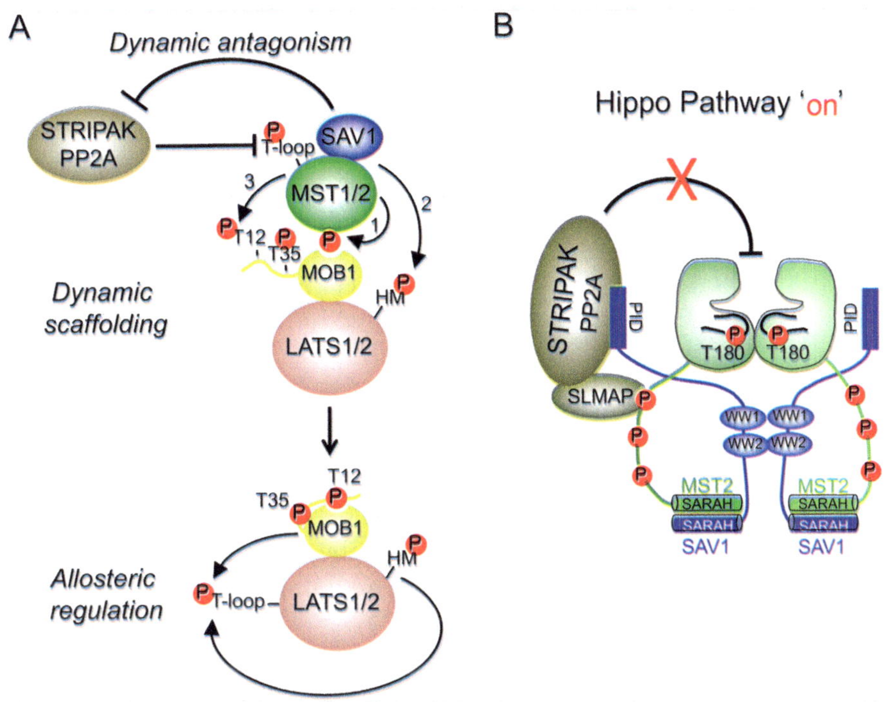

Fig. 2 Activation mechanisms of the core Hippo kinase cascade. (**a**) Model for MST–LATS activation during Hippo signaling. (**b**) Model for SAV1-dependent MST2 activation when Hippo pathway is on. SLMAP, the adaptor protein for STRIPAK PP2A. *PID* phosphatase inhibition domain

and T35 and LATS1 at residue T1079 in the HM motif (Fig. 2a). Once MOB1 is phosphorylated by MST2, the phosphorylated N-terminal tail binds to the pT binding site of MOB1 and competes with MST2 pTM for binding to the same site, thus releasing pMOB1-LATS1 from MST2. pMOB1 maintains the open conformation and stays in complex with LATS1. Furthermore, pMOB1 allosterically stimulates LATS1 autophosphorylation at S909 in its T-loop to allow LATS1 kinase activation. Therefore, MOB1 promotes MST-dependent LATS activation through dynamic scaffolding and allosteric mechanisms [29].

Recombinant MST2 is fully activated in vitro, however, most of MST2 protein remains unphosphorylated at the T-loop in vivo when the Hippo signaling is off [25, 33], indicating that its kinase activation is negatively regulated by phosphatases. We have recently shown that STRIPAK PP2A is a negative regulator of the Hippo pathway and that SAV1 promotes MST2 kinase activation through antagonizing the STRIPAK PP2A phosphatase activity (Fig. 2a) [25]. SAV1 contains a flexible N-terminal region, followed by two WW domains and a C-terminal SARAH domain (Fig. 1). SAV1 SARAH binds to MST2 SARAH to form an MST2–SAV1 heterodimer. The WW2 domain of SAV1 further dimerizes to allow the formation of the MST2–SAV1 heterotetramer. In the context of the MST2-SAV1 heterotetramer, the N-terminal tail of SAV1 directly blocks the STRIPAK PP2A phosphatase activity, thus preventing MST2 dephosphorylation and inactivation by STRIPAK PP2A (Fig. 2b). Therefore, the positive effect of SAV1 on MST2 activation is through counteracting the inhibitory effect of STRIPAK PP2A on MST2.

In summary, our studies demonstrate that two core adaptor proteins SAV1 and MOB1 play critical roles for activation of Hippo kinase cascade through different mechanisms. More structural and functional studies are required to further demonstrate how SAV1 inhibits PP2A and shields MST2 from PP2A-mediated dephosphorylation and inactivation. For future biochemical and structural studies, large amounts of purified recombinant core components of the Hippo pathway are needed. The production of active recombinant LATS1 kinase will be useful for high throughput screening of chemical inhibitors of the Hippo pathway, which may have therapeutic value in tissue regeneration. In this chapter, we describe the detailed procedures of expression, purification, and crystallization of recombinant human MST2–SAV1 and pMOB1–LATS1 complexes. We also describe the methods for studying the SAV1-dependent inhibition of PP2A, and for our in vitro reconstitution assay for understanding the activation mechanism of the LATS1 kinase.

2 Materials

All solutions are prepared using Milli-Q water (18.2 MΩ cm resistivity at 25 °C) and ACS or molecular biology-grade reagents.

1. Miller's LB broth powder.
2. Terrific broth modified capsules.
3. Kanamycin stock solution (50 mg/mL).
4. Ampicillin stock solution (100 mg/mL).
5. Isopropyl-β-D-thiogalactoside (IPTG) stock solution (1 M).
6. 4-(2-Aminoethyl)-benzenesulfonyl fluoride hydrochloride (AEBSF) stock solution (100 mM).
7. Tris (2-carboxyethyl) phosphine (TCEP) stock solution (1 M).
8. Thermo Scientific Pierce universal nuclease (Fisher #PI88702).
9. Chemical competent *E. coli* strains: BL21(DE3)-T1 (Sigma) for protein expression and DH5α (Invitrogen) for general cloning and subcloning.
10. LB-agar plates with relevant antibiotics.
11. Ni-NTA agarose.
12. Glutathione Sepharose 4B resin.

13. Homogenizer (for example Avestin EmulsiFlex-C5).
14. Gravity columns (for example BIO-RAD #7321010).
15. Lysis buffer 1 (for the MST2–SAV1 complex): 20 mM Tris–HCl, pH 8.0, 200 mM NaCl, 5 mM $MgCl_2$, 1 mM DTT (*see* **Note 1**).
16. Tris buffered saline and Tween (TBST): 50 mM Tris–HCl, pH 7.5, 150 mM NaCl, 0.05% Tween-20 (v/v).
17. Lysis buffer 2 (for GST-fusion protein): 20 mM Tris–HCl, pH 8.0, 500 mM NaCl, 1 mM DTT.
18. Elution buffer 2 (for GST-fusion protein): 50 mM Tris–HCl, pH 8.0, 500 mM NaCl, 1 mM DTT, 10–20 mM reduced glutathione (*see* **Note 2**).
19. Lysis buffer 3 (for 6×His-tagged protein): 20 mM Tris–HCl, pH 8.0, 500 mM NaCl, 5% glycerol, 10 mM imidazole, 2 mM 2-mercaptoethanol.
20. Wash buffer 3 (for 6×His-tagged protein): 20 mM Tris–HCl, pH 8.0, 500 mM NaCl, 5% glycerol, 25 mM imidazole, 2 mM 2-mercaptoethanol.
21. Elution buffer 3 (for 6×His-tagged protein): 20 mM Tris–HCl, pH 8.0, 500 mM NaCl, 5% glycerol, 250 mM imidazole, 2 mM 2-mercaptoethanol.
22. Lysis buffer 4 (for denaturing purification): 100 mM NaH_2PO_4, 10 mM Tris–HCl, 8 M urea (adjust pH to 8.0 using NaOH) (*see* **Note 3**).
23. Wash buffer 4 (for denaturing purification): 100 mM NaH_2PO_4, 10 mM Tris–HCl, 8 M urea (adjust pH to 6.3 using HCl).
24. Elution buffer 4 (for denaturing purification): 100 mM NaH_2PO_4, 10 mM Tris–HCl, 8 M urea (adjust pH to 4.5 using HCl).
25. SA buffer for RESOURCE S column: 20 mM MOPS, pH 6.8, 100 mM NaCl, 1 mM DTT.
26. SB buffer for RESOURCE S column: 20 mM MOPS, pH 6.8, 1 M NaCl, 1 mM DTT.
27. Cleavage buffer for 6×His-tagged LATS1: 20 mM Tris–HCl, pH 8.0, 300 mM NaCl, 5% glycerol, 2 mM $MgCl_2$, 1 mM DTT.
28. SA buffer for Mono S column (for LATS1): 20 mM HEPES, pH 7.0, 150 mM NaCl, 5% glycerol, 2 mM $MgCl_2$, 1 mM DTT.
29. SB buffer for Mono S column (for LATS1): 20 mM HEPES, pH 7.0, 1 M NaCl, 5% glycerol, 2 mM $MgCl_2$, 1 mM DTT.

30. Gel filtration buffer: 20 mM Tris–HCl, pH 8.0, 100 mM NaCl, 5 mM $MgCl_2$, 1 mM TCEP.
31. Centrifugal filter units (for example, EMD Millipore Centriprep™ Millipore).
32. Ultra centrifugal filter units (for example, EMD Millipore Amicon™ Millipore).
33. AKTA-FPLC (GE Healthcare).
34. RESOURCE S (1 mL) column (GE Healthcare).
35. Mono S (1 mL) column (GE Healthcare).
36. Superdex 75 10/300 GL gel filtration column (GE Healthcare).
37. Superdex 200 10/300 GL gel filtration column (GE Healthcare).
38. Kinase reaction buffer: 50 mM Tris–HCl, pH 7.5, 150 mM NaCl, 5 mM $MgCl_2$, 1 mM DTT.
39. Refolding buffer: 20 mM Tris–HCl, pH 8.0, 100 mM NaCl, 1 mM DTT.
40. Phosphatase reaction buffer: 20 mM Tris–HCl, pH 8.0, 100 mM NaCl, 5 mM $MgCl_2$, 1 mM DTT.
41. 2× SDS-PAGE sample buffer
42. Adenosine-5′-triphosphate (ATP) stock solution (100 mM, adjust pH to 7.0 by NaOH).
43. Adenylyl-imidodiphosphate (AMP-PNP) stock solution (100 mM).
44. L-[^{35}S]-Methionine (10.2 mCi/mL).
45. Coupled Transcription/Translation System (for example Promega #L2080).
46. Nitrocellulose membranes.
47. Trans-Blot SD semi-dry transfer cell (or equivalent equipment).
48. Biomolecular imager (for example Typhoon FLA 7000).
49. Imaging System (for example Odyssey CLx, LI-COR Biosciences).
50. INTELLI-PLATE 96 with 3 wells (art Robbins instruments).
51. Crystal Gryphon LCP (Art Robbins Instruments).
52. NanoDrop 1000 Spectrophotometer (Thermo Scientific).

3 Methods

3.1 Purification and Crystallization of the MST2–SAV1 Complex

1. Clone the cDNA sequence encoding human MST2$^{\Delta L/D146N}$ (ΔL: Linker-deletion, residues 9–313 and 428–491) into the BamHI and AscI restriction sites of a modified pET29 vector that includes an N-terminal 6×His affinity tag [21]. Clone the cDNA sequence encoding the C-terminal fragment of human SAV1 (SAV1$^{\Delta 198}$, SAV1$^{\Delta 267}$, SAV1$^{\Delta 280}$, SAV1$^{\Delta 290}$, and SAV1$^{\Delta 320}$, respectively) into the FseI and AscI restriction sites of a modified pGEX-6P vector that contains a tobacco etch virus (TEV) protease cleavage site at the N terminus.
2. Co-transform pET29-MST2$^{\Delta L/D146N}$ (300 ng) and pGEX-6P-SAV1 (300 ng) plasmids into 100 μL BL21(DE3)-T1 competent cells with heat shock at 42 °C for 45 s (*see* **Note 4**).
3. Add 200 μL LB and incubate the cells with gentle shaking at 37 °C for 1 h. Spread the entire transformation mix on a single LB-agar plate containing 100 μg/mL ampicillin and 50 μg/mL kanamycin. Incubate the plate at 37 °C overnight.
4. Inoculate 100 mL LB medium containing 100 μg/mL ampicillin and 50 μg/mL kanamycin with a single colony from the overnight plate, and grow the cultures at 37 °C overnight with vigorous shaking.
5. Add 10 mL overnight culture to 500 mL TB medium containing 100 μg/mL ampicillin and 50 μg/mL kanamycin, and grow the cells at 37 °C with vigorous shaking for 3–4 h until the OD_{600} is 2.0–3.0 (*see* **Note 5**).
6. Lower the temperature of the incubator to 16 °C. Wait for 1 h, then add 200 μL IPTG to 500 mL of culture to a final concentration of 0.4 mM to induce the target protein expression. Grow the cultures at 16 °C for an additional 16–20 h.
7. Harvest the cells by centrifugation at 4000 × *g* at 4 °C for 20 min, and discard supernatants.
8. Resuspend the cell pellets in 60 mL cold lysis buffer 1 per 500 mL of culture. Add AEBSF and nuclease to a final concentration of 0.2 mM and 10 U/mL, respectively. Lyse cells by passing through a homogenizer at a pressure of 10,000–15,000 psi 3–4 times.
9. Centrifuge the lysate at 30,000 × *g* at 4 °C for 1 h, and transfer the supernatant to a fresh 50-mL conical tube.
10. Equilibrate Glutathione Sepharose 4B beads with lysis buffer 1.
11. Add 1 mL of the 50% Glutathione Sepharose 4B slurry to 50 mL cleared lysate. Gently rotate end-over-end at 4 °C for 2 h.

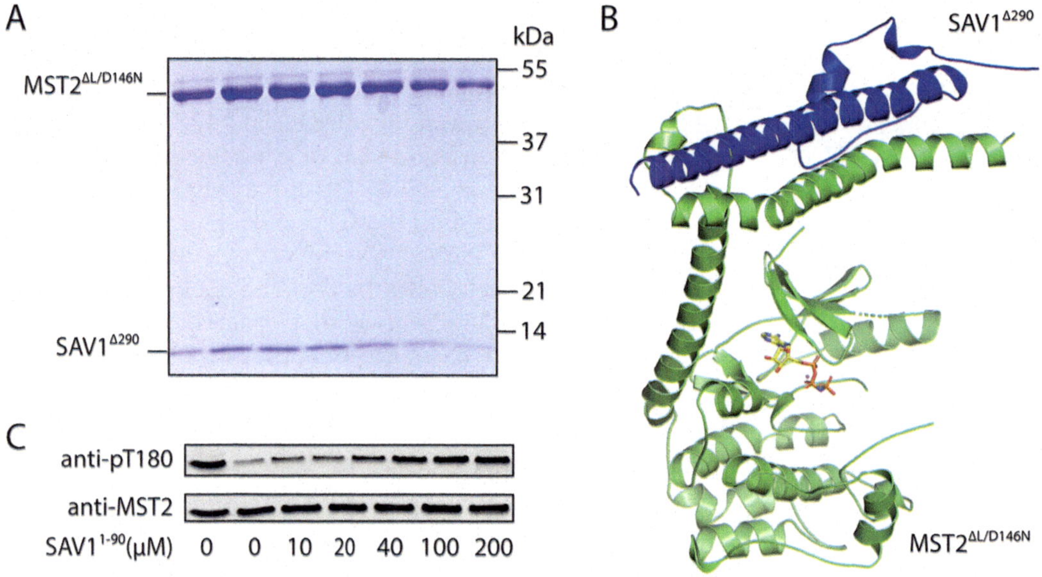

Fig. 3 Structural and functional studies of SAV1 in Hippo pathway. (**a**) The fractions of the MST2$^{\Delta L/D146N}$–SAV1$^{\Delta 290}$ complex from the Superdex 200 gel filtration column that are used for crystallization. The purify of the complex in each fraction is analyzed by SDS-PAGE. (**b**) Cartoon drawing of the crystal structure of MST2$^{\Delta L/D146N}$–SAV1$^{\Delta 290}$ complex (PDB ID: 6A05). MST2 and SAV1 are colored in green and blue, respectively. (**c**) The N-terminal fragment of SAV1 directly inhibits dephosphorylation of MST2 pT180 by PP2A. Quantitative MST2 pT180 and total MST2 immunoblotting of phosphatase reactions containing SAV1$^{1-90}$ at the indicated concentrations

12. Load the mixture of the lysate and the beads into an empty column and collect the column flow through.
13. Wash the Glutathione Sepharose 4B beads with 50× bead-volume of cold lysis buffer 1. Resuspend the beads with 8 mL cold lysis buffer 1 in the same column. Add TEV protease (3 mg/mL) (*see* **Note 6**) and gently rotate the beads at 4 °C overnight (*see* **Note 7**). Check the completeness of TEV cleavage by SDS-PAGE.
14. Elute the cleaved MST2$^{\Delta L/D146N}$–SAV1 complex from the Glutathione Sepharose 4B beads with 8 mL cold lysis buffer 1. Centrifuge the elution at 16,000 × *g* for 20 min.
15. Load half of the elution (4 mL) onto a Superdex 200 column pre-equilibrated with lysis buffer 1 (*see* **Note 8**). Repeat the step with the second 4 mL of the elution. The purity of the final complex is validated by SDS-PAGE (Fig. 3a).
16. Combine the fractions containing the pure complex and add AMP-PNP to a final concentration of 100 μM. Use a centrifugal filter to concentrate the complex to 8 mg/mL. Add additional AMP-PNP to the concentrated sample to a final

concentration of 5 mM (*see* **Note 9**). Incubate on ice for 1 h before setting up crystal screening.

17. Carry out the crystal screening in 96-well plates using sitting-drop vapor-diffusion method on Gryphon LCP. Add 50 μL crystallization solution in reservoir, and mix 0.3 μL protein complex with 0.3 μL reservoir solution in sitting well. Among the five MST2–SAV1 complexes, only the $MST2^{\Delta L/D146N}$–$SAV1^{\Delta 280}$ and $MST2^{\Delta L/D146N}$–$SAV1^{\Delta 290}$ complexes produce diffracting crystals. The crystals of the $MST2^{\Delta L/D146N}$–$SAV1^{\Delta 290}$ complex appear after 24 h at 20 °C. They diffract at much higher resolution and are further optimized for structure determination (Fig. 3b) [25].

3.2 Assaying SAV1-Depedent PP2A Phosphatase Inhibition

3.2.1 Denaturation Purification of the N-Terminal Fragment of $SAV1^{1-90}$

1. Clone the cDNA sequence encoding human $SAV1^{1-90}$ into a modified pET29 with 6×His tag at the N terminus.
2. Grow the cells in TB medium containing 50 μg/mL kanamycin at 37 °C until OD_{600} reaching 2.0–3.0. Add IPTG to a final concentration of 1 mM. Induce protein expression at 37 °C for 4 h.
3. Harvest the cells and resuspend the pellet in 70 mL lysis buffer 4 for denaturing purification per 500 mL cultures (*see* **Note 10**). Gently shake at room temperature overnight for thorough protein denaturation.
4. Centrifuge the lysis solution at 30,000 × *g* for 1 h. Collect the supernatant and mix with pre-equilibrated 50% Ni-NTA slurry with lysis buffer 4 (1 mL slurry per 50 mL supernatant). Rotate at room temperature for 2 h.
5. Pour the supernatant and the Ni-NTA resin mixture onto an empty column. Wash the beads with 100 mL wash buffer 4 for denaturing purification.
6. Elute with 20 mL elution buffer 4 for denaturing purification (*see* **Note 11**).
7. Refold $SAV1^{1-90}$ by dialysis extensively against kinase reaction buffer at 4 °C overnight.
8. Concentrate $SAV1^{1-90}$ to 3 mg/mL. Aliquot and store at −80 °C.

3.2.2 In Vitro PP2A Inhibition Assay

1. The PP2A A-C complex is purified from insect cells using previously published protocol as described [34].
2. Dilute PP2A A-C in phosphatase reaction buffer to a concentration of 1 μM. Dilute the $SAV1^{1-90}$ fragment in phosphatase reaction buffer into relevant concentrations.
3. Mix 2 μL PP2A A-C with 2 μL $SAV1^{1-90}$ at appropriate concentration for each reaction. Incubate the reaction mixtures at room temperature for 15 min.

4. Dilute $MST2^{KD}$ (kinase domain alone) in phosphatase reaction buffer at room temperature to a concentration of 1.25 μM. Add 16 μL $MST2^{KD}$ to each tube containing the PP2A A-C and $SAV1^{1-90}$ mixture. Incubate at room temperature for 30 min.
5. Add 20 μL 2× SDS-PAGE sample buffer to each reaction mixture and boil the samples in 100 °C water bath for 5 min to quench the reaction. The remaining pT180 of MST2 is analyzed by quantitative Western blot using a specific antibody against pT180 of MST2 (Fig. 3c).

3.3 Purification and Crystallization of the pMOB1–LATS1 Complex

3.3.1 Purification of pGST-MOB1 and pMOB1

1. Clone the full-length gene of human MOB1 into a modified pGEX-6P vector containing an N-terminal TEV cleavage site.
2. Grow and induce the cultures as described for the MST–SAV1 complex in Subheading 3.1.
3. Harvest the cells and resuspend pellet in lysis buffer 2 for GST-fusion protein.
4. Lyse the cells and add the Glutathione Sepharose 4B beads pre-equilibrated with lysis buffer 2 to the cleared lysate. Gently rotate the beads and lysate mixture at 4 °C for 2 h.
5. Elute GST-MOB1 from the Glutathione Sepharose 4B beads with 20 mL elution buffer 2 for GST-fusion protein.
6. Dialyze the elution against 2 L kinase reaction buffer at 4 °C overnight. Determine the concentration of GST-MOB1 by UV absorbance at A_{280}.
7. Based on a molar ratio of 1:20 for kinase and substrate, mix the appropriate amount of purified active $MST2^{KD}$ (kinase domain alone) with GST-MOB1. Add ATP to a final concentration of 2 mM to start the kinase reaction. Incubate the reaction at room temperature for 30 min to fully phosphorylate GST-MOB1.
8. Add a new batch of Glutathione Sepharose 4B beads pre-equilibrated with lysis buffer 2 to the reaction mixture. Collect the beads after rotating at 4 °C for 2 h.
9. Wash with 100 mL lysis buffer 2 to remove ATP and $MST2^{KD}$.
10. Split the beads into two parts. Elute half of the Glutathione Sepharose 4B beads with elution buffer 2 to obtain pGST-MOB1.
11. Add TEV protease to the second half of the Glutathione Sepharose 4B beads. Rotate at 4 °C overnight. Elute pMOB1 from the beads.
12. Concentrate purified pGST-MOB1 and pMOB1 to 20 mg/mL, respectively. Aliquot and store at −80 °C.

3.3.2 Identification the MOB1-Binding Domain of LATS1

1. Clone the coding region of human LATS1 ($LATS1^{FL}$, $LATS1^{566-1130}$, $LATS1^{602-1130}$, $LATS1^{631-1130}$, and $LATS1^{651-1130}$, respectively) into a modified pCS2 vector with six copies of N-terminal Myc tags.
2. To obtain in vitro translated ^{35}S-labeled Myc-LATS1 proteins, mix 3 μL of each pCS2-LATS1 plasmid (250 ng/μL) with 20 μL TNT Quick Master Mix (*see* **Note 12**). Add 1 μL ^{35}S-labeled methionine and 1 μL nuclease-free water. Incubate the reaction mixtures in 30 °C water bath for 90 min.
3. Equilibrate Glutathione Sepharose 4B beads with TBST.
4. For each bait reaction, add 5 μL purified pGST-MOB1 protein (50 μM) to 10 μL Glutathione Sepharose 4B beads plus 300 μL TBST. For each control reaction, add 5 μL purified GST protein (50 μM) to 10 μL Glutathione Sepharose 4B beads plus 300 μL TBST. Incubate at 4 °C for 1 h with shaking.
5. Spin down the beads at 300 × *g* for 2 min and discard the supernatant. Wash the beads once with 1 mL TBST.
6. For 25 μL reaction mixture from each in vitro translation reaction, add 3 μL reaction mixture to 47 μL SDS-PAGE sample buffer as 10% input, add 9 μL reaction mixture and 300 μL TBST to the GST-bound beads as control, and add another 9 μL reaction mixture and 300 μL TBST to the pGST-MOB1-bound beads. Incubate at 4 °C for 2 h with shaking.
7. Wash the beads with 1 mL TBST three times. Add 30 μL SDS-PAGE sample buffer to each 10 μL of beads (*see* **Note 13**). Boil in 100 °C water bath for 5 min.
8. Spin at 300 × *g* for 2 min to pellet the beads. Resolve 8 μL of each sample on SDS-PAGE. Dry the gel and scan with the flatbed scanner to show the equal loading of the baits.
9. Expose the dried gel to the erased imaging plate in a phosphoimager cassette. Quantify the ^{35}S signals by typhoon FLA 7000 biomolecular imager. Based on this pull-down assay, the minimal MOB1-binding domain (MBD) of LATS1 was mapped to residue 632–699 [29].

3.3.3 Purification and Refolding of the MOB1-Binding Fragments of LATS1

1. Clone the cDNA fragment encoding human $LATS1^{566-704}$, $LATS1^{602-704}$, and $LATS1^{631-704}$ into the modified pET29 with a 6×His tag at the N terminus, respectively.
2. Purify the LATS1 fragments by denaturation method as described in Subheading 3.2.1.
3. Dialyze the protein elution extensively against refolding buffer at 4 °C overnight.
4. Concentrate each LATS1 sample to 2 mg/mL. Aliquot and store at −80 °C.

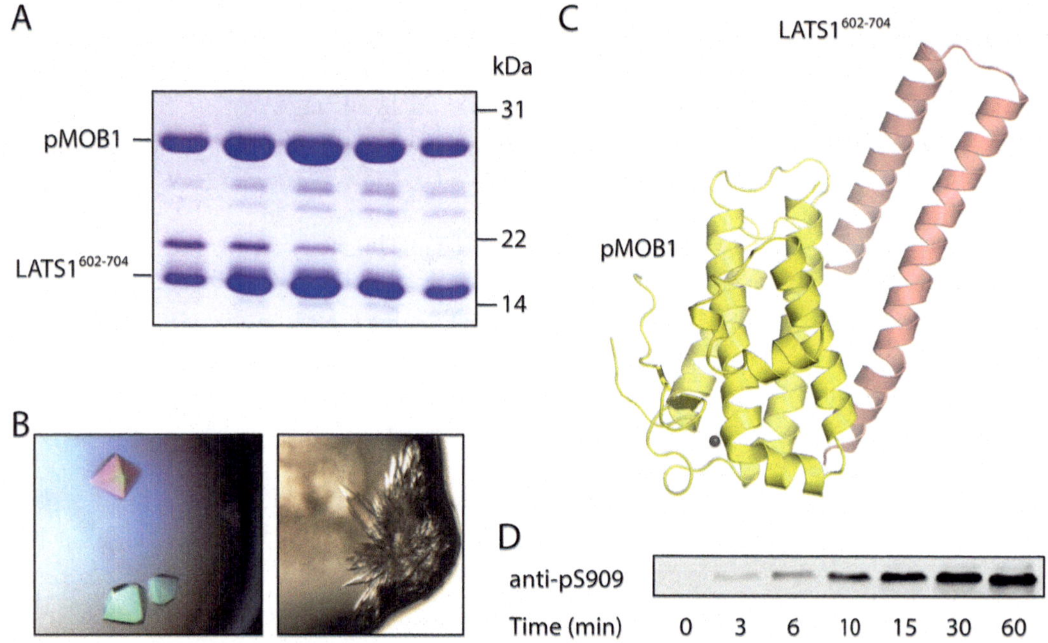

Fig. 4 Structure of pMOB1–LATS1 and reconstitution of LATS1 activation. (**a**) SDS-PAGE of the pMOB1–LATS1$^{602–704}$ complex fractioned on a Superdex 75 gel filtration column. The displayed fractions are combined for crystallization. (**b**) Two different crystal forms of the pMOB1–LATS1$^{602–704}$ complex. Left: pyramid shape appeared in most of the drops. Right: plate clusters only formed in a few drops and used for data collection. (**c**) Cartoon drawing of the crystal structure of the pMOB1–LATS1$^{602–704}$ complex (PDB ID: 5BRK) with pMOB1 colored in yellow and LATS1 colored in salmon, respectively. (**d**) LATS1-pS909 blot of kinase reaction containing LATS1$^{602–1060}$ and the pHM fragment at the indicated time points

3.3.4 Purification of the pMOB1–LATS1 Complex

1. Mix pMOB1 and each LATS1 fragment (LATS1$^{566–704}$, LATS1$^{602–704}$, and LATS1$^{631–704}$, respectively) with a molar ratio of 1:10 and dialyze the mixture against SA buffer for RESOURCE S column at 4 °C overnight.
2. Load the overnight mixture onto a RESOURCE S (1 mL) column to separate the pMob1–LATS1 complex from the extra LATS1 fragment.
3. Verify the fractions by SDS-PAGE. Combine the fractions containing the pure complex, and load onto a Superdex 75 gel filtration column pre-equilibrated with gel filtration buffer. Check the purity and molar ratio of the complex by SDS-PAGE (Fig. 4a).
4. Concentrate the complex to 10 mg/mL. Set up the initial crystal screening in 96-well plate by the sitting drop method at 20 °C. Among the three pMOB1–LATS1 complexes, only the pMOB1-LATS1$^{566–704}$ and pMOB1-LATS1$^{602–704}$ complexes produce crystals under the same crystallization condition (0.2 M diammonium citrate, pH 5.0, and 20% PEG 3350

(w/v)). Crystals from pMOB1–LAT1$^{602-704}$ are further optimized.

5. There are two crystal forms of pMOB1–LAT1$^{602-704}$ appeared in the final optimized crystallization condition in separate drops. Crystals with pyramid shape appear in most of the drops during optimization (Fig. 4b). Crystals of plate cluster appear only in a few drops (Fig. 4b), which give the best diffraction data for structure determination (Fig. 4c) (*see* **Note 14**).

3.4 *In Vitro* Reconstitution of LATS1 Activation

3.4.1 Purification of LATS1$^{602-1060}$

1. Clone the coding region of human LATS1$^{602-1060}$ (residues 602 to 1060) into the MCS 1 site and the cDNA sequence of full-length human MOB1 into the MCS 2 site of a modified pRSF-Duet vector (*see* **Note 15**).
2. Grow BL21(DE3)-T1 cells transformed with pRSF-Duet-LATS1$^{602-1060}$-MOB1 in 12 L of LB medium with relevant antibiotic at 37 °C until OD_{600} reaching 0.8–1.0 (*see* **Note 16**). Lower the temperature of the incubator to 10 °C and continue to grow the cells for 1 h. Add IPTG to a final concentration of 0.2 mM to induce protein expression for 24 h at 10 °C.
3. Harvest the cells and resuspend pellet with 25 mL cold lysis buffer 3 for 6×His-tagged protein per liter of culture. Lyse the cells with the C5 homogenizer. Mix the cleared supernatant with 6 mL 50% Ni-NTA slurry pre-equilibrated with lysis buffer 3 (0.5 mL slurry per 1 L of culture). Incubate at 4 °C for 2 h with shaking.
4. Pour the beads and supernatant mixture into an empty column and collect the flow through.
5. Wash the Ni-NTA resins with a total of 100 mL of cold wash buffer 3 for 6×His-tagged protein (*see* **Note 17**).
6. Elute with 20 mL cold elution buffer 3 for 6×His-tagged protein.
7. Add TEV protease (3 mg/mL) to the elution at a molar ratio 1:30. Dialyze against cleavage buffer for 6×His-tagged LATS1 overnight.
8. Dialyze against SA buffer for Mono S column for another 4–6 h (*see* **Note 18**).
9. Load the cleaved product onto a Mono S (1 mL) column (*see* **Note 19**).
10. Verify the fractions by SDS-PAGE.
11. Combine the fractions containing pure LATS1$^{602-1060}$, and concentrate to 5 mg/mL. Aliquot and store at −80 °C.

3.4.2 Purification of the HM and pHM Fragments of LATS1

1. Clone the cDNA sequence encoding the C-terminal fragment of LATS1$^{1069-1130}$ (residues 1069–1130) containing the HM motif into a modified pET29 vector with a 6×His tag at the N-terminus.
2. Purify the HM fragment of LATS1$^{1069-1130}$ by denaturation method as described in Subheading 3.2.1 (*see* **Note 20**).
3. Refold the HM fragment by dialysis against kinase reaction buffer (without DTT) overnight (*see* **Note 21**).
4. Concentrate the HM fragment to 4 mg/mL. Aliquot and store at −80 °C as non-phosphorylated HM fragment.
5. To obtain the phosphorylated HM (pHM) fragment, add MST2^{KD} (kinase domain) to the HM fragment with a molar ratio of 1:20 in kinase reaction buffer (without DTT) (*see* **Note 22**). Incubate at room temperature for 30 min.
6. To remove MST2^{KD} and ATP, add the second batch of Ni-NTA resins pre-equilibrated with kinase reaction buffer (without DTT) to the pHM fragment and MST2^{KD} mixture. Incubate at 4 °C for 2 h.
7. Collect the elution from the Ni-NTA resin.
8. Concentrate the pHM fragment to 4 mg/mL. Aliquot and store at −80 °C.

3.4.3 In Vitro Reconstitution of LATS1 Kinase Activation

1. Dilute LATS1$^{602-1060}$ in kinase reaction buffer without ATP to 40 μM. Dilute the pHM fragment in kinase reaction buffer without ATP to 80 μM.
2. Mix 10 μL of LATS1$^{602-1060}$ with 10 μL of the pHM fragment. Incubate the reaction at room temperature for 15 min.
3. Add 80 μL of kinase reaction buffer containing 1.25 mM ATP to each reaction to start kinase phosphorylation.
4. To quench the reaction at the corresponding time intervals, take 15 μL reaction mixture and add 15 μL 2× SDS-PAGE sample buffer, and boil in 100 °C water bath for 5 min.
5. Resolve the samples on SDS-PAGE. Transfer the proteins to nitrocellulose membrane using Trans-Blot Semi-dry transfer cell.
6. Incubate the nitrocellulose membrane blot with a specific antibody against pS909 of LATS1 (1:5000 dilution in TBST supplemented with 5% BSA) at 4 °C overnight by gently agitation.
7. Wash the blot with 10 mL TBST three times.
8. Incubate with a fluorescently-labeled secondary antibody in TBST supplemented with 5% milk at 4 °C for 1 h.
9. Rinse the membrane with 10 mL TBST 4–5 times.
10. Analyze quantitatively the fluorescent signals on the membrane blot using an adequate imaging system (Fig. 4d).

4 Notes

1. The pH of Tris-containing buffer is sensitive to temperature. The pH value of all Tris-containing buffers is adjusted at room temperature.
2. Because reduced glutathione may change the pH of the buffer, check the pH of elution buffer 2 to make sure that the pH is in the relevant range.
3. Due to the dissociation of urea, the pH of urea-containing buffers for denaturing purification in this method should be adjusted right before use. Keep the urea-containing buffers at room temperature.
4. In order to successfully co-transform pGEX-6P and pET29 plasmids into a single BL21 competent cell, add 300 ng or more of each plasmid into the transformation mix. Use competent BL21 cells with high transformation efficiency. Failure in either condition will result in no colony grown.
5. For TB rich medium, high cell density with OD_{600} of 2.0–3.0 is needed to have a higher yield for protein expression.
6. Add TEV protease at 1:30 molar ratio of protease vs. protein. Because TEV cleavage reaction is very inefficient at 4 °C, more TEV might be needed.
7. Rotate the column with gentle agitation to prevent frothing.
8. The $MST2^{\Delta L/D146N}$–SAV1 complex is not very soluble and cannot be further concentrated at this purification step. Add a small volume of lysis buffer 1 to the TEV cleavage mix to prevent the precipitation of the final cleaved complex if needed. Load the elution of the cleaved complex directly onto the gel filtration column to avoid the precipitation caused by further concentrating of the complex.
9. Adding AMP-PNP increases the solubility of the $MST2^{\Delta L/D146N}$–SAV1 complex so it can be further concentrated. The $MST2^{\Delta L/D146N}$–SAV1 complex is prone to aggregation and precipitation near the membrane surface during the concentrating step. In this case, Centriprep centrifugal filters have much less precipitation of the complex than Amicon centrifugal filters.
10. In this method, carry out the entire denaturation purification at room temperature unless specified.
11. The $SAV1^{1-90}$ fragment eluted from the Ni-NTA resins has a purity greater than 90% pure on SDS-PAGE.
12. Dissolve the plasmid in nuclease-free water in the last step of the plasmid mini prep. High-quality plasmid is essential for efficient in vitro transcription/translation.

13. To load 10% input, 3 μL reaction mixture is added to 47 μL SDS-PAGE sample buffer which makes a total volume of 50 μL for input. 4 μL of the 50 μL input is used for SDS-PAGE. Assume that the in vitro translated ^{35}S-labeled Myc-LATS1 protein from 9 μL reaction mixture is fully bound to the 10 μL glutathione Sepharose 4B beads used in each reaction. 30 μL SDS-PAGE sample buffer is used to strip all proteins from the beads. 8 μL of the 30 μL stripped protein sample is used for SDS-PAGE. The following formula is used to calculate the 10% input: $\frac{\frac{3}{50}\times 4}{\frac{9}{30}\times 8} = 10\%$.
14. Crystals in pyramid shape diffract to the highest resolution of 3.0 Å, and crystals of plate cluster diffract to the highest resolution of 2.3 Å.
15. LATS1 has an N-terminal 6×His tag. MOB1 has no tag.
16. Use LB medium instead of TB medium to grow cells transformed with pRSF-Duet-LATS1$^{602-1060}$-MOB1. Cells grow in TB medium give very low yield of LATS1.
17. The expression level of LATS1$^{602-1060}$ alone is extremely low and is very difficult for further purification. When co-expressed with MOB1, both yield and solubility of LATS1 are greatly improved. However, the expression level of MOB1 during co-expression is too low to be detected by SDS-PAGE after purification by Ni-NTA resin. Therefore, the vast majority (>99%) of purified protein from co-expression is LATS1.
18. LATS1$^{602-1060}$ eluted from Ni-NTA resins should not be dialyzed directly against SA buffer for Mono S column because His$_6$-LATS1$^{602-1060}$ is not stable in SA buffer. It also takes much longer time to achieve complete cleavage at pH 7.0.
19. LATS1$^{602-1060}$ can only bind to Mono S column, but not RESOURCE S column.
20. The HM fragment eluted from Ni-NTA resins has a purity greater than 90%.
21. Adding DTT in the buffer will decrease the binding efficiency of His6-tagged protein to Ni-NTA resin.
22. Using a molar ratio of 1:20 for kinase vs. substrate in order to get the complete phosphorylation of the HM fragment within 30 min.

Acknowledgements

This work was supported by the National Institutes of Health (GM107415 to XL) and the Robert A. Welch Foundation (I-1932 to XL).

References

1. Harvey K, Tapon N (2007) The Salvador-Warts-Hippo pathway—an emerging tumour-suppressor network. Nat Rev Cancer 7 (3):182–191
2. Pan D (2010) The hippo signaling pathway in development and cancer. Dev Cell 19 (4):491–505. https://doi.org/10.1016/j.devcel.2010.09.011
3. Johnson R, Halder G (2014) The two faces of Hippo: targeting the Hippo pathway for regenerative medicine and cancer treatment. Nat Rev Drug Discov 13(1):63–79. https://doi.org/10.1038/nrd4161
4. Yu FX, Zhao B, Guan KL (2015) Hippo pathway in organ size control, tissue homeostasis, and cancer. Cell 163(4):811–828. https://doi.org/10.1016/j.cell.2015.10.044
5. Badouel C, Garg A, McNeill H (2009) Herding Hippos: regulating growth in flies and man. Curr Opin Cell Biol 21(6):837–843
6. Staley BK, Irvine KD (2012) Hippo signaling in Drosophila: recent advances and insights. Dev Dyn 241(1):3–15. https://doi.org/10.1002/dvdy.22723
7. Barry ER, Camargo FD (2013) The Hippo superhighway: signaling crossroads converging on the Hippo/Yap pathway in stem cells and development. Curr Opin Cell Biol 25 (2):247–253. https://doi.org/10.1016/j.ceb.2012.12.006
8. Zhao B, Lei QY, Guan KL (2008) The Hippo-YAP pathway: new connections between regulation of organ size and cancer. Curr Opin Cell Biol 20(6):638–646
9. Harvey KF, Zhang X, Thomas DM (2013) The Hippo pathway and human cancer. Nat Rev Cancer 13(4):246–257. https://doi.org/10.1038/nrc3458
10. Halder G, Johnson RL (2011) Hippo signaling: growth control and beyond. Development 138(1):9–22. https://doi.org/10.1242/dev.045500
11. Yu FX, Guan KL (2013) The Hippo pathway: regulators and regulations. Genes Dev 27 (4):355–371. https://doi.org/10.1101/gad.210773.112
12. Chan EH, Nousiainen M, Chalamalasetty RB, Schafer A, Nigg EA, Sillje HH (2005) The Ste20-like kinase Mst2 activates the human large tumor suppressor kinase Lats1. Oncogene 24(12):2076–2086. https://doi.org/10.1038/sj.onc.1208445
13. Praskova M, Xia F, Avruch J (2008) MOBKL1A/MOBKL1B phosphorylation by MST1 and MST2 inhibits cell proliferation. Curr Biol 18(5):311–321. https://doi.org/10.1016/j.cub.2008.02.006
14. Hansen CG, Moroishi T, Guan KL (2015) YAP and TAZ: a nexus for Hippo signaling and beyond. Trends Cell Biol 25(9):499–513. https://doi.org/10.1016/j.tcb.2015.05.002
15. Huang J, Wu S, Barrera J, Matthews K, Pan D (2005) The Hippo signaling pathway coordinately regulates cell proliferation and apoptosis by inactivating Yorkie, the Drosophila Homolog of YAP. Cell 122(3):421–434
16. Zhao B, Wei X, Li W, Udan RS, Yang Q, Kim J, Xie J, Ikenoue T, Yu J, Li L, Zheng P, Ye K, Chinnaiyan A, Halder G, Lai ZC, Guan KL (2007) Inactivation of YAP oncoprotein by the Hippo pathway is involved in cell contact inhibition and tissue growth control. Genes Dev 21(21):2747–2761
17. Luo X (2010) Snapshots of a hybrid transcription factor in the Hippo pathway. Protein Cell 1(9):811–819. https://doi.org/10.1007/s13238-010-0105-z
18. Zhao B, Ye X, Yu J, Li L, Li W, Li S, Yu J, Lin JD, Wang CY, Chinnaiyan AM, Lai ZC, Guan KL (2008) TEAD mediates YAP-dependent gene induction and growth control. Genes Dev 22(14):1962–1971
19. Avruch J, Zhou D, Fitamant J, Bardeesy N, Mou F, Barrufet LR (2012) Protein kinases of the Hippo pathway: regulation and substrates. Semin Cell Dev Biol 23(7):770–784. https://doi.org/10.1016/j.semcdb.2012.07.002
20. Jin Y, Dong L, Lu Y, Wu W, Hao Q, Zhou Z, Jiang J, Zhao Y, Zhang L (2012) Dimerization and cytoplasmic localization regulate Hippo kinase signaling activity in organ size control. J Biol Chem 287(8):5784–5796. https://doi.org/10.1074/jbc.M111.310334
21. Ni L, Li S, Yu J, Min J, Brautigam CA, Tomchick DR, Pan D, Luo X (2013) Structural basis for autoactivation of human Mst2 kinase and its regulation by RASSF5. Structure 21 (10):1757–1768. https://doi.org/10.1016/j.str.2013.07.008
22. Xiong S, Couzens AL, Kean MJ, Mao DY, Guettler S, Kurinov I, Gingras AC, Sicheri F (2017) Regulation of protein interactions by Mps One binder (MOB1) phosphorylation. Mol Cell Proteomics 16(6):1111–1125. https://doi.org/10.1074/mcp.M117.068130
23. Couzens AL, Xiong S, Knight JDR, Mao DY, Guettler S, Picaud S, Kurinov I, Filippakopoulos P, Sicheri F, Gingras AC (2017) MOB1 mediated phospho-recognition

in the core mammalian hippo pathway. Mol Cell Proteomics 16(6):1098–1110. https://doi.org/10.1074/mcp.M116.065490

24. Cairns L, Tran T, Kavran JM (2017) Structural insights into the regulation of hippo signaling. ACS Chem Biol 12:601. https://doi.org/10.1021/acschembio.6b01058
25. Bae SJ, Ni L, Osinski A, Tomchick DR, Brautigam CA, Luo X (2017) SAV1 promotes Hippo kinase activation through antagonizing the PP2A phosphatase STRIPAK. eLife 6:e30278. https://doi.org/10.7554/eLife.30278
26. Meng Z, Moroishi T, Guan KL (2016) Mechanisms of Hippo pathway regulation. Genes Dev 30(1):1–17. https://doi.org/10.1101/gad.274027.115
27. Hoa L, Kulaberoglu Y, Gundogdu R, Cook D, Mavis M, Gomez M, Gomez V, Hergovich A (2016) The characterisation of LATS2 kinase regulation in Hippo-YAP signalling. Cell Signal 28(5):488–497. https://doi.org/10.1016/j.cellsig.2016.02.012
28. Hergovich A (2016) The roles of NDR protein kinases in hippo signalling. Genes 7(5):21. https://doi.org/10.3390/genes7050021
29. Ni L, Zheng Y, Hara M, Pan D, Luo X (2015) Structural basis for Mob1-dependent activation of the core Mst-Lats kinase cascade in Hippo signaling. Genes Dev 29(13):1416–1431. https://doi.org/10.1101/gad.264929.115
30. Hergovich A (2012) Mammalian Hippo signalling: a kinase network regulated by protein-protein interactions. Biochem Soc Trans 40(1):124–128. https://doi.org/10.1042/BST20110619
31. Hergovich A (2013) Regulation and functions of mammalian LATS/NDR kinases: looking beyond canonical Hippo signalling. Cell Biosci 3(1):32. https://doi.org/10.1186/2045-3701-3-32
32. Kim SY, Tachioka Y, Mori T, Hakoshima T (2016) Structural basis for autoinhibition and its relief of MOB1 in the Hippo pathway. Sci Rep 6:28488. https://doi.org/10.1038/srep28488
33. Praskova M, Khoklatchev A, Ortiz-Vega S, Avruch J (2004) Regulation of the MST1 kinase by autophosphorylation, by the growth inhibitory proteins, RASSF1 and NORE1, and by Ras. Biochem J 381(Pt 2):453–462. https://doi.org/10.1042/BJ20040025
34. Cho US, Xu W (2007) Crystal structure of a protein phosphatase 2A heterotrimeric holoenzyme. Nature 445(7123):53–57. https://doi.org/10.1038/nature05351

Chapter 19

Isothermal Titration Calorimetry Assays to Measure Binding Affinities In Vitro

Kui Lin and Geng Wu

Abstract

In the study of the Hippo signal transduction pathway, protein-protein interactions are often explored, because various proteins such as MOB1, NDR1/NDR2, and LATS1/LATS2 are very important members in this complicated signaling pathway. The transduction of signals from upstream to downstream is largely dependent on the mutual recognition of proteins and the formation of specific non-covalent complexes between them. In general, protein-protein associations, protein–DNA associations, or protein-small molecule associations cause the release or absorption of heat during the association reaction. The isothermal titration calorimetry (ITC) assay is a convenient and widely used approach to directly measure the amount of heat released or absorbed during association processes of biomolecules (such as protein-protein, protein-DNA, or protein-small molecules) in solution and to quantitatively estimate the interaction affinity.

Key words Isothermal titration calorimetry, ITC, Protein–protein interaction, Dissociation constant, Association constant, Binding affinity

1 Theoretical Background

When two molecules, molecule A (e.g., a protein such as MOB1) and molecule B (e.g., another protein such as NDR2), interact to form a molecular complex, AB, the reaction can be written as:

$$A + B = AB$$

The association constant K_a between molecules A and B can be written as:

$$K_a = [AB]/\{[A][B]\} \quad (1)$$

([AB], [A], and [B] stand for the concentrations of AB, A, and B, respectively.)

On the other hand, the dissociation constant K_d can be written as:

$$K_d = \{[A][B]\}/[AB] \quad (2)$$

Alexander Hergovich (ed.), *The Hippo Pathway: Methods and Protocols*, Methods in Molecular Biology, vol. 1893, https://doi.org/10.1007/978-1-4939-8910-2_19, © Springer Science+Business Media, LLC, part of Springer Nature 2019

Therefore, the dissociation constant K_d is the reciprocal of the association constant K_a:

$$K_d = 1/K_a \tag{3}$$

According to the principle of thermodynamics, the change in Gibbs free energy ΔG during the interaction between A and B is as following:

$$\Delta G = \Delta H - T\Delta S \tag{4}$$

(ΔH, change of enthalpy; T, absolute temperature, in Kelvin; ΔS, change of entropy)

On the other hand, the relationship between the change in Gibbs free energy ΔG and association constant K_a can be described as:

$$\Delta G = -RT * \mathrm{Ln}\, K_a \tag{5}$$

(R, universal gas constant; T, absolute temperature, in Kelvin; Ln, natural logarithm)

Combining Eqs. (4) and (5), we obtain the following equation for the association reaction between A and B:

$$\Delta H - T\Delta S = -RT * \mathrm{Ln} K_a \tag{6}$$

From the isothermal titration calorimetry (ITC) assay, we can measure the change of enthalpy ΔH from the heat released or absorbed during the association reaction between A and B. In addition, we can also obtain the association constant K_a from the fitting of the ITC data. Hence, we can calculate ΔS from Eq. (6). From the association constant K_a, the dissociation constant K_d can be calculated from Eq. (3), which is more commonly used than K_a. As a rule of thumb, K_d on the nanomolar (nM, 10^{-9} M) level describes a strong association, K_d on the micromolar (μM, 10^{-6} M) level describes a medium-strength association, and K_d on the millimolar (mM, 10^{-3} M) level describes a weak association.

2 Operational Principle of ITC

The basic components of the ITC instrument are a sample cell and a reference cell, both of which contain a highly efficient thermal conductor surrounded by an adiabatic jacket. The two cells are connected by sensitive thermocouple circuits, which detect temperature difference between the two cells and between the cells and the jacket. Heaters are located in both cells which are activated when necessary to maintain identical temperatures between all components (Fig. 1).

In an ITC experiment, two solutions of the reactants are prepared and are put into the temperature-controlled sample cell

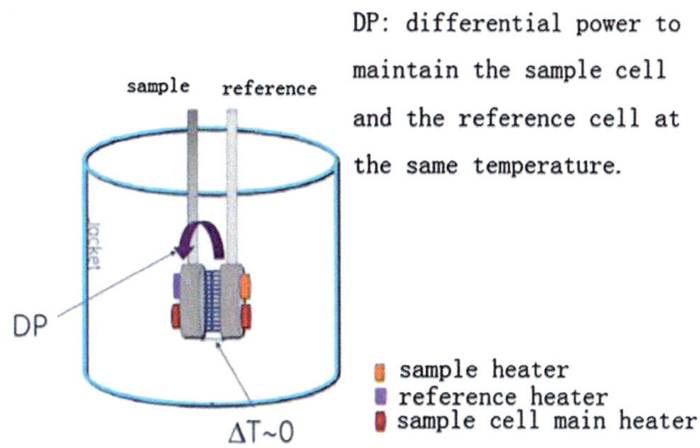

Fig. 1 The basic components of the ITC instrument

and the injection syringe, respectively. Buffer or water without any reactants is added to the reference cell. Prior to the injection of the reactant, a constant power is applied to the reference cell. This signal directs the feedback circuit to activate the heater located on the sample cell, and this represents the baseline signal. During the injection of the titrant into the sample cell, heat is taken up or released depending on whether the association reaction is endothermic or exothermic. The feedback circuit will either increase or decrease power to the sample cell to maintain equal temperature with the reference cell. For example, before the start of the titration, 5 μW of power is applied on both the sample cell and the reference cell to maintain the baseline (Fig. 2). After the start of the titration, the titration reaction releases 3 μW of heat. The feedback circuit perceives this energy change and reduces the power in the sample cell to 2 μW to keep the sample cell and the reference cell at the same temperature (Fig. 3).

In the raw ITC data, the abscissa axis shows the time, and the vertical axis shows the change of heat per unit time (in cal/sec). The 3 μW of heat generated by the exothermic reaction in the sample cell results in the decrease of 3 μW of power on the sample cell as compared with the power applied on the sample cell before titration begins. Therefore, a downward peak is produced in the ITC data (Fig. 4). If the association reaction is an endothermic reaction, the direction of the peak is just the opposite in the original ITC data. When the effect of heat produced or absorbed by a reaction disappears, the power applied on the sample cell and the reference cell will return to 5 μW, and the signal will go back to baseline. This is when a second titration can begin. Each titration reaction would produce a peak (Fig. 4).

For convenience, we would call the reactant in the sample cell as the macromolecule and the reactant in the syringe as the ligand (Fig. 5). The heat released or absorbed during a calorimetric

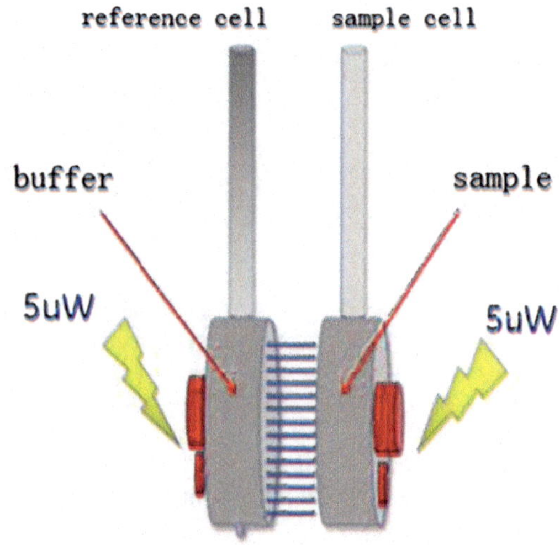

Fig. 2 Before the start of the titration, 5 μW of power is applied on both the sample cell and the reference cell to maintain the baseline

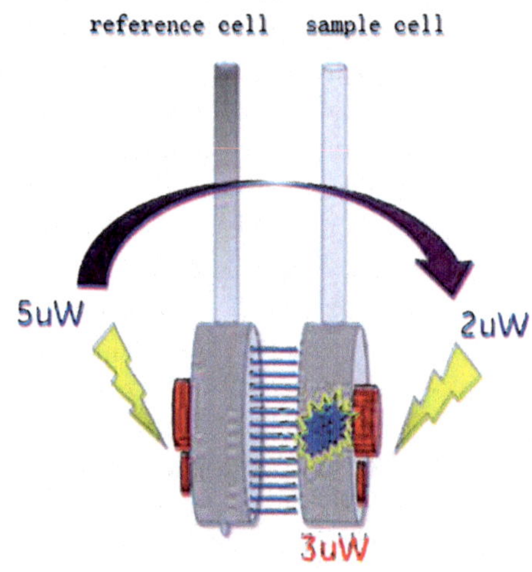

Fig. 3 After the start of the titration, the titration reaction releases 3 μW of heat. The feedback circuit perceives this energy change and reduces the power in the sample cell to 2 μW to keep the sample cell and the reference cell at the same temperature

titration is in direct proportion to the amount of binding that occurs. At the beginning of the titration, all or most of the ligands interact with the receptors in the sample cell to produce large endothermic or exothermic signals (Fig. 6). As the titration proceeds, the receptors in the sample cell are gradually saturated with the increasing amount of the ligand. Therefore, the heat produced or absorbed in the endothermic or exothermic reaction gradually decreases until the receptors in the sample cell are fully saturated (Fig. 7). At the end of the titration, if more ligand is added, there is

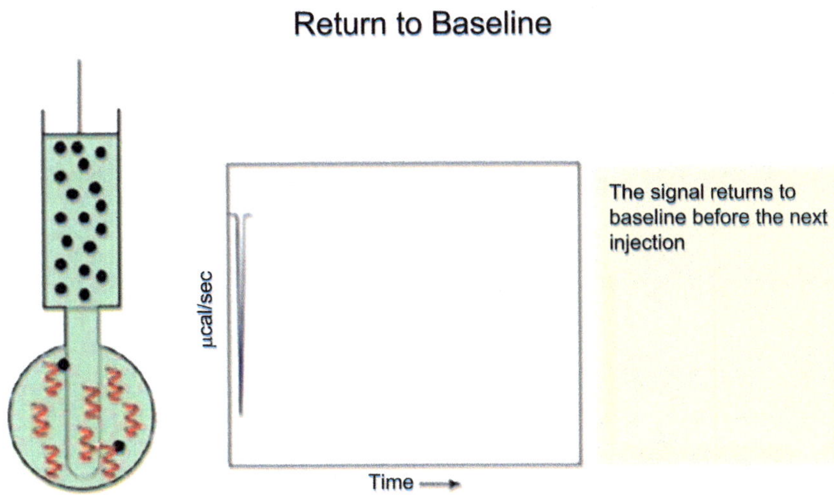

Fig. 4 Because heat is generated by the exothermic reaction in the sample cell during the ITC titration assay, a downward peak is produced in the ITC data

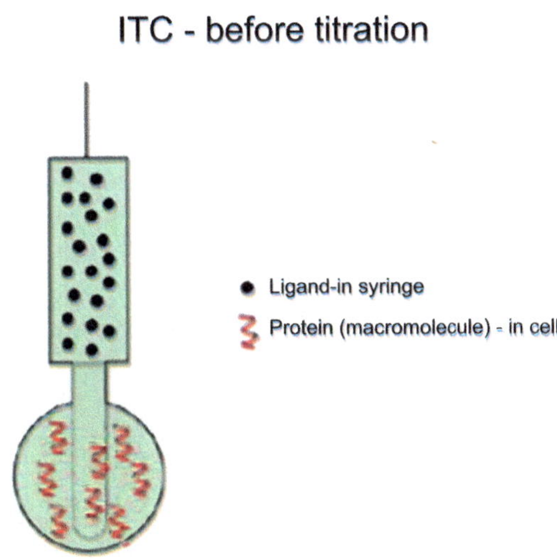

Fig. 5 For convenience, we would call the reactant in the sample cell as the macromolecule and the reactant in the injection syringe as the ligand

no receptor available to interact with it, and the heat produced or absorbed is very little. If the titration continues, the heat produced or absorbed comes only from the dilution of the ligand by the receptor solution in the sample cell (Fig. 8).

The original ITC data can be processed by the Origin software, which is usually installed in the ITC instrument. The abscissa axis is the molar ratio of the ligand to the macromolecule, and the vertical axis is the amount of released heat caused by the injection of ligand per mol into the sample cell. By fitting the interaction pattern of two reactants by the Origin software, we can obtain the change of

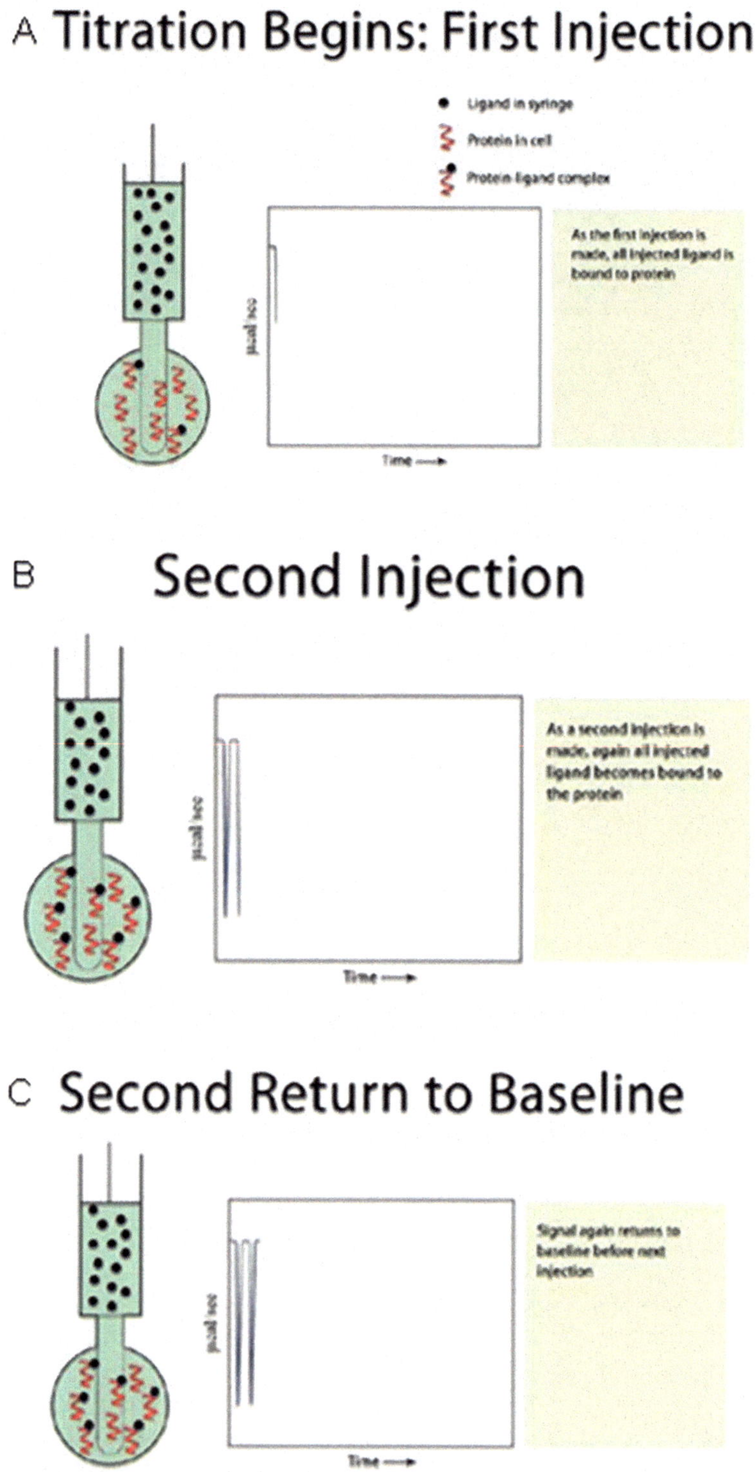

Fig. 6 At the beginning of the titration, all or most of the ligands interact with the receptors in the sample cell to produce large endothermic or exothermic signals. (**a**) First injection. (**b**) Second injection. (**c**) Two downward peaks are produced after the first two injections have finished

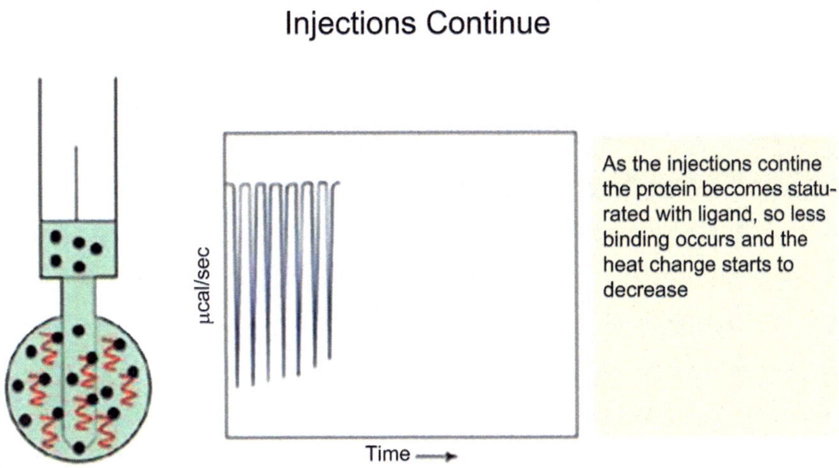

Fig. 7 The heat produced or absorbed in the endothermic or exothermic reaction gradually decreases until the receptors in the sample cell are fully saturated

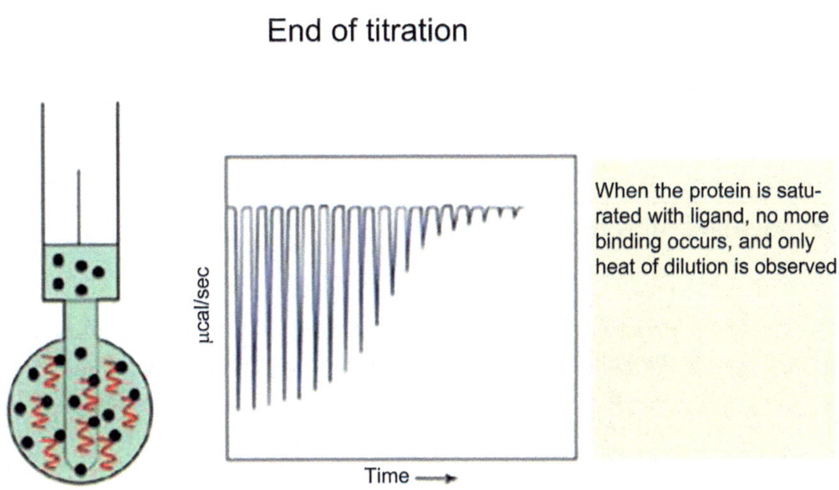

Fig. 8 At the end of the titration, if more ligand is added, there is no receptor available to interact with it, and the heat produced or absorbed is very little

enthalpy ΔH during the reaction, the binding stoichiometry N (the number of binding sites), and the dissociation constant K_d. From Eq. (6), we can obtain the change of entropy ΔS by calculation (Fig. 9).

At present, ITC is the only technique that can determine all the binding parameters at the same time in a single experiment. It provides complete thermodynamic information about the interaction of molecules. In addition, ITC determines the affinity between the reactants in a natural state, so there is no need to modify the reactants by fluorescence labeling or immobilization. Furthermore, the samples are not destroyed during the experiment, so that it can be used for further biochemical analysis.

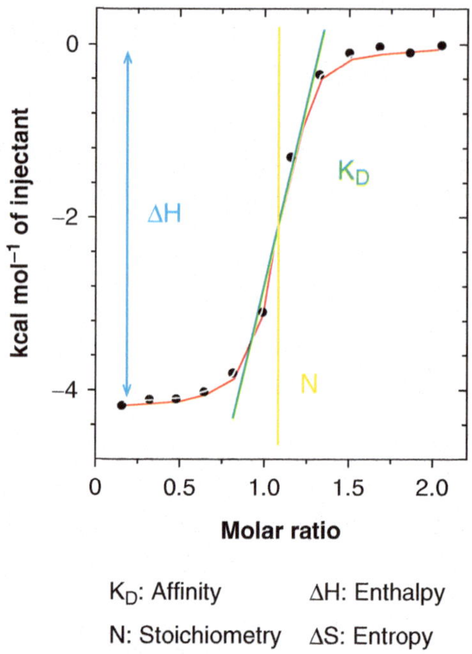

Fig. 9 By fitting the interaction pattern of two reactants by the Origin software, we can obtain the change of enthalpy ΔH, the binding stoichiometry N, the dissociation constant K_d, and the change of entropy ΔS

3 ITC Experimental Procedure

Here, we will use MicroCal ITC200 as an example to describe the procedure for an ITC experiment. The ITC instrument consists of the cell unit (including the sample cell and the reference cell), the injection syringe (which can be in three states: washing, loading, and running), the washing module, and the computer (Fig. 10).

All the experimental steps using the MicroCal ITC200 system are carried out under the computer control. The user only needs to input the experimental parameters such as the temperature, the number of titration drops, and the volume of each titration drop. In order to allow the ITC system to run properly, the computer needs to be turned on first to the start the Windows operating system. The next step is to turn on the power of the ITC instrument and to open the ITC200 controlling software. Now a real-time copy of Origin will automatically pop up (Fig. 11).

The basic experimental steps of an ITC experiment include sample preparation, instrument washing (Fig. 12), parameter setup, sample loading (Fig. 13), titration (Fig. 14), titration data processing, and instrument washing again. Here we will emphasize on some of most important details deserving attention for an ITC experimentalist.

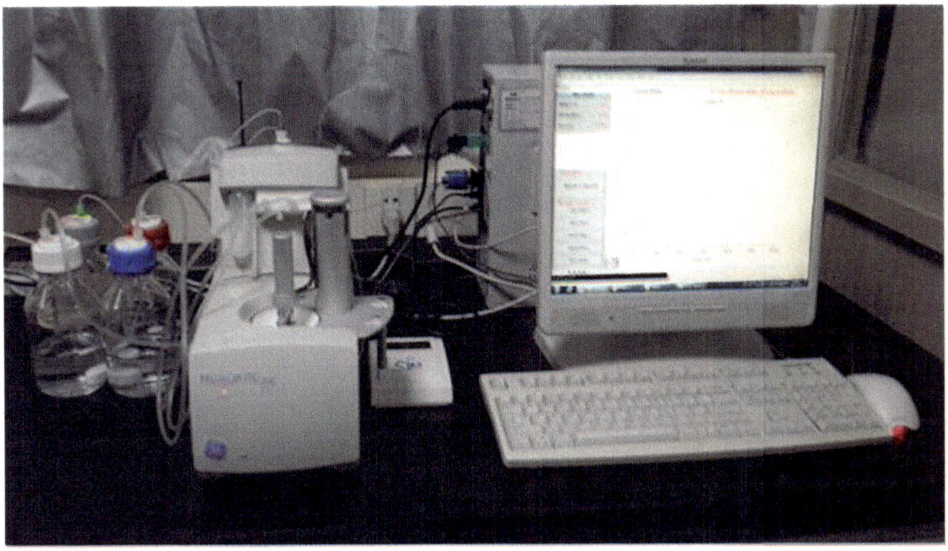

Fig. 10 The ITC instrument setup (using MicroCal ITC200 as an example)

3.1 Sample Preparation

As to the preparation of the sample, the first consideration is that which buffer should be used. The two reactants should be soluble and stable in the buffer used. What is more, the buffer itself should not lead to any change of enthalpy (ΔH) of the reaction system. Some buffers will release hydrogen ions into the aqueous solution and release heat. If the interaction between two reactants can also release hydrogen ions, it is possible for these hydrogen ions to interact with the buffer conjugate base, resulting in the change of enthalpy of the solution, which would interfere with the ΔH produced by the interaction between the reactants. Therefore, some buffers, such as Tris, are not suitable for ITC experiments. The buffers suitable for ITC experiments require ΔH to be close to 0 in aqueous solution. Therefore, phosphate, HEPES, acetate, formate, citrate, sulfate, cacodylate, glycine, etc. are candidate buffers suitable for using in ITC experiments. We commonly used phosphate and HEPES in our laboratory.

Another important principle is that the buffer components used to dissolve the two reactant molecules, including salt and reducing agents, need to be exactly the same. In order to achieve this, the two reactants are usually prepared in the same buffer when doing dialysis, gel filtration, or dilution. Some researchers also put two reactants in the same beaker for dialysis or use the dialysate from preparation of the macromolecule solution as the "solvent" for preparation of the ligand. If the buffer used by the two reactants is not completely the same, the change of enthalpy produced by mixing or diluting different components may interfere with the change of enthalpy produced by the interaction of the two reactants.

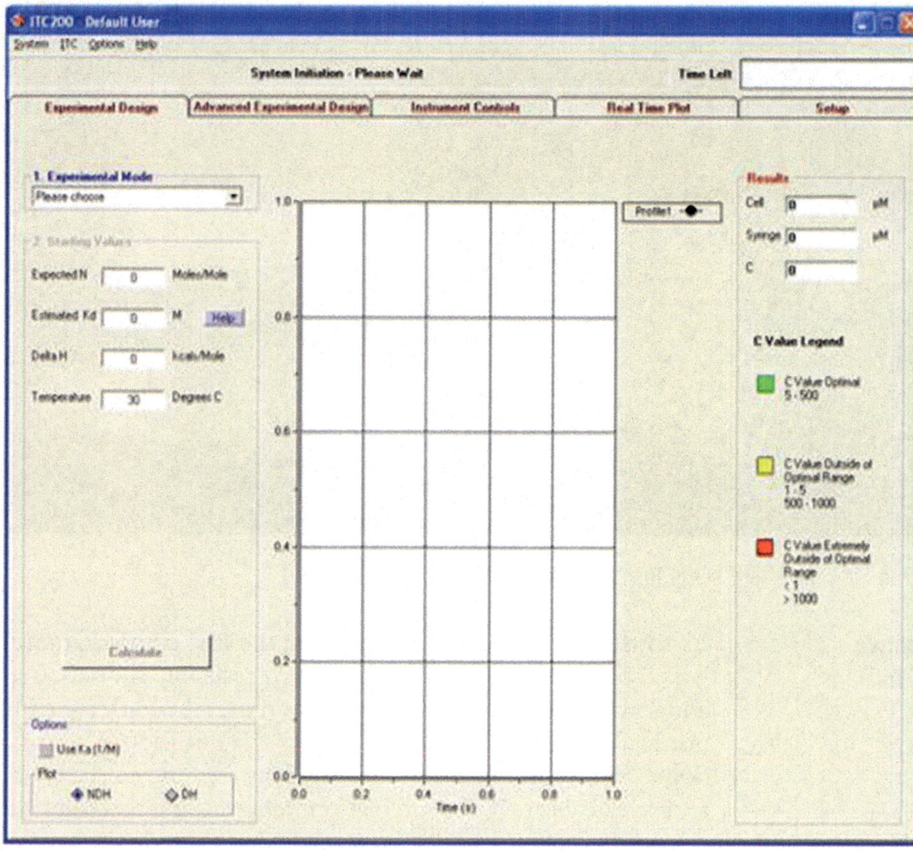

Fig. 11 The interface of the Origin software on ITC200

For some proteins, it is necessary to add a reductant to the buffer to keep the proteins stable and active or avoid protein aggregation. This reducing agent should not be dithiothreitol (DTT) because DTT is unstable and is easily oxidized in the air, which affects the baseline of the ITC. If a reducing agent is a must, beta-mercaptoethanol or TCEP is recommended. Our experience showed that in the 40 mM HEPES, pH 7.5 buffer system, beta-mercaptoethanol up to 5 mM or TCEP up to 2 mM has little effect on the baseline.

3.2 Choosing the Concentrations of the Two Reactants

In an ITC experiment, the choice of concentrations of the two reactants depends on the binding capacity between the two reactants according to the formula:

$$C = N * K_a * [M]_T \tag{7}$$

where C is the binding capacity, K_a is the association constant, $[M]_T$ is the total macromolecular concentration in the cell, and N is the binding stoichiometry. When the binding stoichiometry (N) and the concentration of the reactants in the sample cell are fixed, the

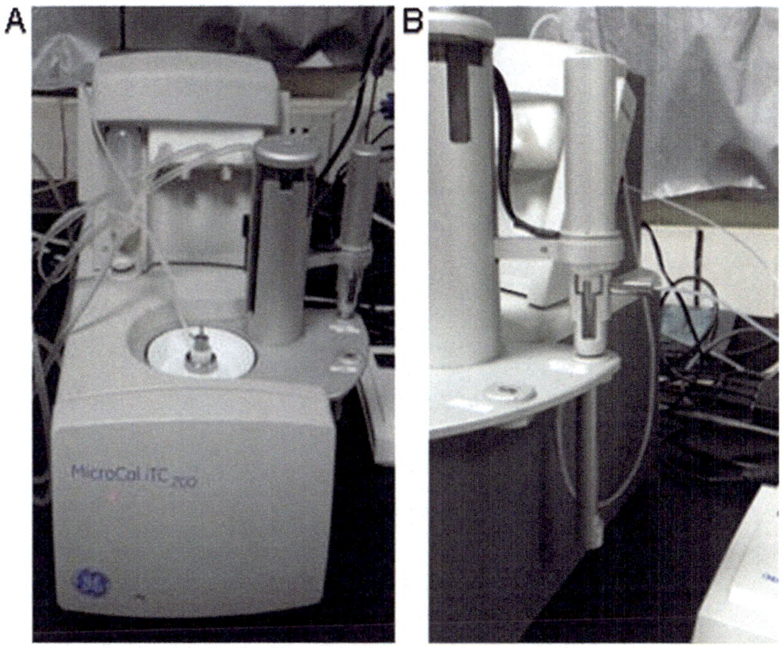

Fig. 12 Washing the cells and the injection syringe under the computer control. (**a**) Washing the cells. (**b**) Washing the injection syringe

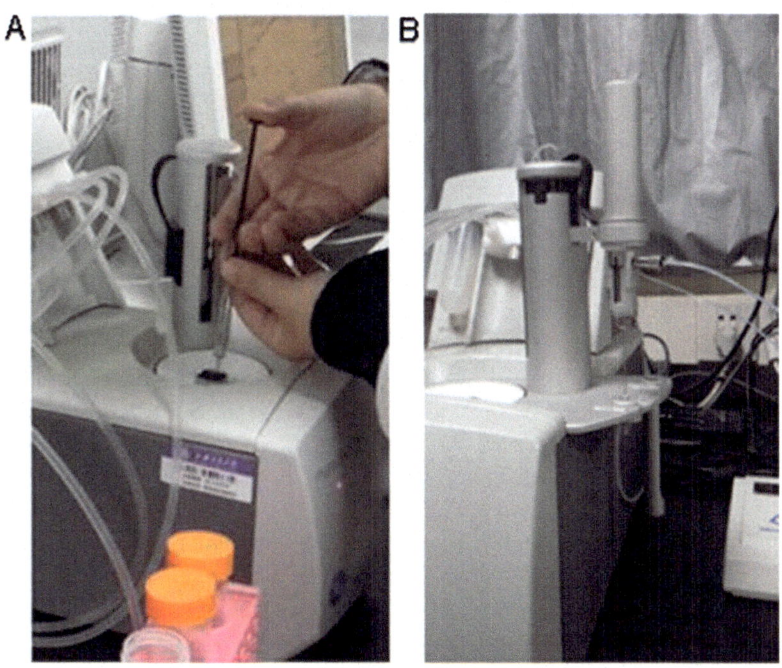

Fig. 13 Loading the sample into the cell and into the injection syringe. (**a**) Loading the sample into the cell. (**b**) Loading the sample into the injection syringe

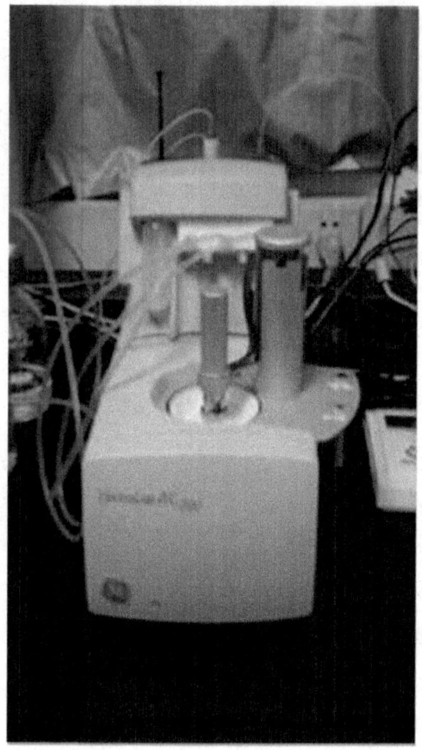

Fig. 14 Titration of one reactant in the sample cell by another reactant in the injection syringe

greater the C is, the stronger the binding capacity between the two reactants will be. When C is too big, accurate K_a cannot be obtained because the curve is too steep and is devoid of data points in the transition region. However, the N and ΔH values are accurate under this condition. Therefore, when C is too large, that is, when the two reactants bind strongly, the concentration of the reactant in the sample cell should be reduced. On the other hand, the smaller the C is, the weaker the binding ability between the two reactants is. When C is too small, correct N, ΔH, and K_a cannot be obtained because the curve is too flat. So, for the weak association, it is necessary to increase the concentration of the reactant in sample cell. When the C value is between 10 and 100, very reliable N, K_a, and ΔH values can be obtained. C value of 5–500 is generally considered as a good range to get the above parameters, and C value of 1–5 or 500–1000 is also acceptable (Fig. 15).

In addition, it is hoped that the molar ratio of the molecules from the injection syringe to the molecules from the sample cell is 2–3 by the end of an ITC experiment (for 1:1 binding stoichiometry), and between 7 and 13 injections, the molar ratio of the two reactant molecules is 1:1. Considering the volume of the sample cell, the injection volume, and the total number of injections, the

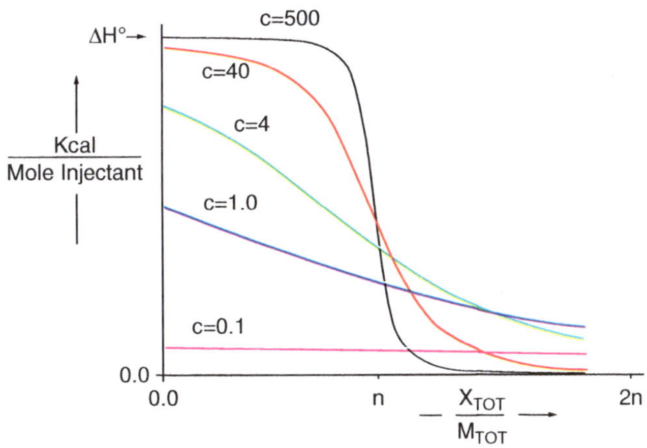

Fig. 15 The binding capacity C should be in the range of 5–500 to get sufficient ITC data points. X_{TOT}, the concentration of injected solute in the cell before each injection. M_{TOT}, the concentration of macromolecule in the cell before each injection after correction for volume displacement

concentration of the reactant in the syringe is usually 10–20 times the concentration of the reactant in the sample cell. Usually, during an actual ITC experiment, if the dissociation constant K_d of the two reactants is unknown, it is suggested that the concentration of the reactant is 20 μM in the sample cell and 200 μM in the injection syringe to get started. Moreover, it is particularly worth to pay attention to the correct determination of the concentration of two reactants for it is essential to get reliable N, K_d, and ΔH in ITC.

It is also very important that before the ITC experiment, the two reactants need to be degassed, because even very small air bubbles may interfere with the feedback circuit and the baseline.

3.3 Cleaning of the ITC Instrument

The cleaning of the ITC instrument is very important, and lack of cleaning can affect the stability of the baseline, the proper data acquisition, and the consistence of the data. Generally speaking, the cleaning of the control cell is relatively simple, and once a week is recommended. Just take out the water with syringe, rinse it with water or buffer several times, and replace with fresh water or buffer. To be cautious, the control cell should be cleaned before the experiment in case the control cell is polluted for various reasons by other people. Even if the sample cell and the injection syringe have been cleaned by the previous person using the instrument, the liquid in the sample cell must be sucked out, and the sample cell and the syringe should be cleaned under the control of the software. If the titration data of water by water is not ideal, then detergent is recommended to be included in the washing buffer to thoroughly clean the sample cell and the injection syringe. Once one titration run is finished, the sample cell and the injection syringe must be cleaned immediately, otherwise proteins or salts will precipitate and increase the difficulty for washing.

3.4 Parameter Setup

The parameters we commonly used for the ITC200 instrument in our laboratory are:

Total number of injections: 20

Measurement temperature: 25 °C (298 K)

Reference power: 5 μcal per second

Initial delay (the time between the start and the first injection): 60 s

Reactant concentration in the sample syringe: millimolar (mM)

Reactant concentration in the sample cell: millimolar (mM)

Stirring speed: 1000 rpm

Feedback mode/gain: high (high sensitivity and small response time)

Injection volume: 2 μl

Duration for injection: 0.8 s (This setting is automatically assigned according to the injection volume.)

Spacing between injections: 90 s

Filter period (the signal-average time period): 5 s

Volume of the first injection: 0.4 μl (The heat associated with the first injection is not included in the data analysis.)

This setting can be used in the first experiment and then adjusted according to the result. For example, the total number of injections can be increased by 4–5 drops after the macromolecules in the cell are saturated by the ligand in order to determine the ΔH produced by the dilution of the ligand. The reference power can be modified if a large exothermic or endothermic reaction is expected. The stirring speed of the syringe needle can be adjusted so that the two reactants can be mixed quickly and evenly, and bubbles produced by viscous macromolecules (such as proteins) need to be avoided in the sample pool. The spacing time between injections should be set to ensure that the signal of the previous injection returns to the baseline before the new injection starts. The volume of the first injection is usually less than 3 μl. The ΔH produced by this titration is usually discarded and is not used for subsequent data processing because of interdiffusion of the solutions during the insertion of the syringe or the equilibration stage. The measurement temperature should also be set properly. The ΔH produced by the interaction of the two reactants varies with the temperature. Therefore, if the ΔH is close to 0 at a set temperature, it is necessary to try other temperatures with about 5–10 °C difference. In some cases, the data can be optimized by changing the temperature without changing the concentrations of the two reactants. The dissociation constant K_d could change as the temperature changes, but the binding stoichiometry N does not change with the temperature. If the ΔH generated by the reaction is very large or very small, the concentrations of the reactants are usually adjusted to get optimal ΔH, instead of changing the volume of the reactants in the sample pool and syringe.

3.5 Sample Loading

Loading the sample is the process that tests the patience of people using ITC. This is due to the fact that it is easy to generate bubbles when very thick liquid such as protein is added into the sample pool. In order to avoid the bubbles, first of all, the concentration of protein or other viscous liquid should be reduced as much as possible. In addition, the total volume of the sample cell and the reservoir of ITC200 is around 400 μl. We usually pull more than 400 μl of reactant into a Hamilton syringe, and this should also be done slowly to avoid too many bubbles. We usually then turn the syringe upside down, so the needle is pointing upward, and tap the syringe gently so that all the bubbles are at the top, and then push all the air out of the needle and make sure that the reactant is around 400 μl. To load the sample into the cell, the needle of the Hamilton syringe should be inserted into the bottom of the cell, and the reactant should be pushed out very slowly. When the liquid level can be seen, gently raise the needle tip bit by bit until the filling is finished. If 400 μl of reactant can be all put into the sample cell and reservoir and there is no overflow, then there should be no air bubble in the sample cell, and the procedure of loading the sample into the cell is complete. Otherwise, the bubbles must be driven out by moving the needle up and down slowly along the wall of the sample cell. If this does not work, then all the reactant in the sample cell need to be removed, and the sample cell should be cleaned before doing the loading again. Loading the sample into the injection syringe also requires avoiding bubbles. However, it is relatively easy compared with loading the sample into the cell, because sometimes the reactant in the syringe is small molecule with less viscosity and it is software controlled.

3.6 Control of ITC Experiments

The first control that needs to be done before using real samples is the titration of water by water. In this experiment, ΔH should be less than 0.02 μcal per second, and DP values should be between 4.5 and 5.5 which indicates that the cells and the syringe are clean without any contaminants, and the ITC should run normally. Another important control experiment that may need to be done is to titrate the buffer without the reactant in the cell using the ligand in the syringe. In our ITC setting, the volume of each injection of the ligand is 2 μl, while the volume of the reactant in the cell is around 200 μl. That is to say, the ligand is diluted by 100 times for the first injection. If 20 injections are performed in the whole run, then the total volume of the ligand added into the cell is 40 μl, which means that the ligand is diluted six times. Since the dilution might release heat and produce ΔH, it might be necessary to do this control. The control data should be subtracted from that of sample before data processing.

When the ligand is titrated into the sample cell, the dilution of the reactant in the sample cell is relatively small compared with that of the ligand. The ΔH caused by this dilution should be small and negligible.

3.7 Troubleshooting

When the ITC graphics is abnormal, such as large baseline fluctuation or peak pattern reversal, the first thing to consider is whether the sample cell and the injection syringe are clean. Although after each run, the sample cell and the injection syringe are cleaned using methanol and water as directed by the software, it is still necessary to use detergent from time to time to clean them. In our experiments, almost all the problems can be solved by intensive cleaning. It is also necessary to consider whether the water or buffer in the reference cell is contaminated. If that is the case, cleaning the reference cell using flesh water or buffer should work. In addition, attention should be paid to whether there are air bubbles in the sample cell or syringe. If there are no problems in these three aspects, it is possible that the failure of the instrument causes a graphic anomaly, and it is necessary to seek professional and technical assistance.

Bibliography

1. Freyer MW, Lewis EA (2008) Isothermal titration calorimetry: experimental design, data analysis, and probing macromolecule/ligand binding and kinetic interactions. Methods Cell Biol 84:79–113
2. GE MicroCal™ ITC200 Training manual, GE Healthcare
3. Han JC, Han GY (1994) Anal Biochem 220:5–10
4. http://structbio.vanderbilt.edu/wetlab/designing.itc.v7.expts.pdf, Center for Struct Biol, Vanderbilt University
5. ITC expert users manual, GE Healthcare
6. Liang Y (2008) Applications of isothermal titration calorimetry in protein science. Acta Biochim Biophys Sin Shanghai 40:565–576
7. Myszka DG, Abdiche YN, Arisaka F, Byron O, Eisenstein E, Hensley P, Thomson JA, Lombardo CR, Schwarz F, Stafford W, Doyle ML (2003) The ABRF-MIRG'02 study: assembly state, thermodynamic, and kinetic analysis of an enzyme/inhibitor interaction. J Biomol Tech 14:247–269
8. Pierce MM, Raman CS, Nall BT (1999) Isothermal titration calorimetry of protein-protein interactions. Methods 19:213–221
9. Wiseman T, Williston S, Brandts JF, Nin LN (1989) Rapid measurement of binding constants and heats of binding using a new titration calorimeter. Anal Biochem 179:131–137
10. Vel'azquez-Campoy A, Ohtaka H, Nezami A, Muzammil S, Freire E (2004) Isothermal titration calorimetry. Curr Protoc Cell Biol Chapter 17:Unit 17.8

Chapter 20

GST Pull-Down Assay to Measure Complex Formations

Sun-Yong Kim and Toshio Hakoshima

Abstract

The GST pull-down assay is an intuitive and fast in vitro method for analyzing protein–protein or protein–ligand interactions and is comprised of a "bait" which is a GST-fused protein expressed in *E. coli* host or a baculovirus expression system and a "prey" which comprises putative binding partner protein(s) or other ligand molecule(s). This method is suitable for examining the direct interaction between two purified proteins and estimating the extent of the affinity.

Key words GST pull-down assay, Glutathione S-transferase, Glutathione-conjugated resin, Direct protein–protein interaction, In vitro assay, Screening of binding partners

1 Introduction

GST (glutathione S-transferase) binds strongly to glutathione-conjugated resin with high specificity given the enzyme-substrate relationship. The high binding affinity, high solubility, and effective expression as a recombinant protein make GST the most useful and popular tag for protein purification and binding assays via generation of a GST fusion protein. Since first being presented as an affinity chromatographic method for the purification of foreign proteins expressed in *E. coli* [1], many laboratories now apply this method not only for purifying recombinant protein but also for verifying direct protein interactions in an effort to identify putative binding partners. Although the yeast two-hybrid system remains a powerful tool since its introduction for the screening of protein–-protein interactions [2], indirect interactions may sometimes result from endogenous "adaptor" molecules present in the host cell, thereby complicating analysis of the results. Especially in the case of protein biochemical studies, confirmation of a direct interaction between proteins is fundamental in determining whether the interaction is functional for the generation of a protein complex, which should be stable and not short-lived since the interaction must be maintained during the course of the investigation. The GST

Alexander Hergovich (ed.), *The Hippo Pathway: Methods and Protocols*, Methods in Molecular Biology, vol. 1893,
https://doi.org/10.1007/978-1-4939-8910-2_20,

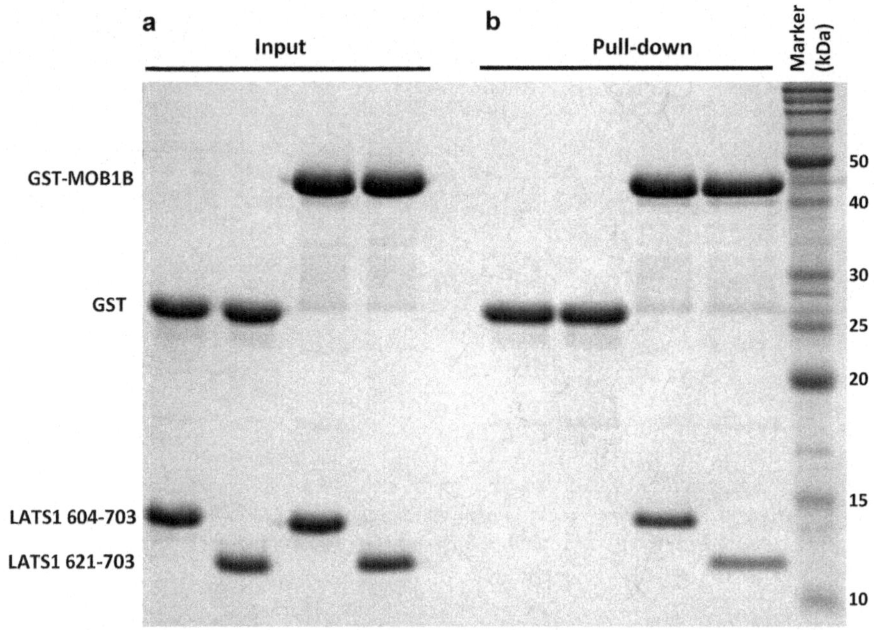

Fig. 1 GST pull-down binding assays of the NTR domains of LATS1 with GST-MOB1B show a typical result of the assay. (**a**) shows input of protein samples in pull-down binding assays of two NTR domain constructs of LATS1 (604–703 and 621–703) as preys with GST-MOB1B of the N-truncated form (33–126) as a bait. (**b**) shows pull-down binding assays of the NTR domain constructs of LATS1 with the GST-N-truncated MOB1B. GST is adopted as a negative control

pull-down assay is an intuitive and fast in vitro method for analyzing direct protein–protein interactions and is comprised of a "bait" which is a GST-fused protein expressed in *E. coli* host or a baculovirus expression system and a "prey" which comprises putative binding partner protein(s) or other ligand molecule(s). The bait, which is immobilized onto glutathione-conjugated resin, traps the prey via specific interactions. The prey can be a single purified protein or unknown protein from a crude extract that possesses the ability to interact with the bait. Therefore, the GST pull-down assay can be applied to the screening of binding partner protein (s) from crude extracts and to investigating the extent of an interaction between two proteins. Here we introduce an example that utilizes a GST-MOB1 (Mps one binder 1) fusion protein to pull down the NTR (N-terminal regulatory) domains of LATS1 (large tumor suppressor 1) [3–5] (Fig. 1).

2 Materials

1. Reaction and washing buffer: PBS (phosphate-buffered saline; 8 g of NaCl, 0.2 g of KCl, 1.44 g of Na_2HPO_4, and 0.24 g of KH_2PO_4 in 1 L of ultrapure H_2O, pH 7.4) or TBS (Tris-

buffered saline; 8 g of NaCl, 0.2 g of KCl, and 3 g of Tris base in 1 L of ultrapure H_2O, pH 7.4) (*see* **Note 1**).

2. Elution buffer: The reaction buffer containing 10 mM glutathione (*see* **Note 2**).
3. Buffers for SDS-PAGE (sodium dodecyl sulfate-polyacrylamide gel electrophoresis): Sample buffer and running buffer which are commonly used in your laboratory.
4. Commercial glutathione-conjugated resin: Agarose- or Sepharose™-based resin.
5. 1.5 ml microcentrifuge tubes.
6. Micropipettes and disposable tips.
7. Molecular weight standard for SDS-PAGE.
8. Polyacrylamide gel and gel electrophoresis apparatus.
9. Vortex mixer.
10. Spectrophotometer for measuring protein concentration.
11. Microcentrifuge equipped with a cooling device.
12. GST-fused protein (GST from *Schistosoma japonicum* is generally adopted for preparation of the GST-tagged protein. In this section, we will not describe the method of the GST-tagged protein expression and purification because the method has been well described in numerus papers. Importantly, if glutathione in the buffer of purified GST-tagged protein still remains, it must be removed by dialysis against the reaction buffer (*see* **Note 3**).

3 Methods

3.1 GST Pull-Down Assay for Examining Direct Protein–Protein Interactions

This method is suitable for examining protein–protein direct interactions between two purified proteins and estimating the extent of the affinity. In the case of structural biology, use of this method may provide decisive data in terms of whether the study is suitable for structure determination.

3.1.1 General Considerations

- Protein sample

  Concentrate the protein solution as much as possible, since the pull-down assay generally requires concentrations from 5 to 50 μM that can be clearly visualized by Coomassie Brilliant Blue staining with SDS-PAGE. Therefore, the required concentration of each prey and bait should be at least 10 μM. When using more sensitive staining methods such as silver staining or Western blotting, lower concentrations of proteins should suffice. However, the dissociation constants for the specific interactions should be relatively low (usually in the nanomolar or micromolar range), as low concentrations of protein might show

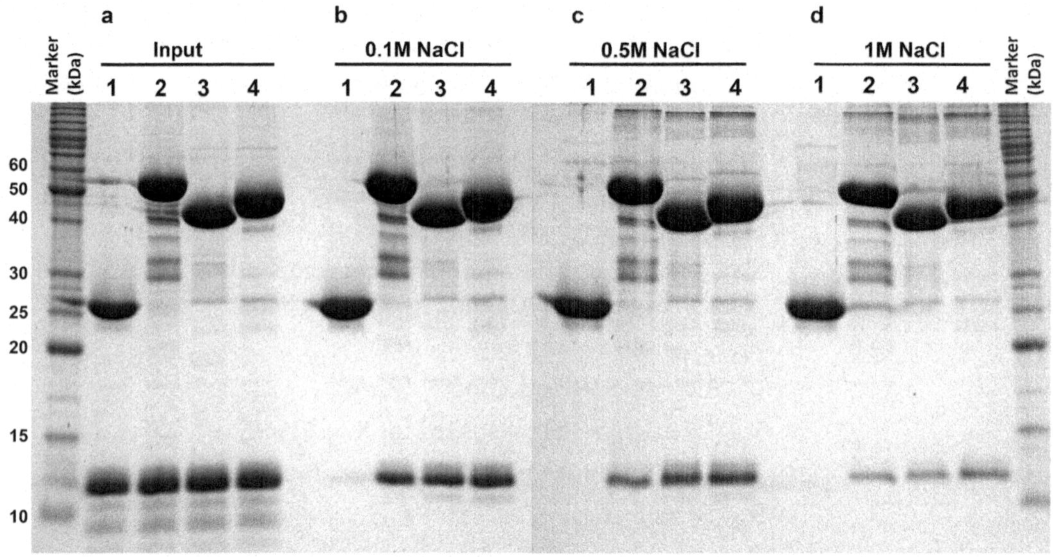

Fig. 2 GST pull-down assay was performed under various salt concentrations for affinity comparison. SDS-PAGE analysis of pull-down assays of the NTR domain of LATS1 as a prey with GST-MOB1B constructs as baits and GST as a negative control (lanes labelled *1*). GST-MOB1 (wild-type) (lanes labelled *2*), GST-MOB1B the N-terminal 32-residue truncated form (33–216) (lanes labelled *3*), and GST-MOB1B double mutant (T12/35D, phosphorylation mimic replacement of Thr12 and Thr35 with Asp) of full-length forms (lanes labelled *4*) were adopted as bait. Influence of NaCl concentration for pull-down assay was examined by varying NaCl concentrations at 0.1 M (**b**), 0.5 M (**c**), and 1 M (**d**) in washing buffers comprised of 20 mM Tris–HCl (pH 7.5) and 1 mM DTT. Input of protein samples are (**a**)

false-negatives. Additionally, the molecular sizes shown as a result of SDS-PAGE should be sufficiently different to distinguish between bait and prey proteins. Careful design of fusion protein sizes will ensure a clear result.

- Assay buffer composition
  The pH and salt concentration are critical factors for protein–-protein interactions. We examined the effect of salt concentration in the assay buffer for pull-down assays with GST-MOB1 derivatives and the LATS1 NTR domain. Although the interaction comprised electrostatic and hydrophobic effects, it is noteworthy that different NaCl concentrations in the assay buffer influenced the binding affinity between the two proteins (Fig. 2). A protein interaction may sometimes result in precipitation. In such cases, the assay buffer should be reconsidered since precipitants formed will remain in the resin and give rise to false-positives.
- Assay temperature
  The assay is performed at room temperature (*see* **Note 4**).

3.1.2 Procedure

1. Aliquot 50 μl of GST resin bed volume (100 μl of 50% slurry in preservative solution) into 1.5 ml microcentrifuge tubes as needed to accommodate samples and controls.
2. Centrifuge at 500 × *g* for 5 min and discard the supernatants using a micropipette (*see* **Note 5**).
3. Add ten bed volumes of the assay buffer into each tube and tap the tubes gently but thoroughly (*see* **Note 6**). The resin should be completely dispersed in the solution for exchanging the trace preservative solution with the buffer effectively (*see* **Note 7**).
4. Centrifuge at 500 × *g* for 5 min and remove the supernatants using a micropipette.
5. Repeat **steps 3** and **4** another two times for exchanging the reservoir solution with the assay buffer completely (*see* **Note 8**).
6. Prepare a mixture of the bait protein (GST-fused protein) and prey protein and that of GST and prey protein as a negative control in 1.5 ml microcentrifuge tubes. The total reaction volume is 100 μl. Incubate these solutions if required (*see* **Note 9**).
7. Add 90 μl of control or protein sample into the tubes prepared using **steps 1–5**, and tap each tube gently but thoroughly, to facilitate binding of the bait and GST to the resin. Incubation of the tubes for the time being is optional (*see* **Note 10**). The remaining 10 μl of each mixture will represent the "input."
8. Centrifuge at 500 × *g* for 5 min and remove the supernatants, which represent the "flow-through."
9. Add 5 bed volumes of the assay buffer into each tube and tap gently but thoroughly.
10. Centrifuge at 500 × *g* for 5 min and remove the supernatants using a micropipette.
11. Repeat washing **steps 8** and **9** another two times to wash away protein that did not bind to the bait (*see* **Note 11**).
12. Add 75 μl elution buffer into the tubes and tap the tubes to facilitate release of the protein from the resin.
13. Centrifuge at 500 × *g* for 5 min and collect the supernatants, which contain bait and prey protein, using a micropipette.
14. Analyze the results using SDS-PAGE. If you find the prey band which is essentially not found in the GST control, it means the proteins directly interact each other. Furthermore, the relative band intensity of preys implies the affinity between the proteins.

3.2 GST Pull-Down Assay for Screening of the Binding Partner from Crude Extract

This method may be more suitable for screening putative binding protein(s) from cell lysates (*see* **Note 12**).

3.2.1 Immobilization of Bait onto Prepared Resin

1. Aliquot 100 μl of GST resin bed volume (200 μl of 50% slurry in preservative solution) into two 1.5 ml microcentrifuge tubes.
2. Centrifuge at 500 × *g* for 5 min and discard the supernatants using a micropipette (*see* **Note 5**).
3. Add ten bed volumes of the assay buffer into each tube and tap the tubes gently but thoroughly. The resin should be completely dispersed in the solution.
4. Centrifuge at 500 *g* for 5 min and remove the supernatants using a micropipette.
5. Repeat **steps 3** and **4** another two times exchanging the reservoir solution with the assay buffer.
6. Add GST and the bait GST-fused protein into each tube, and tap the tubes gently but thoroughly to facilitate binding to the resin.
7. Centrifuge at 500 × *g* for 5 min and remove the supernatants, containing unbound GST and bait protein, using a micropipette.
8. Add five bed volumes of assay buffer into each tube and tap the tubes gently but thoroughly.
9. Centrifuge at 500 × *g* for 5 min and remove the supernatants using a micropipette.

3.2.2 Pull-Down Screening

1. Add cell lysate which includes the putative endogenous or bacterial expressed exogenous prey into each tube prepared above using the immobilization steps, and shake the tubes gently for 1 h at 4 °C (*see* **Note 13**).
2. Centrifuge at 500 × *g* for 5 min and remove the supernatants using a micropipette (*see* **Note 14**).
3. Add 5 bed volumes of the assay buffer into each tube and tap gently but thoroughly.
4. Centrifuge at 500 × *g* for 5 min and remove the supernatants using a micropipette.
5. Repeat washing **steps 4** and **5** another two times to wash away proteins that did not bind to the bait (*see* **Note 11**).
6. Add 75 μl the elution buffer into the tubes, and tap the tubes to facilitate removal of the protein from the resin. Incubation is not required.

7. Centrifuge at 500 × *g* for 5 min and collect the supernatants, which contain bait and prey protein, using a micropipette.
8. Analyze the results using SDS-PAGE. If you find the prey band which is essentially not found in the GST control, it means the proteins directly interact each other (*see* **Note 15**).

4 Notes

1. PBS or TBS may be suitable for the first trial. Additionally, reducing agent and nonionic detergent can be used as necessary.
2. Glutathione reduced form will change pH of the buffer. The final pH should be adjusted to 7.4 with 1 N NaOH solution.
3. Alternatively, we recommend to purify the protein with another purification step(s), ion-exchange chromatography, and/or gel filtration chromatography, which will make the protein higher purity and remove the glutathione by itself.
4. If your protein is unstable at room temperature, the assay at 4 °C or in an ice bath is strongly recommended.
5. The centrifugal force employed should follow that stipulated by the resin manufacturer's instructions.
6. Vortexing is not recommended for the fragile agarose-based resin.
7. Ethanol from the preservative solution may have a possibility of causing protein precipitation or aggregation. In this step, trace ethanol should be removed completely.
8. This step is also required for equilibration of the resin with assay buffer.
9. Certain types of protein interactions require a relatively long incubation time to generate a stable complex, such as 30 min or a couple of hours.
10. The incubation in this step is not required as long as PBS, TBS, or ordinary buffers are used for the reaction buffer. However, if the reaction buffer contains the material which may effect on binding of GST to GST-conjugated resin, the incubation time should be extended.
11. The final buffer wash flow-through solution of GST control should have no absorbance at 280 nm (OD280) when all unbound protein has been removed from the resin. If the OD280 > 0, then simply repeat the washing steps.
12. It should be noted that this method gives rise to diffusion effects and a potent competitive effect derived from endogenous proteins that engage in partial non-specific bait

interactions from host cells. As a result, it is more difficult to detect the binding compared to using an assay with purified proteins. Consideration should also be given to potential non-specific interactions that yield false-positives.

13. To protect proteins from endogenous protease activity, supplement of protease inhibitor cocktail in the lysis buffer is recommended.
14. If the amount of putative prey protein in the cell lysate is very low, i.e., low expression level or endogenous protein for prey, consider repeating **steps 1** and **2** for more clear result.
15. Comparing with the assay with purified proteins, minor protein bands derived from non-specific interaction to the resin will be seen in the SDS-PAGE. Careful reconsideration of lysis buffer and reaction buffer will reduce the bands.

References

1. Fields S, Song OK (1989) A novel genetic system to detect protein–protein interactions. Nature 340:245–246
2. Kim S-Y, Tachioka Y, Mori T, Hakoshima T (2016) Structural basis for autoinhibition and its relief of MOB1 in the Hippo pathway. Sci Rep 6:28488
3. Kohler RS, Schmitz D, Cornils H, Hemmings BA, Hergovich A (2010) Differential NDR/LATS interactions with the human MOB family reveal a negative role for human MOB2 in the regulation of human NDR kinases. Mol Cell Biol 30:4507–4520
4. Ni L, Zheng Y, Hara M, Pan D, Luo X (2015) Structural basis for Mob1-dependent activation of the core Mst-Lats kinase cascade in Hippo signaling. Genes Dev 29:1416–1431
5. Smith DB, Johnson KS (1988) Single-step purification of polypeptides expressed in Escherichia coli as fusions with glutathione S-transferase. Gene 67:31–40

Chapter 21

Determining the Phosphorylation Status of Hippo Components YAP and TAZ Using Phos-tag

Rui Chen, Steven W. Plouffe, and Kun-Liang Guan

Abstract

Protein phosphorylation is one of the most important posttranslational modifications in cell signaling and regulation. Protein phosphorylation can be detected by site-specific phospho-antibodies, but generating these antibodies can be costly, time-consuming, and difficult. Recently, Phos-tag technology has been developed to detect protein phosphorylation. Here, we describe our method for using Phos-tag gels to compare the phosphorylation status of YAP and TAZ, the most important downstream effectors of the Hippo pathway.

Key words Hippo pathway, Phos-tag gel, Phosphorylation, YAP, TAZ

1 Introduction

The Hippo pathway is a highly conserved signaling pathway which controls cell growth, tissue homeostasis, organ size, and tumorigenesis [1]. Core components of Hippo pathway include mammalian Ste20-like kinases 1/2 (MST1/2) and the mitogen-activated protein kinase kinase kinase kinase (MAP4Ks), which phosphorylate and activate the large tumor suppressor 1/2 kinases (LATS1/2). Activated LATS1/2 phosphorylate the downstream effectors of the Hippo pathway, Yes-associated protein (YAP), and transcriptional coactivator with PDZ-binding motif (TAZ), to induce YAP/TAZ cytoplasmic localization and inactivation of their transcriptional activity. The phosphorylation status of YAP/TAZ is commonly used as an indicator for Hippo pathway activity. Researchers often use phospho-specific antibodies to detect YAP/TAZ phosphorylation; however, this data is not quantitative, and different batches of commercial antibodies may vary significantly.

Phos-tag technology was developed to detect protein phosphorylation [2]. It can slow the migration of proteins in a phosphorylation-dependent manner [3, 4]. The phosphate-affinity site is a polyacrylamide-bound di-nuclear Mn^{2+} complex (Mn^{2+}-

Alexander Hergovich (ed.), *The Hippo Pathway: Methods and Protocols*, Methods in Molecular Biology, vol. 1893,
https://doi.org/10.1007/978-1-4939-8910-2_21,

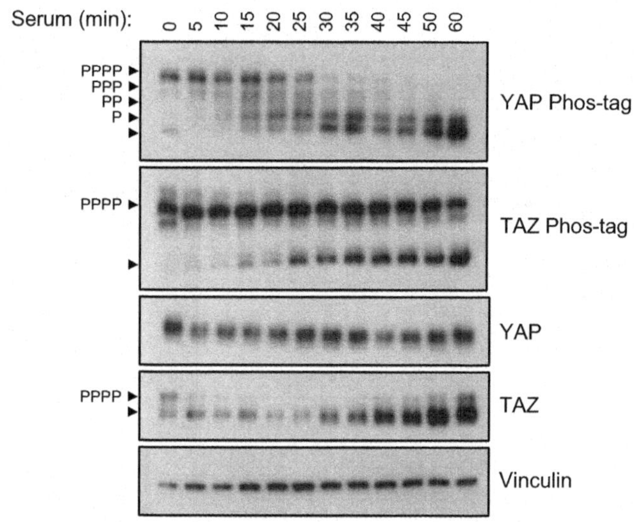

Fig. 1 Serum stimulation induces YAP and TAZ dephosphorylation. HEK293A cells were serum starved for 24 h and then stimulated with 10% serum for indicated times. YAP phosphorylation status is detected using Phos-tag gel, and TAZ phosphorylation status is detected using Phos-tag and regular gel

Phos-tag) that interacts with phosphate, thus decreasing the mobility of phosphorylated proteins in an otherwise regular polyacrylamide gel. A huge advantage of the Phos-tag gel is that it provides quantitative information regarding the phosphorylation level [5]. Another advantage is that it does not require any additional information regarding phosphorylation sites or specific phospho-antibodies. However, one limitation is that Phos-tag cannot provide information regarding which specific phosphorylation sites are phosphorylated, unless the protein in question only has one phosphorylation site. Here, we describe the Phos-tag gel method we routinely use to detect and compare the phosphorylation status of YAP and TAZ. Examples are provided in Fig. 1 and 2.

2 Materials

Prepare all solutions using ultrapure water and analytical grade reagents. Prepare and store all reagents at room temperature (unless indicated otherwise).

2.1 Phos-tag SDS Polyacrylamide Gel

1. Separating gel buffer: 1.5 M Tris–HCl, pH 8.8. Weigh 181.8 g Tris–HCl and transfer to a clean glass beaker. Add water to a volume of 800 ml. Mix and adjust the pH with HCl to pH 8.8. Bring up to 1 L with water. Store at 4 °C (*see* **Note 1**).
2. Stacking gel buffer: 1 M Tris–HCl, pH 6.8. Weigh 121.2 g Tris–HCl and prepare a 1 L solution as in the previous step. Store at 4 °C.

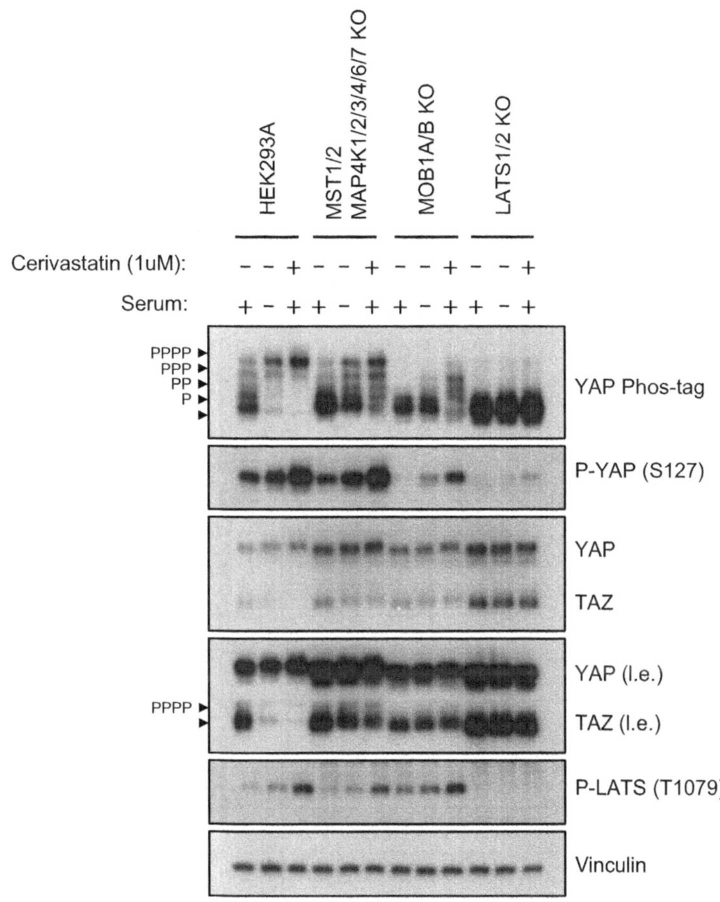

Fig. 2 Serum starvation and cerivastatin (1 μM) treatment both induce Hippo pathway activation and YAP/TAZ phosphorylation. HEK293A wild-type or various knockout cells were serum starved or treated with cerivastatin for 24 h as indicated. Phosphorylation status as detected by Phos-tag gel is also consistent with phosphorylation status detected using YAP phosphor-specific antibodies. When core components of the Hippo pathway are deleted, YAP/TAZ phosphorylation is severely compromised, as observed by Phos-tag. Essentially, all YAP phosphorylation depends on LATS1/2 under these experimental settings

3. 40% acrylamide (e.g., Fisher Scientific; # BP14081) and store at 4 °C.
4. Phos-tag solution: Dissolve 10 mg Phos-tag (purchased from Wako Pure Chemicals Industrials, Ltd) in 100 μl methanol; shake, and then add 3.2 ml H_2O. The stock solution tube should be wrapped with aluminum foil because the Phos-tag may be light-sensitive.
5. 100 mM $MnCl_2$ buffer: Dissolve 198 mg $MnCl_2{\cdot}4\ H_2O$ in 10 ml water (*see* **Note 2**).
6. 4× SDS-PAGE sample buffer: 0.2 M Tris–HCl, pH 6.8, 8% SDS, 0.1% bromophenol blue, 40% glycerol, 20%

β-mercaptoethanol. Leave one aliquot at 4 °C for current use, and store the remaining aliquots at −20 °C (*see* **Note 3**).

7. 10× SDS running buffer: 0.25 M Tris–HCl, pH 8.0, 2 M glycine, 1% SDS.

2.2 Immunoblotting

1. Polyvinylidene difluoride (PVDF) membrane.
2. 1× transfer buffer: 0.025 M Tris–HCl pH 8.0, 0.2 M glycine, 20% methanol. Store at 4 °C (*see* **Note 4**).
3. 10× TBS: 1.7 M NaCl, 20 mM Tris–HCl pH 8.0.
4. 1× TBST: Dilute 100 ml 10× TBS in 800 ml water, and add Tween-20 to 0.1% final concentration and bring up with water to 1 L.
5. Block buffer: 5% milk in TBST (*see* **Note 5**).
6. Antibody dilution solution: 5% BSA in TBST (*see* **Note 5**). Store at −20 °C.
7. Mini-PROTEAN® glass plate size (W × L), short plate 10.1 × 7.3 cm, spacer plate 10.1 × 8.2 cm.
8. Plastic combs (1.0 mm thickness).
9. Medium binder clips (1¼ in.).
10. Chromatography paper (e.g., Fisherbrand® #05-714-4).
11. Mini Trans-Blot® Cell system.

3 Methods

Carry out all procedures at room temperature unless otherwise specified.

Phos-tag SDS gel running and immunoblotting of YAP or TAZ proteins

1. Cell treatment, cell lysate preparation: Seed cells in 12-well plates. When cells reach 70–80% confluence, treat cells with desired stimulation or compound. Remove medium as completely as possible (*see* **Note 6**). Harvest cells by adding 150 μl 1× SDS sample buffer (*see* **Note 6**) per well, and shake the plate using a plate shaker. Collect each sample and heat at 95 °C for 5–10 min, vortex, and centrifuge. Samples can be stored at −20 °C for 1 year. Vortex and centrifuge each sample before loading 12 μl of sample per well onto Phos-tag SDS gel. Make sure that the sample does not contain EDTA or EGTA.
2. Separating gels: Prepare a 7.5% Phos-tag separating gel, 10 ml for two gels. Add TEMED last. Once mixed, pour the separating gel and then gently add isopropanol to the top of the gel to ensure the top of the gel is level. Wait 20–30 min for the gel to polymerize. If you have problems with the gel leaking, try the

Table 1
Formula for the preparation of 7.5% Phos-tag gels

	Separating gel	Stacking gel
H_2O	5.4 ml	3.2 ml
40% acrylamide	1.9 ml	0.49 ml
1.5 M Tris (pH 8.8)	2.5 ml	
1 M Tris (pH 6.8)		1.25 ml
10% SDS	100 μl	50 μl
Phos-tag	30 μl	
100 mM $MnCl_2$	5 μl	
10% APS	100 μl	50 μl
TEMED	10 μl	15 μl

following: before adding TEMED, take 200 μl of the separating gel and put into a microfuge tube, add 2 μl TEMED, mix, and add the mixture to the bottom of glass plate quickly. Try to avoid any bubbles. This will seal the bottom of the glass plates. Then you may proceed with the above procedure.

3. Stacking gel: Preparing the stacking gel, 5 ml for two gels. After pouring, insert the comb (15 well, 1 mm) gently to avoid producing bubbles (*see* **Note 7**) (Table 1).
4. Loading sample: Aspirate any remaining gel in the wells before loading the samples. Load 12 μl of sample per well (*see* **Note 8**).
5. Running: Start running at 60–70 V for 30 min, until all the samples enter the Phos-tag separating gel. Then run at 110–120 V until the 55 kDa marker is around 2 cm from the bottom of the gel (YAP protein is above the 70 kDa marker). This takes about 3 h (*see* **Note 9**).
6. Transferring: The PVDF membrane should be soaked in methanol for about 5 min (*see* **Note 10**). Assemble the transferring system, and transfer for 1.5–2 h in the cold room. Set constant current (around 400 mA) (*see* **Note 11**).
7. Wash the membrane one time with TBST.
8. Block with 5% milk in TBST for 1 h at room temperature (*see* **Note 12**).
9. Dilute the primary antibody (dilute according to the antibody manufacturer's instructions) in 5% BSA in TBST. Incubate overnight at 4 °C (*see* **Note 13**).
10. Wash 4× with TBST for 10 min each (*see* **Note 14**).
11. Incubate with secondary antibody (diluted in 5% milk) for 1 h at room temperature.

12. Wash 4× with TBST for 5 min each.
13. Prepare a working solution of the Millipore Corporation Immobilon Western Chemiluminescent HRP Substrate (#WBKLS0500) by mixing equal parts of luminol reagent and peroxide solution. Incubate the membrane in the working solution for 1 min at RT. Remove the membrane from the working solution, and place it in a plastic page protector or plastic wrap. Remove any excess liquid or bubbles with an absorbent tissue, and then expose the protected membrane to the X-ray film. The exposure time may vary. A good starting point is an exposure time of 1 min.

4 Notes

1. The separating gel buffer can be stored at 4 °C for 1 year. The pH of the separating gel buffer will increase slowly over time, but this has no obvious effect on gel quality.
2. It is better to prepare 1 M $MnCl_2$ buffer and then dilute into 100 mM $MnCl_2$ buffer.
3. The 4× SDS-PAGE running buffer should not contain any EDTA or EGTA. EDTA or EGTA binds with the Mn^{2+} and may affect the protein shift by Phos-tag.
4. Prepare 1× transfer buffer. Make 10× transfer buffer: 0.25 M Tris–HCl pH 8.0, 2 M glycine, then dilute 2 L of 10× native buffer to 14 L water, and add 2 L of methanol. Bring up to 20 L with water. Store at 4 °C.
5. It is better to prepare fresh blocking buffer and antibody diluent solution, because the blocking buffer and antibody dilution solution contain milk and BSA, which supports microbe growth.
6. There is no need to add protease and phosphatase inhibitors to the sample buffer.
7. Regular gels (non-Phos-tag gels) can be usually stored at 4 °C for 2–3 weeks. However, Phos-tag gels can only be stored at 4 °C for 2–4 days. It is suggested to use freshly made Phos-tag gels.
8. Dilute the protein ladder (catalog number: 26616. Thermo Fisher Scientific) using 1× sample buffer (250 μl protein ladder diluted in 1 ml 1× sample buffer (should be EDTA-free), and add loading buffer to any empty wells so the lanes run evenly.
9. The selection of acrylamide percentage of the Phos-tag and the time of the running depend on the molecular weight of the protein of interest. For example, the molecular weight of YAP is about 70 kDa, so the 7.5% Phos-tag gel is suitable for

separating the phosphorylated form of YAP. If the molecular weight of the protein of interest is about 100–140 kDa, the 6–6.8% Phos-tag gel should be selected. For proteins with a molecular weight more than 150 kDa, it is hard to separate using Phos-tag gels.

10. After the PVDF membrane is soaked in methanol for about 5 min, rinse the membrane with water 2–3 times, and then soak in 1× transfer buffer before transferring.
11. The time and current of transferring depend on the molecular weight of the protein. It takes about 90 min to transfer the YAP protein with 400 mA constant current after the running in a Phos-tag gel.
12. The time of blocking can be 20–60 min.
13. The primary antibody diluted in antibody dilution solution can be repeatedly used and stored at −20 °C. It is recommended to use the YAP (D8H1X; #14074) and TAZ (V386; #4883) antibodies from Cell Signaling Technology.
14. Sometimes, washing the membrane three times with TBST for 15 min each is sufficient. However, if the background of the primary antibody is not clean, the washing times should be increased.

Acknowledgments

This work was supported by grants from the National Institutes of Health (CA196878, CA217642, and GM51586) to K.-L.G. S.W.P. is supported by the UCSD Graduate Training Program in Cellular and Molecular Pharmacology (T32 GM007752). K.-L.G. is a co-founder and has an equity interest in Vivace Therapeutics, Inc. The terms of this arrangement have been reviewed and approved by the University of California, San Diego, in accordance with its conflict of interest policies.

References

1. Yu FX, Zhao B, Guan KL (2015) Hippo pathway in organ size control, tissue homeostasis, and Cancer. Cell 163(4):811–828
2. Zhao B et al (2010) A coordinated phosphorylation by Lats and CK1 regulates YAP stability through SCF beta-TRCP. Genes Dev 24 (1):72–85
3. Kinoshita E, Kinoshita-Kikuta E, Koike T (2009) Separation and detection of large phosphoproteins using Phos-tag SDS-PAGE. Nat Protoc 4(10):1513–1521
4. Kinoshita-Kikuta E, Kinoshita E, Koike T (2009) Phos-tag beads as an immunoblotting enhancer for selective detection of phosphoproteins in cell lysates. Anal Biochem 389(1):83–85
5. Plouffe SW et al (2016) Characterization of hippo pathway components by gene inactivation. Mol Cell 64(5):993–1008

Chapter 22

Quantifying the Kinase Activities of MST1/2

Niamh A. O'Driscoll and David Matallanas

Abstract

The functions of the kinases MST1 and MST2 rely heavily on their ability to phosphorylate and become phosphorylated themselves. Hence, it is important to precisely measure the kinase activities of both isoforms in a reproducible manner. Here, we describe in detail the protocol for an in-gel kinase assay for the quantification of the kinase activity of MST1/2, which involves immunoprecipitation of MST1/2 and the incorporation of radiolabeled phosphate from [γ-^{32}P]-ATP into a substrate immobilized in a polyacrylamide gel. We also include a protocol for indirect measurement of MST1/2 activation status using immunoblotting.

Key words MST1, MST2, Kinase assay, Immunoprecipitation, Radiolabel

1 Introduction

MST1 (also known as KRS2 or STK4) and MST2 (KRS1 or STK3) are serine/threonine kinase members of the STE20 family [1]. MST1 and MST2 present a 77% sequence homology and both contain N-terminal kinase domains and a C-terminal SARAH domain that enable them to interact with Salvador and the RASSF family, in addition to being able to homo- and heterodimerize with each other [2]. Previous work has shown that full-length MST1 and MST2 are activated by homodimerization and autophosphorylation on Threonine 183/180, respectively [1, 3, 4]. These kinases are central nodes of the mammalian MST1/2-Hippo pathway, and several substrates have been identified to date including LATS1/2, AKT, FOXO, and H2B [5]. Although these proteins seem to have a certain level of functional redundancy, they also mediate specific functions [1, 5]. Importantly these kinases are also cleaved by caspases during apoptosis, and the kinase domain becomes an activated kinase that phosphorylates different residues resulting in the activation of cell death [1].

Protein phosphorylation is a key posttranslational modification (PTM) that acts as a molecular switch for many regulatory events in

Alexander Hergovich (ed.), *The Hippo Pathway: Methods and Protocols*, Methods in Molecular Biology, vol. 1893, https://doi.org/10.1007/978-1-4939-8910-2_22,

signal transduction networks, such as proliferation, division, differentiation, and apoptosis, and it is often deregulated in many pathological conditions [6]. Protein kinases catalyze the attachment of phosphates to proteins and therefore play a key role in the regulation of biological functions. Hence, the precise measurement and determination of kinase activity are central in the characterization of the roles of these proteins and the understanding of life itself. Different experimental approaches are used to monitor kinase activity both in direct and indirect manners. Direct measurement of kinase activity is commonly done by using radiolabeled [γ-^{32}P]-ATP and measuring the incorporation of radioactive ^{32}P to appropriate substrates for the specific kinase being studied [7, 8], and this still remains the golden standard for quantification of kinase activities [8]. Chernoff's group developed an in-gel kinase assay for the measurement of MST1/2 kinase activity [9] that, with different adaptations, has been utilized by our group to characterize the central role of these kinases in the mammalian MST1/2-Hippo pathway [10–13]. This is a simple protocol for the detection and quantification of the kinase activity of MST1/2. The method monitors immunoprecipitated (IP) endogenous or exogenously expressed MST1/2 direct kinase activity and can be used to detect the specific activation of full-length (FL) MST1 or MST2 and the activated N-terminal kinase cleaved domains. This assay involves the incorporation of the kinase substrate into an SDS-PAGE gel and the samples being separated by electrophoresis. The SDS can then be removed, and a number of steps are performed, such as denaturation, renaturation, and incubation of the gel with [γ32P]-ATP. Autoradiography is subsequently performed in order to evaluate the phosphorylation activity of the kinases on the substrate that is present in the gel [14]. Indirect determination of kinase activation is often used and is most commonly performed by immunoblotting using specific antibodies that recognize activating phosphorylations in the kinases or incorporation of phosphates in known substrates of these enzymes. At the end of the chapter, we include a simple protocol for the detections of the autophosphorylation of MST1/2 as an indirect readout for the activation of these kinases using specific antibodies.

2 Materials

Prepare all solutions using ultrapure water and analytical grade reagents. Prepare and store all reagents and buffers at room temperature, unless otherwise stated. Follow all standardized waste disposal regulations when discarding waste materials.

2.1 Preparation of IP Samples/Cell Lysates

1. Cells/samples 100 cm plate per condition. The cells can be transfected, untreated, or treated as per experiment (*see* **Note 1**).
2. Phosphate-buffered saline (PBS): 137 mM NaCl, 2.7 mM KCl, 10 mM Na_2HPO_4, and KH_2PO4 1.8 mM.
3. Standard lysis buffer: 20 mM HEPES pH 7.5, 150 mM NaCl, and 1% NP40, supplemented with the following protease and phosphatase inhibitors 10 mM sodium fluoride, 1 mM sodium orthovanadate, 2 mM sodium pyrophosphate, 10 μg/mL leupeptin, 10 μg/mL aprotinin, and 10 mM β-galactosidase.
4. Cell scraper.
5. Refrigerated centrifuge cooled to 4 °C.
6. Protein G beads for endogenous IP, FLAG-beads, GFP-Trap-A beads for exogenous IP (*see* **Note 2**).
7. Antibodies for IP: MST1, anti-STK4, and MST2, Krs-1 (*see* **Note 3**).
8. Antibodies for Western blot (WB): MST1, anti-STK4, and MST2, anti-STK3, anti-Krs1, anti-phospho-MST1/2 threonine 183/threonine 180, anti-GAPDH (*see* **Note 4**).
9. Secondary antibodies: Rabbit IgG, HRP-linked (*see* **Note 5**).
10. NuPAGE LDS sample buffer supplemented with 50 mM dithiothreitol (DTT) (*see* **Note 6**).
11. Eppendorf tube orbital rotator.

2.2 SDS-Polyacrylamide Gel

1. Samples prepared as explained in Subheading 3.1, **step 1** per normal and denatured.
2. SDS-PAGE mini gel casting system (*see* **Note 26**).
3. Medium binder clips.
4. Ammonium persulfate (APS): 10% (w/v) solution in water (*see* **Note 7**).
5. Resolving gel buffer (10%): 3.3 mL (30%) acrylamide, 5 mL water, 1.66 mL Tris–HCl (2 M, pH 8.9), and 56 μL EDTA (400 mM). To polymerize: 66 μL of 10% APS and 14 μL of TEMED.
6. Stacking gel buffer (100 mL): 15 mL (30%) acrylamide, 76.75 mL water, 6.25 mL Tris–HCl (2 M, pH 6.8), 1 mL 10% SDS, 1 mL EDTA (400 mM). To polymerize 4 mL of stacking gel buffer: 30 μL APS and 10 μL TEMED.
7. Precision Color Protein Standard.
8. 10× running buffer: 250 mM Tris–HCl, 1.92 M glycine, and 1% (w/v) SDS.
9. 1× running buffer: 200 mL of 10× running buffer and 1800 mL water.
10. Electrophoresis power supply.

2.3 In-Gel Kinase Assay

1. MBP-containing gel resolving buffer: Use the same recipe shown in Subheading 2.2, **item 5** and incorporate myelin basic protein (MBP) into the gel by adding the reconstituted MBP to the resolving buffer to a final concentration of 0.5 mg/mL and mix thoroughly (*see* **Note 8**).
2. MBP stock (Sigma-Aldrich): Resuspend the lyophilized protein in water to a final concentration of 12.5 mg/mL.
3. Samples prepared as in Subheading 3.1, **step 1**.
4. Buffer to remove SDS: 20% isopropanol in 50 mM Tris–HCl pH 8.0 (*see* **Note 9**).
5. Washing buffer: 50 mM Tris–HCl pH 8.0, 5 mM β-mercaptoethanol (*see* **Note 9**).
6. Denaturing buffer: 6 M guanidine HCl, 50 mM Tris–HCl pH 8.0, 5 mM β-mercaptoethanol. Store at 4 °C (*see* **Note 9**).
7. Renaturing buffer: 50 mM Tris–HCl pH 8.0, 5 mM β-mercaptoethanol containing 0.04% Tween 20. Store at 4 °C (*see* **Note 10**).
8. Re-equilibrium buffer: 40 mM HEPES pH 8.0 and 10 mM magnesium chloride and 2 mM DTT.
9. Phosphorylation reaction buffer: 40 mM HEPES pH 8.0, 10 mM $MgCl_2$, 0.5 mM EGTA, 50 μM ATP, and 25 μCi [γ-^{32}P]-ATP (*see* **Note 11**).
10. Stop reaction wash buffer: 5% trichloroacetic acid (TCA) and 1% sodium pyrophosphate.
11. Methacrylate box (*see* **Note 12**).
12. Bio-Rad gel dryer.
13. 3MM Whatman paper.
14. X-ray film (*see* **Note 13**).
15. X-ray film cassette with intensifying screen (*see* **Note 13**).

2.4 Immunoblotting

1. PVDF membrane (*see* **Note 14**).
2. 3MM Whatman paper.
3. Western blot semidry transfer buffer: 192 mM glycine, 25 mM Tris-base, 20% (v/v) methanol, 2 mL 10% SDS and made up to 2 L with water.
4. 100% methanol (*see* **Note 15**).
5. 10× Tris-buffered saline (TBS): 1.5 M NaCl, 0.1 M Tris–HCl, pH 7.4.
6. 1× TBS containing 0.05% Tween 20 (TBST).
7. Blocking solution and antibody diluent solution: 5% milk in TBST. Store at 4 °C.
8. Electrophoresis tank.

9. Semidry transfer cell (*see* **Note 16**).
10. Benchtop orbital shaker.
11. Enhanced chemiluminescence (ECL) substrate solution (*see* **Note 17**).
12. Medical X-ray films.

3 Methods

Carry out all procedures at room temperature unless otherwise specified.

3.1 In-Gel Kinase Assay

3.1.1 Sample Preparation

1. Use one 10 cm plate per sample. Treatments should be performed prior to sample preparation. Prepare lysis buffer as described above (Subheading 2.1, **item 3**), and keep on ice (*see* **Note 18**).
2. Remove cell culture media by aspiration and wash each plate gently with 10 mL of ice cold PBS.
3. Add 1 mL of lysis buffer to each plate. Keep all plates on ice.
4. Scrape the cells using a cell scraper and transfer all contents to a pre-chilled 1.5 mL Eppendorf tubes.
5. Vortex each tube for 5 s, and spin down the tubes in a pre-cooled 4 °C centrifuge for 10 min at 18,000 × *g* to remove cellular debris.
6. Take off the supernatant and transfer to a labeled, fresh pre-chilled Eppendorf tube, and discard the pellets (*see* **Note 19**).
7. Transfer 50 μL to a clean tube and keep as total lysate samples at −20 °C.
8. Prepare tubes for the IP. Add 10 μL of protein G beads per tube, 2 μL of antibody (anti-MST1 or anti-MST2), and 450 μL cell lysate (*see* **Note 20**).
9. Place the tubes on an orbital wheel at 4 °C. Rotate tubes at 1 × *g* for 2 h.
10. Collect the tubes and spin them down in a centrifuge at 4 °C at 3300 × *g* for 30 s and discard the supernatant (*see* **Note 21**).
11. Wash the beads three times with 0.5 mL standard lysis buffer. Each wash consists of resuspending the beads in lysis buffer and then spinning down the beads (3300 × *g* for 30 s), followed by supernatant removal and addition of fresh lysis buffer (*see* **Note 22**).
12. After the last wash, discard supernatant and keep the bead pellets. Release the bead bound IP by adding 20 μL of 2× NuPAGE LDS sample buffer. Denature the samples by boiling them at 100 °C for 5 min in a dry thermo-block.

3.1.2 Kinase Assay (See Note 23)

1. Prepare 5 mL of the resolving gel in a 15 mL plastic tube by mixing 1.66 mL (30%) acrylamide, 2.5 mL water, 0.83 mL Tris-HCl (2 M, pH 8.9), and 28 μL EDTA (400 mM) and 250 μL of MBP (12.5 mg/mL). To polymerize quickly add 33 μL of 10% (APS) and 7 μL of TEMED and pour this into pre-assembled gel cassette. Allow space at the top (approximately one third) for the stacking gel. Immediately after, overlay the gel with isopropanol and allow the gel to fully polymerize for at least 15–20 min (*see* **Note 24**).
2. Once the separation phase of the gel is visible, remove the isopropanol overlaying the polymerized gel by inverting the gel and rinsing with water and removing any excess water using 3MM Whatman paper. Prepare the stacking gel by taking 4 mL of the stacking gel stock solution prepared in Subheading 2.2, **item 6**. In order to polymerize this solution, add 30 μL 10% APS and 10 μL TEMED and immediately pour on top of the resolving gel. Add the specific Bio-Rad comb gently and allow them to polymerize for 15 min (*see* **Note 25**).
3. Remove the combs gently. Assemble the gel in the Bio-Rad electrophoresis tank, and ensure there are no leakages by pouring the 1× running buffer in between the two gels (*see* **Note 26**).
4. Top up the tank with 1× running buffer to the designated line on the side of the tanks.
5. Load previously prepared samples alongside a molecular weight marker (precision plus dual color protein standard) that can be loaded onto the gel using gel loading tips.
6. Run the gel at 100 V for the entire course (*see* **Note 27**). Ensure the bands are well separated especially in the 50–75 kDa range.
7. Transfer the gels into the container (20 cm × 20 cm × 5 cm with sealing lid to avoid spills and any smell escaping) that is used for the following steps of the kinase assay (*see* **Note 28**).
8. Remove SDS from the gels by performing 3 × 20 min washes with 40 mL of removed SDS buffer rocking at room temperature in buffer as described above (Subheading 2.3, **item 4**). All incubations are done rocking at low rpm to avoid damage to the gels and with 40 mL of buffer.
9. Once SDS is removed, wash 3 × 20 min with wash buffer, rocking at room temperature in 40 mL washing buffer (Subheading 2.3, **item 5**) (*see* **Note 23**).
10. Place the gel in denaturing buffer (Subheading 2.3, **item 6**), for 1 h, rocking at room temperature.
11. Incubate the gel in 50 mL of renaturing buffer (Subheading 2.3, **item 7**) 2 × 45 min, rocking at 4 °C, followed by an

overnight incubation with 50 mL of renaturing buffer at 4 °C (*see* **Note 29**).

12. The following morning, place the gel in 50 mL of fresh renaturing buffer for a further 1 h incubation period at 4 °C.
13. Re-equilibrate the gel by placing it in 40 mL re-equilibrating buffer (Subheading 2.3, **item 8**), for 45 min, rocking at room temperature.
14. Remove as much of the re-equilibrating buffer as possible before the phosphorylation reaction is carried out (*see* **Note 31**).
15. Place the container with the gels within a methacrylate box (min of 0.5 cm thick) to prevent radiation exposure of the operator during the next steps, and keep it behind a protecting screen. All the handling of radioactive material during the reaction has to be done behind the protecting screen with the operator wearing protective gloves and a lab coat. A Geiger counter should be used to monitor the surface and operator for possible contamination after each step. The radioactive ATP must be kept at 4 °C in the container provided by the manufacturer.
16. Prepare reaction buffer behind the methacrylate protection screen (1 cm thickness is recommended). Use the minimum volume necessary to cover the gel (*see* **Note 32**).
17. Add 10 mL of phosphorylation reaction buffer per container to incubate the gels with the solution. If more than one gel is being assayed, we recommend placing one gel per container or, if necessary, using a larger container for more than one gel in order to make sure that the gels do not stack up on top of each other and allow it to rotate in the methacrylate box at room temperature for 2 h (*see* **Note 33**).
18. Discard the tube and pipette tips into a designated container for radiolabeled ^{32}P according to national/institutional guidance.
19. Remove the incubation buffer by carefully pouring the liquid into a 50 mL plastic tube; close the lid and discard in the ^{32}P waste (*see* **Note 34**).
20. Monitor the incorporation of [γ-^{32}P]-ATP into the gel using a Geiger counter with a probe adequate for the detection of γ-radiation. There are no set times for the next steps in the protocol, and all times are based on our experience. The washing of the gels is done in order to get defined bands and reduce background by getting rid of the [γ-^{32}P] ATP that has not been incorporated into the MBP present in the gel. Initially, the gel and the container should give a very strong reading that will decrease when you complete the washes. The container will

eventually not give much reading, and the gel will go from a diffused reading to a more localized reading at the molecular level of MST1/2 or the cleaved kinase domain (~55 kDa, ~37 kDa, respectively) (*see* **Note 35**).

21. Stop the reaction by washing the gel in stop reaction buffer (Subheading 2.3, **item 10**), typically 5–10× using 50 mL of stop reaction wash buffer each time, over a 3-h period or until counts drop to background low levels in most of the gel and the main signal comes from the bands where the kinases are localized. Collect the liquid into 50 mL plastic tubes and discard with radioactive material.
22. Dry the gel onto 3MM Whatman paper using Bio-Rad dryer. Place the gel carefully on the paper and cover with cling film to avoid contamination of the machine. Vacuum seal and dry at 80 °C for 2 h (*see* **Note 36**).
23. After the gel is dry, the reading from the background should be even lower and the area where the kinase is should be easier to identify. Estimate the counts to see how long the exposition has to be (*see* **Note 37**).
24. Perform autoradiography: Place the gels into a film cassette so that the gel is facing the screen, and place the X-ray film between the gel and the screen (*see* **Note 38**). Expose depending on signal (*see* **Note 39**).
25. Bands can be quantified using ImageJ and if you have a phosphor-imager quantify using specific program.

3.2 Immunoblotting

In order to make sure that any difference observed in the kinase activity of the different samples is not due to different amounts of the kinases being IP or expressed in the samples, it is also recommended to monitor MST1/2 levels in the total lysates and the IP by performing WB. The blotting with anti-phospho MST1/2 can be used as a readout for the activation (*see* **Note 4**).

1. The gels are prepared and run as described in Subheading 2.2, **item 5** without adding MBP to the resolving gel solution. The voltage may be increased to 120 V.
2. Immediately after the dye front runs out of the gels, stop the electrophoresis. Carefully remove the gels from their casts by immersing the gel in semidry transfer buffer, and equilibrate the gel rocking for 10 min (*see* Subheading 2.4, **item 3**).
3. Activate the PVDF membrane by dipping it in methanol prior to transfer until the membrane is translucent (~30 s) (*see* **Note 15**).
4. Prepare a "transfer sandwich" by immersing three pieces of 3MM Whatman paper in semidry transfer buffer, placing the methanol activated membrane on top, followed by the gel and

then another three pieces of 3MM Whatman paper that have also been previously soaked in semidry buffer (*see* **Note 40**).

5. The gels are transferred to the activated PVDF membrane using a semidry transfer cell. Time and voltage for transfer are calculated based on the number and size of membranes, using the formula 5.5 mA/cm^2 for one gel and 3 mA/cm^2 for more than one gel, as recommended by Bio-Rad (*see* **Note 41**).
6. Following transfer of the proteins to the membrane, the membrane is left to air-dry at room temperature for 15 min before being reactivated by submerging it in 100% methanol (*see* **Note 42**).
7. Rinse membrane in distilled water, followed by TBST.
8. Block the membranes in 5% milk, for 1 h on a shaker at room temperature (*see* **Note 43**).
9. Primary antibodies are prepared by diluting the antibody in 5% milk (Stk3 and MST1 (1/1000), GAPDH (1/10,000), and pMST1/2 (1/500)). These solutions can be used numerous times and are stored at −20 °C (*see* **Note 44**).
10. Membranes incubated as a whole or cut corresponding to the size of the protein of interest are incubated in primary antibody overnight on a shaker at 4 °C (*see* **Note 45**).
11. Remove the primary antibody the following day. The primary antibodies are poured off into their relevant tubes (stored at −20 °C).
12. Wash the membranes 3 × 10 min in TBST at room temperature shaking.
13. Prepare secondary antibodies by diluting rabbit IgG HRP-linked (1/5000 in TBST for pMST1/2 or 1/10,000 5% milk for MST1/2 and GAPDH) (*see* **Note 46**).
14. Incubate membranes in secondary antibody for 1 h at room temperature.
15. Remove secondary antibody solutions, and wash the membranes 3 × 10 min in TBST at room temperature.
16. Remove TBST and incubate with ECL peroxidase substrate (Subheading 2.4, **item 10**) by covering the membranes with the solution for 2 min (*see* **Note 47**).
17. Drain the ECL and wrap the membrane with cling film. Place the membrane into a film cassette, and expose the membrane to light-sensitive X-ray film (*see* **Notes 48** and **49**).

4 Notes

1. Cell confluence may vary depending on specific experiments, but 80–90% confluency usually gives better results. When possible we recommend to perform endogenous IPs, but epitope-tagged MST1 or MST2 (FLAG or GFP) can be transfected to measure kinase activity. Treatment of cells with staurosporine for 15 min [10] or okadaic acid results in the activation and cleavage of MST1/2 kinase activity and can be used as a positive control. Total lysates may be also loaded, but more than one band can show up due to other kinases that can also phosphorylate MBP in the gel (*see* Fig. 1).
2. We use protein G beads (GE Healthcare) for endogenous IP and FLAG®-beads (Sigma-Aldrich) and GFP-Trap-A® beads (ChromoTek) for exogenous IP. Other antibodies can be used depending on the epitope tagged to exogenous MST1/2.

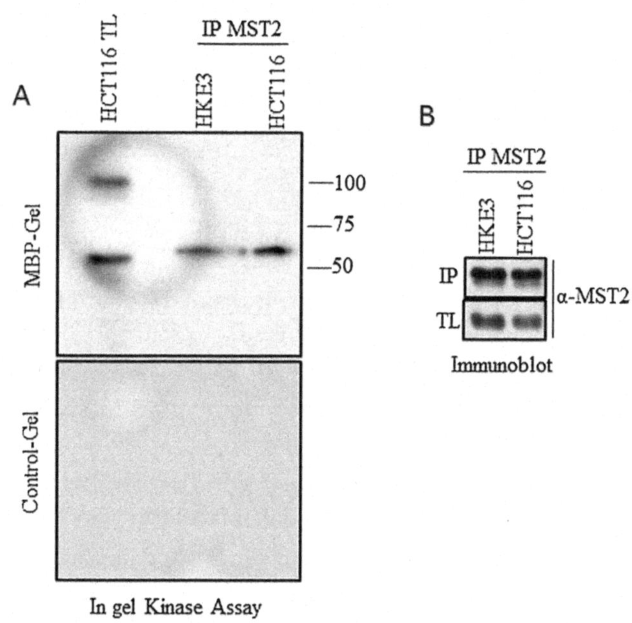

Fig. 1 In-gel kinase assay example: HCT116 and HKE3 colorectal cancer cell lines were serum deprived for 16 h and subsequently lysed using standard lysis buffer. Lysates were split into total lysate and two IP samples. Endogenous MST2 was immunoprecipitated (two IPs per cell line) using anti-STK3 antibody. (**a**) HCT116 extracts (TL) or MST2 IP were loaded in an MBP gel or a control gel without MBP and kinase. HCT116 TL shows a band at ~55 kDa and another at ~100 kDa. Numbers indicate molecular weight in kDa. (**b**) MST2 IPs and extracts were resolved by Western blot and the proteins were monitored by blotting with anti-STK3 antibodies

3. Anti-STK4 (Abcam ab97399) and MST2: Krs-1 (Santa Cruz sc-130405). Other anti-MST1 and anti-MST2 antibodies can be used, but these antibodies specifically IP each isoform and are optimized for the protocols described herein. Some antibodies cannot distinguish between both isoforms.
4. Anti-STK4 (Abcam ab97399), Anti-STK3 (Abcam, ab52641), Anti-Krs1 (Santa Cruz sc-130,405), Anti-phospho-MST1/2 Threonine 183/Threonine 180 (Cell Signaling Technology #3681), Anti-GAPDH (Cell Signaling Technology #2118). The antibody for detection of phosphorylated MST1/2 (Cell Signaling Technology #3681) cannot distinguish between phosphorylated MST1 (Threonine 183) and phosphorylated MST2 (Threonine 180).
5. Antibodies conjugated to other detection systems can also be used.
6. Homemade Laemmli sample buffer can also be used.
7. We find that it is best to prepare this fresh each time for the MBP-containing gels.
8. Mix by flipping the tube and avoid vortexing as we find that this affects the quality of the gel.
9. To prepare 300 mL buffer to remove SDS, make a solution of 60 mL isopropanol and 15 mL of 1 M Tris–HCl pH 8.0 and make up to 300 mL with ultrapure water. For 300 mL of washing buffer, make a solution of 15 mL 1 M Tris-HCl pH 8.0 and 105 mL of β-mercaptoethanol and make up to 300 mL with ultrapure water. These buffers can be stored for up to 3 months. For 100 mL denaturing buffer, accurately weigh out 57.3 g of guanidine HCl and add to 5 mL 1 M Tris–HCl pH 8.0 and 35 mL of β-mercaptoethanol and make up to 100 mL with ultrapure water. Do not store denaturing buffer for more than 1 week and it is better if it is prepared extemporary.
10. To prepare 500 mL, add 25 mL 1 M Tris–HCl pH 8.0 with 175 mL β-mercaptoethanol and 200 mL Tween 20 and make up to 500 mL with ultrapure water. Do not store for more than 1 month since we have noticed that this can affect the intensity of the signal.
11. Do not store and prepare fresh just before the reaction begins.
12. The box is to protect the operator from radiation. If there are no boxes available, the rocker should be placed behind a suitable protective screen.
13. This is not necessary if a phosphor-imager is used.
14. Nitrocellulose membranes can also be used.

15. Do not use methanol when using nitrocellulose membranes. Activate the membrane in transfer buffer.
16. Wet transfer methods can also be used. Transfer times must be adjusted as per operator's protocol.
17. We use handmade ECL prepared by mixing solution 1 Tris–HCl 1 M pH 8.5, p-coumaric acid 0.2 mM, luminol 1.25 mM and water and solution 2 Tris–HCl 1 M pH 8.5, H_2O_2 (30% w/w), and water. Mix solution 1 and solution 2 at a ratio of 1:1. As an alternative to handmade ECL, commercial ECL is also available from standard suppliers and should be used as recommended by the manufacturer.
18. Placing the plates on ice stops the kinase activity and prevents MST1/2 activation by stress stimuli such as change of temperature. Use a flat ice tray to place the plates on.
19. The lysates can be divided into two to perform parallel control experiments. This allows a better control of the experiment. Transfer 400 μL to another tube which typically is sufficient to get the necessary amount of IP MST1/2 for a kinase assay or immunoblotting. Alternatively two 6 cm plates can be treated or transfected in parallel and lysed at the same time to use as experimental controls. Splitting the lysates reduces the number of cell plates and reduces intra-experimental variability.
20. The amount of beads can be decreased to reduce costs as long as the pellets are visible after centrifugation. When using other antibodies, the amount added may vary and should be adjusted.
21. Place the tubes on ice to avoid protein degradation.
22. If possible aspirate the supernatant using a vacuum pump. Use a loading tip as a probe to avoid the beads being aspirated. Using pipettes can disturb the pellets and result in more loss of beads than when using the pump. During the washes there is no need to remove all the washing solution, but in the last step, it is recommended to remove as much as possible without disturbing the beads.
23. MST1/2 can autophosphorylate, and therefore it is possible that some of the signal observed is due to incorporation of [γ-^{32}P]-ATP to the kinases. We rarely observe autophosphorylation, but it should be monitored through the use of experimental controls. In order to monitor MST1/2 autophosphorylation, prepare a control gel without MBP (normal SDS-PAGE gel buffer described above) and run the IPs in parallel with the MBP-containing gel. Subsequently, the in-gel kinase assay is performed with the control gel in parallel to the MBP gel. The control gel that does not contain any MBP should give a much lower reading as there is no substrate,

and if no autophosphorylation has occurred, there will not be any defined band after exposure. Importantly, the control gel can be considered as a readout for background levels when comparing with the MBP gel, and when very low counts are detected, the gels can be considered ready to dry.

24. Other gel resolving buffers recipes containing SDS can be used, but we find that our formula is better for the kinase assay and reduces the washing times with the removed SDS buffer.
25. We recommend 1 mm gels but other spacers can be used. However, thinner or thicker gels can affect the incubation times during the kinase assay and subsequent washes. Ten-well combs are the best to use as they will give better resolution, but 15-well combs may also be used. If there is space, it is beneficial to leave some empty wells between the samples (loaded with NuPAGE sample buffer alone) which will result in a clearer result and can also avoid samples with a very intense signal masking neighboring weaker sample signals.
26. Other electrophoresis systems can be used, but the current protocol has been optimized for the Bio-Rad Mini-PROTEAN® 3 system glass plates.
27. The gels can be run at higher voltages, but we have observed that this results in less defined bands. Running the gels at 4 °C results in defined, clearer bands.
28. We recommend that the containers used for radioactive assays are not used for other experiments. The dimensions described are optimized for our gels and the volumes mentioned in the protocol, but they can be of any size as long as the gels are fully covered during all the steps.
29. Shorter times could be used, but it is then necessary to increase the number of incubations. We do not recommend to change the indicated times since this step is critical for the assay, and we have seen that shorter times can have severe effects on the quality of the experiments and can result in the kinase assay not working.
30. From this step all work must take place in a designated radiation room/area. When working with radioactivity, it is essential that the operator strictly adheres to the national/institutional guidelines and has received the proper training. All work with radioactivity should be carried out in a special radioactive room/area with all of the equipment used being confined to this room and properly decontaminated after use. This helps reduce the risk of contamination.
31. Removing as much of this buffer as possible reduces the time of the kinase assay reaction.

32. Adjusting the volumes to the minimum amount necessary to perform the assay reduces the amount of radioactive waste.
33. Longer periods can result in saturation of the reaction and no difference between different conditions will be observed. The time is optimized for 22–24 °C and may have to be modified for lower or higher temperatures.
34. Most of the [γ-^{32}P]-ATP will still be in the solution so this must be handled with strict care to avoid spills.
35. The control gel can be considered as a readout for background levels when comparing with the MBP gel since no incorporation of [γ-^{32}P]-ATP should have occurred. When very low counts are detected using the Geiger counter, the gels can be considered ready to dry.
36. Other gel dryers can be used. Air-dry protocol is not recommended since it is time-consuming but can be used if there is no access to a gel dryer.
37. Usually the area where the kinase is located gives a reading close to 100 counts per second in the series 800 Geiger counter from Thermo Fisher, while the background will be close to 1 count per second.
38. If a phosphor-imager is available, this increase in the time of exposure can be observed in real time, facilitating this step.
39. First, test a 1 h exposition if the signal is very strong, and depending on the intensity of the bands, you can adjust the exposure time. Usual and overnight exposure is necessary for weaker bands. Detection can be accelerated by placing the cassette at −80 °C. In this case you should see a good signal in 2–4 h in an optimal assay. Longer times are usually recommended and necessary to make sure you see all the bands. For the autophosphorylation control, expose 24 h, and if no band is observed, then the effect of autophosphorylation is irrelevant.
40. If using wet transfer, this has to be adjusted as appropriated.
41. As per manufacturer's recommendations, this may change for other semidry cells.
42. This step is not strictly necessary, but we have observed it helps to preserve the color of the molecular weight marker especially if the membrane is stripped and reblotted several times.
43. 4–5% BSA solutions in TBST can also be used, but for the phospho-MST1/2 antibody for WB, we recommend using milk.
44. Dilutions may be modified depending on experimental conditions and intensity of the signal. In our experience, the phospho-MST1/2 antibody gives a very week signal, and it

does not work after freezing once, so we recommend using a fresh dilution each time.

45. If the membranes are cut, a sufficient amount of surrounding area must be left to monitor any unspecific bands. The membranes can be stripped and reblotted with different antibodies several times, but the signal tends to get weaker. Incubation times may be reduced. For example, a 1-h incubation period is sufficient for the anti-STK3 antibody, but overnight incubation is necessary for the anti-phospho-MST1/2 antibody.
46. These dilutions change depending on suppliers. These dilutions are for the cell signalling secondary antibodies.
47. Follow manufacturer's instructions when using commercial ECL.
48. If using a digital imager, follow your standard procedures.
49. When measuring the MST1/2 activity indirectly with phospho-antibodies, the amount of phosphorylated MST1/2 must be quantified against total levels of MST1/2 to get a percentage of phosphorylation.

Acknowledgments

This work was supported by Science Foundation Ireland grants 15/CDA/3495 and 16/TIDA/4064.

References

1. Radu M, Chernoff J (2009) The DeMSTification of mammalian Ste20 kinases. Curr Biol 19 (10):R421–R425. https://doi.org/10.1016/j.cub.2009.04.022
2. Scheel H, Hofmann K (2003) A novel interaction motif, SARAH, connects three classes of tumor suppressor. Curr Biol 13(23): R899–R900
3. Creasy CL, Ambrose DM, Chernoff J (1996) The Ste20-like protein kinase, Mst1, dimerizes and contains an inhibitory domain. J Biol Chem 271(35):21049–21053
4. Praskova M, Khoklatchev A, Ortiz-Vega S, Avruch J (2004) Regulation of the MST1 kinase by autophosphorylation, by the growth inhibitory proteins, RASSF1 and NORE1, and by Ras. Biochem J 381(Pt 2):453–462. https://doi.org/10.1042/BJ20040025
5. Fallahi E, O'Driscoll NA, Matallanas D (2016) The MST/hippo pathway and cell death: a non-canonical affair. Genes (Basel) 7(6):28. https://doi.org/10.3390/genes7060028
6. Bononi A, Agnoletto C, De Marchi E, Marchi S, Patergnani S, Bonora M, Giorgi C, Missiroli S, Poletti F, Rimessi A, Pinton P (2011) Protein kinases and phosphatases in the control of cell fate. Enzyme Res 2011:329098. https://doi.org/10.4061/2011/329098
7. Hastie CJ, McLauchlan HJ, Cohen P (2006) Assay of protein kinases using radiolabeled ATP: a protocol. Nat Protoc 1(2):968–971. https://doi.org/10.1038/nprot.2006.149
8. Karra AS, Stippec S, Cobb MH (2017) Assaying protein kinase activity with radiolabeled ATP. J Vis Exp 123:e55504. https://doi.org/10.3791/55504
9. Creasy CL, Chernoff J (1995) Cloning and characterization of a human protein kinase with homology to Ste20. J Biol Chem 270 (37):21695–21700
10. O'Neill E, Rushworth L, Baccarini M, Kolch W (2004) Role of the kinase MST2 in suppression of apoptosis by the proto-oncogene product

Raf-1. Science 306(5705):2267–2270. https://doi.org/10.1126/science.1103233

11. Matallanas D, Romano D, Al-Mulla F, O'Neill E, Al-Ali W, Crespo P, Doyle B, Nixon C, Sansom O, Drosten M, Barbacid M, Kolch W (2011) Mutant K-Ras activation of the proapoptotic MST2 pathway is antagonized by wild-type K-Ras. Mol Cell 44 (6):893–906. https://doi.org/10.1016/j.molcel.2011.10.016
12. Matallanas D, Romano D, Yee K, Meissl K, Kucerova L, Piazzolla D, Baccarini M, Vass JK, Kolch W, O'Neill E (2007) RASSF1A elicits apoptosis through an MST2 pathway directing proapoptotic transcription by the p73 tumor suppressor protein. Mol Cell 27 (6):962–975. https://doi.org/10.1016/j.molcel.2007.08.008
13. Romano D, Maccario H, Doherty C, Quinn NP, Kolch W, Matallanas D (2013) The differential effects of wild-type and mutated K-Ras on MST2 signaling are determined by K-Ras activation kinetics. Mol Cell Biol 33 (9):1859–1868. https://doi.org/10.1128/MCB.01414-12
14. Wooten MW (2002) In-gel kinase assay as a method to identify kinase substrates. Sci STKE 2002(153):pl15. https://doi.org/10.1126/stke.2002.153.pl15

Chapter 23

Measuring the Kinase Activities of the LATS/NDR Protein Kinases

Alexander Hergovich

Abstract

The Hippo tissue growth control and regeneration pathway is a main regulator of the YAP/TAZ effectors. In this regard, the LATS/NDR serine/threonine protein kinases can function as central components of the Hippo core module. More specifically, LATS/NDR-mediated phosphorylation of YAP/TAZ on different residues can regulate the subcellular localization and/or stability of YAP/TAZ. Therefore, the assessment of LATS/NDR activities can serve as readout for the activity status of the Hippo pathway. Here, we describe our preferred methodology regarding the measurement of the activities of LATS/NDR kinases.

Key words Hippo pathway, Protein kinases, Kinase assays, Kinase substrates, LATS1, LATS2, NDR1, NDR2, STK38, STK38L

1 Introduction

Research over the past decade and beyond has uncovered that the Hippo pathway serves as a central regulator of diverse cellular processes in order to coordinate tissue growth and regeneration in multicellular organisms [1–6]. In a nutshell, the Hippo pathway is organized into three main levels: a vast array of upstream inputs, a core kinase module, and the co-transcriptional regulators Yorkie and YAP/TAZ as main effectors in flies and mammals, respectively [1–6]. The Hippo core module comprises members of the Ste-20 like and AGC families of serine/threonine protein kinases as well as the scaffolding proteins Salvador (WW45) and Mats (MOB1) [7–11].

In mammalian cells, LATS1/2 and NDR1/2 (aka STK38/STK38L) represent the AGC kinases that can function as members of the Hippo core module [7, 12, 13]. The LATS1/2 kinases phosphorylate YAP and TAZ on five or four different residues, respectively, thereby controlling the subcellular localization and/or stability of YAP/TAZ [14–19]. The NDR1/2 kinases can also phosphorylate YAP on four different residues (that are identical

Alexander Hergovich (ed.), *The Hippo Pathway: Methods and Protocols*, Methods in Molecular Biology, vol. 1893, https://doi.org/10.1007/978-1-4939-8910-2_23,

with the LATS1/2 targeting sites), consequently regulating YAP activity in a similar fashion as LATS1/2 kinases [13, 20]. The direct regulation of TAZ by NDR1/2 has not been reported yet [13]. Nevertheless, these findings cumulatively indicate that LATS/NDR kinases can function as central components of the Hippo core module. Therefore, the measurement of LATS/NDR kinase activities can serve as readout for Hippo pathway activity upstream of the YAP/TAZ effectors.

Here we describe our preferred methodology regarding the quantification of LATS/NDR kinase activities [21–26]. Basically, the LATS/NDR kinase of interest is first purified by immunoprecipitation using specific antibodies, followed by kinase assays using peptides as substrates. Possible alternations of the immunoprecipitation procedure and alternative kinase substrates are discussed as well.

2 Materials

In order to ensure robust and reliable results, all buffers and solutions should be prepared using ultrapure deionized water and molecular biology-grade reagents. Unless otherwise stated, prepare at room temperature. Storage conditions are indicated individually.

2.1 Immunoprecipitation

1. Cell lysis buffer (20 mM Tris–HCl pH 8.0, 150 mM NaCl, 1% NP40, 10% glycerol, 0.5 mM EDTA, 0.5 mM EGTA, 20 mM beta-glycerophosphate, 50 mM NaF, *see* **Note 1**) supplemented freshly with protease and phosphatase inhibitors (*see* **Notes 2** and **3**). Store at 4 °C.
2. Washing buffer containing 1 M NaCl (*see* **Note 1**) supplemented freshly with protease and phosphatase inhibitors (*see* **Notes 2** and **3**). Store at 4 °C.
3. Kinase assay washing buffer (20 mM Tris–HCl pH 7.5) supplemented freshly with protease and phosphatase inhibitors (*see* **Notes 2** and **3**). Store at 4 °C.
4. Suitable antibodies and protein A Sepharose beads (*see* **Notes 4–6**). Here we focus on describing our preferred methodology regarding the analysis of HA-tagged LATS/NDR kinases; however, depending on your needs, this approach can be adjusted accordingly (*see* **Note 4**).

2.2 Kinase Assays

1. Stock solution of 1 M Tris–HCl pH 7.5, stored at room temperature.
2. Stock solution of 0.5 M EDTA pH 8.0, stored at room temperature.

3. Kinase substratepeptide (NH2-KKRNRRLSVA-COOH) (*see* **Notes** 7 and **8**), prepared in sterile ddH_2O as a 10 mM substrate peptide stock solution. Store aliquots at −20 °C.
4. cAMP-dependent protein kinase inhibitor (*see* **Note 9**), prepared in ddH_2O as a 50 μM stock solution. Store aliquots at −20 °C.
5. Protease and phosphatase inhibitors (*see* **Note 2**).
6. Phosphocellulose P81 paper (e.g., obtainable from Whatman). Prepare in approx. 2 × 2 cm (1 × 1 in.) squares (you will need one square per kinase reaction).
7. 1% phosphoric acid (dilute 110 ml of 85% stock solution in 10 l of ddH_2O). Store at room temperature.
8. Prepare 1 mM ATP stock solution in ddH_2O. Store aliquots at −20 °C.
9. Stock solution of 1 M $MgCl_2$. Store at room temperature.
10. [γ-P32]-ATP (3000 Ci/mmol specific activity) (*see* **Note 10**). Store at 4 °C (or as recommended by the manufacturer).
11. Prepare a 10× kinase buffer and store in aliquots at −20 °C. For a total volume of 100 μl, mix 20 μl 1 M Tris–HCl pH 7.5, 1 μl of 1 M benzamidine, 1 μl of 4 mM leupeptin, 3 μl of 1 M DTT, 3 μl of 1 mM microcystin, and 72 μl sterile ddH_2O.
12. Washing beaker unit with stirring bar (*see* **Note 11**).
13. Scintillation vials and scintillation solution (*see* **Note 12**).
14. Access to scintillation counter.
15. Magnetic stirrer unit for the washing beaker unit.
16. Thermomixer allowing you to heat and shake samples simultaneously.
17. Acetone (≥99.5%).

3 Methods

3.1 Immunoprecipitation

Always make sure to keep your samples on ice at 4 °C. Our preference is to perform the entire immunoprecipitation procedure in a cold room at 4 °C. As outlined in Subheading 2.1, we are describing here our preferred methodology regarding the analysis of HA-tagged LATS/NDR kinases; however, depending on your needs, this approach can be adjusted accordingly (*see* **Note 4**). We have successfully used this procedure to purify HA-tagged LATS/NDR kinases for subsequent kinase assays using transiently and stably expressed kinase versions [21–26]. Before actually performing the experiments, please take the time to properly plan your positive and negative controls (*see* **Notes 13** and **14**).

1. Before harvesting the cells, wash them once briefly with ice-cold PBS (containing Mg^{2+}/Ca^{2+}), and gently aspirate off all residual liquid.
2. Depending on the surface area of your tissue culture, add 0.5–1.0 ml cell lysis buffer supplemented freshly with protease and phosphatase inhibitors (*see* **Note 15**).
3. Collect cells with a cell scraper (our preferred model is FB55160 from Fisherbrand), and transfer them to pre-chilled and properly labeled 1.5 ml Eppendorf tubes.
4. Incubate your samples on ice for 30 min, with inverting them gently a few times every 10 min to assist the lysis process.
5. Separate the cell debris from the soluble cell components by centrifugation for 10 min at 20,000 × *g* at 4 °C.
6. Transfer the supernatants containing your soluble kinases of interests into fresh pre-chilled 1.5 ml Eppendorf tubes (cell lysates may be frozen at this point; *see* **Note 1**). Store samples on ice.
7. Collect aliquots of cell lysates as lysate input controls (*see* **Notes 13** and **16**). Store these lysate input controls at −20 °C (or if you have the space at −80 °C) for later analysis by Western blotting.
8. To pre-clear cell lysates from factors that can bind to protein A-sepharose, we add 10 μl of protein A Sepharose (50% slurry) per sample, followed by incubation at 4 °C for 30–60 min on a rolling/rocking shaker.
9. Collect the protein A Sepharose beads by centrifugation for 10 s at 20,000 × *g* at 4 °C.
10. Transfer the supernatants into fresh pre-chilled 1.5 ml Eppendorf tubes. Store samples on ice.
11. Add the anti-HA protein A Sepharose beads (*see* **Note 5**) to the cell lysates. 10 μl of beads are fully sufficient. Keep samples on ice.
12. Incubate the antibody-sepharose beads with the cell lysate at 4 °C for at least 2 h on a rolling/rocking shaker. Optionally, this incubation can be performed overnight.
13. Wash the antibody-sepharose beads twice with 0.5 ml of cell lysis buffer supplemented freshly with protease and phosphatase inhibitors. For each washing step, centrifuge the 1.5 ml tubes for 10 seconds at 20,000 × *g* at 4 °C, followed by very careful aspiration of supernatants (the supernatants can be discarded from this step onward).
14. Wash the antibody-sepharose beads once with 1.0 ml of washing buffer containing 1 M NaCl supplemented freshly with protease and phosphatase inhibitors (*see* **Note 17**).

15. Wash the antibody-sepharose beads once with 1.0 ml of cell lysis buffer supplemented freshly with protease and phosphatase inhibitors (*see* **Note 18**).
16. Wash the antibody-sepharose beads twice with 1.0 ml of kinase assay washing buffer supplemented freshly with protease and phosphatase inhibitors. Optionally, if you have the time, a third wash never hurts at this step.
17. Finally, remove very gently all of the supernatant, and add to the beads 20 μl of kinase assay washing buffer supplemented freshly with protease and phosphatase inhibitors. Keep the samples on ice until you are ready to perform the kinase assays.

3.2 Kinase Assays

Keep your samples on ice until you are ready to add the kinase assay master mix. Prepare the kinase assay master mix on ice. The kinase assay is performed at 30 °C, while the washing steps of the P81 paper are performed at room temperature.

1. Make sure that all required materials are ready (*see* Subheading 2.2 above). Have the thermomixer ready at 30 °C.
2. Prepare the kinase assay master mix as follows (the indicated volumes are per kinase reaction and the components are listed in the order of their addition; *see* **Note 19**):

   1.6 μl ddH_2O, 1.0 μl of 10× kinase buffer, 0.6 μl of 50 μM PKI stock solution, 0.3 μl of 1 M $MgCl_2$, 3 μl of 10 mM substrate peptide stock solution, 3 μl of 1 mM ATP stock solution, and 0.5 μl of [γ-P32]-ATP (3000 Ci/mmol). Mix well (do not vortex, simply flick the tube several times, and then spin down the liquid by centrifugation). Keep on ice.
3. To start the kinase reactions, add 10 μl of master mix solution per reaction to the samples from **step 17** (*see* Subheading 3.1). Keep samples on ice.
4. Incubate the kinase reactions for 60 min at 30 °C with shaking (*see* **Note 20**).
5. In the meantime properly label your squares of P81 paper with a pencil (*see* **Note 21**).
6. Terminate the kinase reactions by centrifuging the 1.5 ml tubes containing your samples for 10 s at 20,000 × *g* at room temperature, followed by addition of 3 μl of 0.5 M EDTA per reaction.
7. Mix the samples by flicking them gently a few times, before centrifuging them once more for 10 s at 20,000 × g at room temperature.
8. Carefully collect by pipetting 20 μl per reaction, and spot the samples onto the pre-labeled squares of P81 paper at room temperature (*see* **Note 22**). Spot the radioactive sample onto the labeled side.

9. Allow the P81 paper squares to dry for at least 10 min at room temperature.
10. Fill the washing beaker unit containing a stirring bar with about 500 ml of 1% phosphoric acid solution, and place the unit onto a magnetic stirrer.
11. Use a forceps to place the P81 paper squares individually into the washing beaker unit. Prevent to pill them on top of each other.
12. Start washing the P81 paper squares by turning on the magnetic stirrer. The ideal stirring intensity is achieved when the P81 paper squares are slight rotating along with the stirrer while still floating in solution. In case P81 paper squares start to pill up on top of each other, separate them using a forceps.
13. Wash the P81 paper squares four times for 10 min in 1% phosphoric acid solution at room temperature. Every 10 min turn off the magnetic stirrer, then gently decant the used 1% phosphoric acid solution, and add about 500 ml of fresh 1% phosphoric acid solution, followed by turning on the magnetic stirrer again.
14. For the final wash, turn off the stirrer; then decant as much as possible of the used 1% phosphoric acid solution, followed by addition of about 500 ml of acetone; and turn on the stirrer for another 5 min.
15. Turn off the stirrer. Using forceps take out the P81 paper squares from the washing beaker unit, and place the P81 paper squares upside up on a clean surface for drying (the labeled side should face upward). Let them dry at room temperature for at least 10 min.
16. In the meantime prepare per kinase reaction (per P81 paper square) one scintillation vial containing scintillation liquid (*see* **Note 12**). Make sure to have the vials properly labeled.
17. Using forceps place each completely dry P81 paper square into its individual scintillation vials and close the vials.
18. Measure in a suitable scintillation counter the number of counts per minute (cpms) for the used P32 isotope.
19. Calculate the relative kinase activities (*see* **Note 23**).

4 Notes

1. The cell lysis buffer and washing buffer containing 1 M NaCl are prepared using the following stock solutions: 1 M Tris–HCl pH 8.0, 5 M NaCl, 10% NP40-substitute, 0.5 M EDTA pH 8.0, 0.5 M EGTA pH 8.0, and 1 M NaF. Beta-glycerophosphate is dissolved as powder upon preparation of

buffers. Glycerol is added to allow optional storage of cell lysate samples at −80 °C, before processing for immunoprecipitation (we prefer to process cell lysates immediately for subsequent immunoprecipitation; however, sometimes the timing of sample collection makes it difficult to commence immediately, in which case samples can be frozen at −80 °C until subsequent processing). In general, we prefer to prepare 500 to 1000 ml of each buffer.

2. In general, any reliable protease and phosphatase inhibitors should be sufficient to protect your kinase of interest. We prefer to add inhibitors at the following final concentrations: 0.5 mM PMSF, 4 μM leupeptin, 1 mM benzamidine, 1 mM DTT, 1 μM microcystin, and 1 mM Na3VO4 using the following stock solutions, 0.5 M PMSF (prepared in DMSO), 4 mM leupeptin (dissolved in ddH_2O), 1 M benzamidine (prepared in ddH_2O), 1 mM microcystin (dissolved in EtOH; our preferred source is ALX-350-012 from Enzo Life Sciences), 1 M DTT (prepared in ddH_2O), and 200 mM Na3VO4 (*see* **Note 3** below). Except for Na3VO4, stock solutions are stored in aliquots at −20 °C.
3. The 200 mM Na_3VO_4 stock solution is prepared as follows: dissolve Na3VO4 in autoclaved ddH_2O; adjust pH to 10, upon which the solution will turn yellow; then heat (boil) until the solution is colorless again; allow the solution to cool down to room temperature, followed by another adjustment of the pH to 10; and continue to repeatedly heat, cool, and pH adjust the solution until the Na3VO4 solution remains colorless at pH 10 at room temperature. Finally store the solution at 4 °C.
4. While we focus on describing our preferred methodology regarding the analysis of HA-tagged LATS/NDR kinases, it is of course possible to adjust the protocol accordingly. Actually, if you already have a preferred immunoprecipitation protocol that works fine for you, then we only suggest to make the following adjustments: firstly, include also a washing step with buffer containing 1 M NaCl to ensure that unspecifically bound proteins are removed from your immunoprecipitates, and, secondly, make sure to wash your immunoprecipitates with kinase assay washing buffer as outlined in Subheading 3.1. In case you want to measure the activities of differently tagged LATS/NDR or maybe even endogenous kinases, you will first need to optimize your immunoprecipitation conditions, taking into account that depending on the antibody subclass, possibly alternative sepharose beads work better than protein A Sepharose and that the specificity of your immunoprecipitating antibody will need to be determined.
5. The preparation of protein A Sepharose beads covalently coupled to anti-HA antibodies is performed as follows: incubate

10 ml of concentrated anti-HA 12CA5 hybridoma supernatant with 1 ml 10% NP-40 solution and 1.5 ml protein A Sepharose beads (50% slurry) on a roller shaker overnight at 4 °C; the next day wash the beads twice with ice-cold PBS (without Mg^{2+}/Ca^{2+}), making sure that none of the beads are lost; wash another three times with 0.2 M Na-borate (pH 9.0) (*see* **Note 6**); remove supernatant and fill up to total volume of 5 ml with 0.2 M Na-borate (pH 9.0); dissolve 51.8 mg of DMP in 5 ml of 0.2 M Na-borate (pH 9.0) and add the dissolved DMP solution to the 5 ml antibody-beads slurry, so that the total volume is about 10 ml; incubate for 1–2 h on roller shaker at room temperature; wash the beads twice with 0.2 M ethanolamine (pH 8.0); aspirate the supernatant and add 10 ml of 0.2 M ethanolamine (pH 8.0), followed by incubation for 2–4 h at room temperature; finally wash the beads three times with ice-cold PBS (without Mg^{2+}/Ca^{2+}); aspirate the supernatant and store the antibody-beads solution as a 50% slurry in ice-cold PBS (without Mg^{2+}/Ca^{2+}) containing 0.2% sodium azide. Store at 4 °C. In our experience it works well to perform the incubation and washing steps in 15 ml tubes, with performing at least two couplings in parallel (which ensures that one has a counter balance for centrifugations, while also saving time, since it is basically the same amount of work). For the washing steps, we centrifuge 15 ml tubes for 1 min at 200 × *g* at room temperature, followed by very careful aspiration of the supernatant.

6. To prepare 50 ml 0.2 M Na-borate (pH 9.0), you should mix 45 ml of 0.05 M Borax with 4.5 ml of 0.2 M boric acid (pH 4.0). Measure the pH before using the Na-borate buffer. Make sure to store stock solutions at 4 °C.

7. Of course, also alternative peptide substrates should work, as long as the core targeting motif of RXXS is taken into consideration. However, it is more important to consider a phosphoacceptor mutant peptide (Ser/Thr to Ala change) as a negative control. As alternative to peptide substrates, one can also consider recombinant proteins or protein fragments as substrates (*see* **Note 8** below).

8. In case you want to use recombinant proteins or protein fragments as alternative substrates, you need to consider additional controls in order to ensure that your recombinant substrate preparation does not contain any internal kinase activity that might interfere with your interpretation of your kinase assays. When using radiolabeled ATP, you should incubate your substrate alone (without any immunoprecipitated kinase) in the kinase assay buffer as outlined in Subheading 3.2, expecting to detect no incorporation of radioactivity in case your substrate preparation is free of any kinase activity. When using phospho-

specific antibodies in combination with Western blotting, you need to make sure that your phospho-specific antibody is indeed highly specific for your phosphorylated substrate. In this regard, we recommend comparing wild-type vs. phospho-acceptor (Ser/Thr to Ala change) versions. Furthermore, when using recombinant kinases as substrates, you will need to consider the preparation of kinase-dead versions in order to exclude any interference with your kinase assays due to the possible autophosphorylation activity of your substrate. In general, you need to be aware that the analysis of recombinant proteins as kinase substrates is more labor and equipment intensive, since you will need to separate your substrate using SDS-PAGE, before performing autoradiography (in which case you need access to adequate gel imaging equipment) or Western blotting (in which case you need reliable phospho-specific antibodies in addition to Western blotting equipment).

9. The cAMP-dependent protein kinase (PKA) can be quite active and abundant in cell lysates; hence possibly a fraction of PKA might be co-immunoprecipitated in large-scale experiments. Moreover, PKA can very efficiently phosphorylate many substrates with a RXXS motif. Consequently, to rule out that any residual PKA activity might interfere with your kinase assay readouts, we strongly recommend including a PKA inhibitor in the kinase assay reactions. Our preferred inhibitor peptide is ALX-151-014 from Enzo Life Sciences.
10. Always be very careful when handling any form of radioactive substance. Never use radiolabeled ATP without prior training in how to handle and discard radioactive substances. Our preferred provider of [γ-P32]-ATP is Hartmann Analytic (product code: SRP-301).
11. The washing beaker is prepared as follows: cut off the top of a 1000 ml centrifuge bottle (e.g., Thermo Scientific 3120–1010) to broaden the top opening for easier access; cut (best to burn with a hot metal object) about 5- to 6-mm-wide holes into the cut bottle (make holes every 3 to 4 cm on each side); finally assemble the washing beaker unit by placing a stirring bar (about 50 to 70 mm in length) in a 1000 ml glass beaker, followed by placing the cut bottle on top of the stirring bar.
12. At the end you can use your scintillation vials and liquids of choice; however, our preferred products are 6 ml vials from Sarstedt (73.680) or Perkin Elmer (6000292) combined with Liquid Scintillation Ready Safe from Beckman Coulter (141349).
13. We recommend that you include controls for your immunoprecipitation and kinase assays. For the immunoprecipitation

we suggest to include at least two controls: an immunoprecipitation using anti-HA-coupled protein A Sepharose beads in cell lysates derived from empty vector (EV) expressing cells (negative control) and the analysis of all input lysates (to verify the expression levels of HA-tagged proteins). For the subsequent kinase assay, we suggest the following controls in addition to your samples of interest: anti-HA immunoprecipitates from EV samples undergoing kinase assays without radiolabeled ATP (a negative control that should not give you any readings in the scintillation counter), anti-HA immunoprecipitates from EV samples undergoing kinase assays in the presence of radiolabeled ATP (a further negative control that will inform you about any background levels that you will need to take into account when calculating relative kinase activities; normally, you can expect readings of a few to a several hundred cpm), and last, but not least, anti-HA immunoprecipitates from cells expressing HA-LATS/NDR wild-type vs. kinase-dead with prior treatment with okadaic acid (OA; *see* **Note 14** below) subjected to kinase assays with radiolabeled ATP (OA-treated wild-type samples should give you readings of at least several ten thousands cpms as positive controls, while OA-treated kinase-dead samples should only give readings of a few to a several hundred cpm like the EV negative control described above).

14. Okadaic acid (OA) treatments are performed as follows: first prepare a 1 mM stock solution of OA in DMSO (our preferred source is ALX-350-003 from Enzo Life Sciences; store aliquots at −20 °C), and then treat cells at 37 °C for 1 h with 1 μM OA (final concentration; dilute the 1 mM stock solution 1/1000 in the appropriate tissue culture medium), before processing cells for analysis.
15. When using 6 cm tissue culture plates, we add 0.5 ml cell lysis buffer. In the case of 10 cm tissue culture plates, we add 1.0 ml cell lysis buffer. Please make sure to adjust your volumes to match your preferred tissue culture materials.
16. Normally we collect 20 μl per 0.5 ml cell lysate. In order to be on the safe side, we tend to collect two input lysate aliquots per sample, so that even if one aliquot series is lost or compromised due to technical errors (which can happen), we still can analyze a second round of samples.
17. To properly expose the antibody-sepharose beads to the washing buffer containing 1 M NaCl, we add the washing buffer to the beads and subsequently incubate the beads for 5 min on ice, before moving to the next washing step. Keep samples on ice.

18. Normally at this step, we also incubate the beads for 5 min with the cell lysis buffer, before we continue with the final washes in kinase assay washing buffer. In our experience the prolonged incubation times can dramatically reduce the background signals in the kinase assays (i.e., get rid of unspecific kinase activity contaminations).
19. To prevent undesired surprises when distributing the kinase assay master mix between samples, we always prepare the master mix with a 10% excess (e.g., when we are performing 20 kinase reactions, we are preparing a master mix for 22 reactions). Make sure to prepare the kinase assay master mix on ice and add the radiolabeled ATP at the very end.
20. Our preference is to incubate samples in an Eppendorf Thermomixer at 600–700 rpm at 30 °C.
21. Good labeling of the P81 squares is essential to ensure that samples are easily identified. We recommend labeling every corner of the squares, so that they are always identifiable, even if a corner gets damaged during the washing process.
22. Make sure to avoid the pipetting of any beads onto the P81 paper, in order to prevent any measurement of autophosphorylation activities of LATS/NDR kinases. Alternatively to spotting 20 μl onto one P81 paper square, one can also pipette two times 10 μl onto two separate P81 paper square (if you wish to have duplicates of each single kinase reaction).
23. When you included the proper controls (*see* **Note 13** above), you can easily calculate the relative kinase activities as follows: the wild-type kinase sample cpms minus the EV cpms gives you the actual cpms of your wild-type kinase; this value is then compared to the values in the other samples (for which you of course also have to subtract the EV cpms). For example, you have 400 cpms in your EV as background signal, 5000 cpms for your wild-type sample and 9600 cpms for your sample of interest. So, you calculate 5000–400 = 4600 and 9600–400 = 9200. When then 4600 is set as 1, we get 9200/4600 = 2. In other words, you have a relative kinase activity increase of twofold in your sample of interest in this case.
24. Please keep in mind that in order to obtain statistically useful numbers, you will need to perform at least three independent experiments. In this regard, we prefer to perform each experiment in two independent replicates, so that for three independent experiments, we will have at least six values in total to analyze.
25. Last, but not least, we would like to mention that the described setup of immunoprecipitations followed by kinase assays is ideally suited to also incorporate the analysis of the

phosphorylation status of your LATS/NDR kinase of interest. In order to achieve this, you need to adjust step 17 in Subheading 3.1. Briefly, instead of adding 20 μl of kinase assay washing buffer supplemented freshly with protease and phosphatase inhibitors, you add 1 ml of kinase assay washing buffer; mix the sample gently by flicking and inverting the tube, followed by transferring 500 μl to a fresh, pre-chilled and properly labeled. Subsequently, you process one 500 μl aliquot for kinase assays as outlined in Subheading 3.2, while the other 500 μl aliquot is processed for Western blotting using specific anti-phospho antibodies as outlined in [21–26].

26. In general, all experimental procedures described in Subheadings 3.1 and 3.2 can be completed within 1 working day. However, our preference is to separate the protocol into two parts split over 2 days. On day 1, the immunoprecipitation is performed up to **step 12** described in Subheading 3.1 and the antibody-sepharose beads are incubated with the cell lysate at 4 °C overnight on a rolling/rocking shaker. On the second day, the immunoprecipitation and the subsequent kinase assays are completed as outlined in Subheadings 3.1 and 3.2 above.

Acknowledgments

The Hergovich laboratory was supported by the Wellcome Trust (090090/Z/09/Z), BBSRC (BB/I021248/1), Worldwide Cancer Research (AICR; 11-0634), UCL Cancer Research UK Centre, and the National Institute for Health Research University College London Hospitals Biomedical Research Centre.

References

1. Harvey KF, Zhang X, Thomas DM (2013) The Hippo pathway and human cancer. Nat Rev Cancer 13(4):246–257. https://doi.org/10.1038/nrc3458
2. Johnson R, Halder G (2014) The two faces of Hippo: targeting the Hippo pathway for regenerative medicine and cancer treatment. Nat Rev Drug Discov 13(1):63–79. https://doi.org/10.1038/nrd4161
3. Meng Z, Moroishi T, Guan KL (2016) Mechanisms of Hippo pathway regulation. Genes Dev 30(1):1–17. https://doi.org/10.1101/gad.274027.115
4. Yu FX, Zhao B, Guan KL (2015) Hippo pathway in organ size control, tissue homeostasis, and cancer. Cell 163(4):811–828. https://doi.org/10.1016/j.cell.2015.10.044
5. Irvine KD, Harvey KF (2015) Control of organ growth by patterning and hippo signaling in Drosophila. Cold Spring Harb Perspect Biol 7(6):a019224
6. Pan D (2010) The hippo signaling pathway in development and cancer. Dev Cell 19(4):491–505
7. Hergovich A, Stegert MR, Schmitz D, Hemmings BA (2006) NDR kinases regulate essential cell processes from yeast to humans. Nat Rev Mol Cell Biol 7(4):253–264
8. Pearce LR, Komander D, Alessi DR (2010) The nuts and bolts of AGC protein kinases. Nat Rev Mol Cell Biol 11(1):9–22
9. Avruch J, Zhou D, Fitamant J, Bardeesy N, Mou F, Barrufet LR (2012) Protein kinases of the Hippo pathway: regulation and substrates. Semin Cell Dev Biol 23(7):770–784

10. Hergovich A (2011) MOB control: reviewing a conserved family of kinase regulators. Cell Signal 23(9):1433–1440. https://doi.org/10.1016/j.cellsig.2011.04.007
11. Sudol M, Harvey KF (2010) Modularity in the Hippo signaling pathway. Trends Biochem Sci 35(11):627–633
12. Manning G, Whyte DB, Martinez R, Hunter T, Sudarsanam S (2002) The protein kinase complement of the human genome. Science 298(5600):1912–1934
13. Hergovich A (2016) The roles of ndr protein kinases in hippo signalling. Genes 7(5):21
14. Dong J, Feldmann G, Huang J, Wu S, Zhang N, Comerford SA, Gayyed MF, Anders RA, Maitra A, Pan D (2007) Elucidation of a universal size-control mechanism in Drosophila and mammals. Cell 130(6):1120–1133
15. Hao Y, Chun A, Cheung K, Rashidi B, Yang X (2008) Tumor suppressor LATS1 is a negative regulator of oncogene YAP. J Biol Chem 283 (9):5496–5509
16. Lei Q-Y, Zhang H, Zhao B, Zha Z-Y, Bai F, Pei X-H, Zhao S, Xiong Y, Guan K-L (2008) TAZ promotes cell proliferation and epithelial-mesenchymal transition and is inhibited by the hippo pathway. Mol Cell Biol 28 (7):2426–2436
17. Liu C-Y, Zha Z-Y, Zhou X, Zhang H, Huang W, Zhao D, Li T, Chan SW, Lim CJ, Hong W (2010) The hippo tumor pathway promotes TAZ degradation by phosphorylating a phosphodegron and recruiting the SCF-β-TrCP E3 ligase. J Biol Chem 285 (48):37159–37169
18. Zhao B, Li L, Tumaneng K, Wang C-Y, Guan K-L (2010) A coordinated phosphorylation by Lats and CK1 regulates YAP stability through SCFβ-TRCP. Genes Dev 24(1):72–85
19. Zhao B, Wei X, Li W, Udan RS, Yang Q, Kim J, Xie J, Ikenoue T, Yu J, Li L (2007) Inactivation of YAP oncoprotein by the Hippo pathway is involved in cell contact inhibition and tissue growth control. Genes Dev 21 (21):2747–2761
20. Zhang L, Tang F, Terracciano L, Hynx D, Kohler R, Bichet S, Hess D, Cron P, Hemmings BA, Hergovich A (2015) NDR functions as a physiological YAP1 kinase in the intestinal epithelium. Curr Biol 25 (3):296–305
21. Cook D, Hoa LY, Gomez V, Gomez M, Hergovich A (2014) Constitutively active NDR1-PIF kinase functions independent of MST1 and hMOB1 signalling. Cell Signal 26 (8):1657–1667
22. Hergovich A, Bichsel SJ, Hemmings BA (2005) Human NDR kinases are rapidly activated by MOB proteins through recruitment to the plasma membrane and phosphorylation. Mol Cell Biol 25(18):8259–8272
23. Hergovich A, Schmitz D, Hemmings BA (2006) The human tumour suppressor LATS1 is activated by human MOB1 at the membrane. Biochem Biophys Res Commun 345(1):50–58
24. Hoa L, Kulaberoglu Y, Gundogdu R, Cook D, Mavis M, Gomez M, Gomez V, Hergovich A (2016) The characterisation of LATS2 kinase regulation in Hippo-YAP signalling. Cell Signal 28(5):488–497. https://doi.org/10.1016/j.cellsig.2016.02.012
25. Kohler RS, Schmitz D, Cornils H, Hemmings BA, Hergovich A (2010) Differential NDR/LATS interactions with the human MOB family reveal a negative role for human MOB2 in the regulation of human NDR kinases. Mol Cell Biol 30(18):4507–4520
26. Vichalkovski A, Gresko E, Cornils H, Hergovich A, Schmitz D, Hemmings BA (2008) NDR kinase is activated by RASSF1A/MST1 in response to Fas receptor stimulation and promotes apoptosis. Curr Biol 18(23):1889–1895

Chapter 24

MST1/2 Kinase Assays Using Recombinant Proteins

Marta Gomez, Yavuz Kulaberoglu, and Alexander Hergovich

Abstract

The Hippo tumor suppressor pathway is fundamental to the coordination of death, growth, proliferation, and differentiation on the cellular level. At the molecular level, a highly conserved Hippo core cassette is central for the regulation of effector activities such as the co-transcriptional activity of YAP. In particular, the mammalian MST1/2 serine/threonine protein kinases (termed Hippo kinase in *Drosophila melanogaster*) can act as central signal transducers as part of the Hippo core cassette. In this chapter we describe in vitro kinase assays using recombinant MST1/2 kinases and recombinant MST1/2 kinase substrate.

Key words The Hippo pathway, Serine/threonine protein kinases, Kinase assays, Recombinant kinase substrates, MST1, STK4, MST2, STK3, Hippo kinase

1 Introduction

In flies and mammals, the Hippo pathway controls organ growth, tissue homeostasis, and regeneration [1–5]. The Hippo pathway can be affected by numerous external and internal factors, which influence the Hippo core kinase cassette [1–4]. Importantly, the deregulation of the Hippo pathway has been linked to various human diseases, in particular cancers [6, 7], and hence has attracted research interests from various angles.

The best understood Hippo core kinase cassette comprises the MST1/2 (aka STK4 and STK3) and LATS1/2 protein kinases. Upon activation the MST1/2 kinases can phosphorylate the LATS1/2 protein kinases and the MOB1 signal transducer protein, thereby supporting the formation of an active MOB1/LATS complex [3, 8–13], which appears to be essential for development and tissue growth control [14]. In contrast, stable MOB1/MST complex formation does not seem to be essential in this regard [14, 15].

Whatever the case, active LATS1/2 kinases can phosphorylate the transcriptional co-regulators YAP/TAZ, thereby promoting the cytoplasmic retention and/or degradation of YAP/TAZ [3, 16]. Notably, the NDR1/2 kinases (aka STK38 and STK38L)

Alexander Hergovich (ed.), *The Hippo Pathway: Methods and Protocols*, Methods in Molecular Biology, vol. 1893, https://doi.org/10.1007/978-1-4939-8910-2_24,

can also phosphorylate and thereby regulate the YAP protein [17, 18]. NDR1/2, like LATS1/2, can form stable interactions with MOB1 [19] and also serve as direct effector substrates of the MST1/2 kinases [17, 20]. Consequently, the Hippo pathway signals through distinct kinases, such as MST1/2, LATS1/2, and NDR1/2, in concert with the fundamental signal adaptor MOB1. To complicate things further, MST1/2 also phosphorylate MOB1 [9]; hence, LATS1/2, NDR1/2, and MOB1 have been established as direct substrates of the MST1/2 kinases [17, 19, 20]. However, the list of MST1/2 substrates appears to expand beyond components of the Hippo core cassette [21, 22], making it crucial to accurately define and validate direct MST1/2 substrates using in vitro kinase assays.

Here, we describe the in vitro phosphorylation of recombinant full-length MOB1 by full-length recombinant MST1/2 kinases. More specifically, we describe first the purification of the recombinant substrate, followed by the definition of in vitro kinase assays, which are finally analyzed by Western blotting using highly specific anti-phospho antibodies. In more general terms, we define here a nonradioactive in vitro kinase assay procedure that is reliable, robust, and reproducible. While our protocol is focused on describing MST1/2 as kinases and MOB1 as phospho-acceptor, our protocol may also be applied to other protein kinases and other (putative) substrates of interest.

2 Materials

Prepare all buffers and solutions using ultrapure deionized water analytical grade reagents. All buffers, solutions, and reagents are to be kept at 4 °C unless indicated otherwise.

2.1 Recombinant Protein Production and Purification

1. pMAL™ protein fusion and purification system (#E8200S, New England Biolabs) (*see* **Note 1**).
2. *Escherichia coli* BL21 (DE3) competent cells (#C2527I, New England Biolabs) (*see* **Note 2**).
3. Amylose resin (#E8021L, New England Biolabs).
4. Access to cell sonicator system.
5. Beckman JS-4.2 and JA-17 rotors (or equivalent centrifugation systems).
6. 40 μm cell strainer (our preferred model is #352340 from Corning Inc.).
7. Glass columns (our preferred model is #7374152, Econo columns from Bio-Rad).
8. Centrifugal filter units (our preferred model is #UFC901008, Merck) (*see* **Note 3**).

9. Dialysis tubing (our preferred model is #132676, Spectrum) (*see* **Note 4**).
10. Magnetic stirrer and stirring bars in 2 l glass beakers.
11. 10% SDS-PAGE gels (*see* **Note 5**) and the corresponding Western blotting equipment.
12. Phosphate-buffered saline (PBS; 10 mM Na2HPO4, 2 mM KH2PO4, 137 mM NaCl, 2.7 mM KCl at pH 7.4).
13. 20 mM Tris–HCl pH 7.5.
14. Glycerol. Store at room temperature.
15. BSA standard (or another protein standard for protein quantifications).
16. LB medium (*see* **Note 6**).
17. LB agar plates (*see* **Note 7**).
18. 100 mM IPTG (Isopropyl β-D-1-thiogalactopyranoside). Store in aliquots at −20 °C.
19. 200 mM column washing buffers (*see* **Note 8**).
20. 1 M column washing buffer (*see* **Note 9**).
21. Coomassie staining solution (*see* **Note 10**). Keep at room temperature.
22. Destaining solution (*see* **Note 11**). Keep at room temperature.
23. 5× Laemmli SDS-PAGE loading buffer (*see* **Note 12**). Store at room temperature.

2.2 Kinase Assays with Recombinant Proteins, Followed by SDS–PAGE and Western Blotting

1. Recombinant Mal-tagged protein (*see* Subheding 3.1). Store at −80 °C.
2. Recombinant GST-MST1 and GST-MST2 kinases (*see* **Note 13**). Store at −80 °C.
3. 1 mM ATP stock solution prepared in sterile ddH2O. Store aliquots at −20 °C.
4. Stock solution of 1 M $MgCl_2$. Store at room temperature.
5. 10× kinase assay buffer (50 mM Tris–HCl pH 7.5, 25 mM beta-glycerophosphate, 10 mM EGTA, 10 mM Na_3VO_4, 40 mM $MgCl_2$).
6. Kinase assay master mix. Per kinase reaction mix together 2 μl of 10× kinase assay buffer, 2 μl of 1 mM ATP, 0.2 μl 100 mM DTT, and 0.8 μl sterile ddH_2O. Prepare freshly before the assay and store on ice until use.
7. Thermomixer to heat and shake samples simultaneously.
8. 10% SDS-PAGE gels (*see* **Note 5**) and the corresponding Western blotting equipment, materials, and solutions (*see* **Note 14**).

9. 5× Laemmli SDS-PAGE loading buffer (*see* **Note 12**). Store at room temperature.
10. Membrane blocking solution (TBS-T (*see* **Note 14**) with 5% (w/v) milk powder or BSA). Store at 4 °C.
11. Protein marker (*see* **Note 15**). Store as recommended by the manufacturer.
12. Primary antibodies (*see* **Note 16**). Store as recommended by the manufacturer.
13. Secondary antibodies (*see* **Note 17**). Store as recommended by the manufacturer.
14. ECL (enhanced chemiluminescence) solutions (*see* **Note 18**) and the corresponding standard equipment for the detection of chemiluminescence. Store at 4 °C.

3 Methods

3.1 Production and Purification of Recombinant MAL-Tagged Protein

1. Transform DNA plasmids allowing Mal-tagged protein expression into BL21 (DE3) competent bacteria (*see* **Notes 1**, **2**, and **19**).
2. Before producing a large batch of your protein, confirm protein expression on a smaller scale (*see* **Note 20**).
3. Once you have confirmed protein expression (*see* **Note 20**), start the large-scale production by inoculating 5 ml of LB medium containing antibiotics with a single bacteria colony in the morning. Incubate at 37 °C for at least 8 h in a shaking incubator.
4. Allow 95 ml of fresh LB medium with antibiotics to adjust to room temperature during the day, and in the evening, add the 5 ml culture grown during the day.
5. Incubate the 100 ml culture overnight at 37 °C in a shaking incubator.
6. Prepare 900 ml of fresh LB medium with antibiotics and allow it to adjust to room temperature overnight.
7. The next morning, add the 100 ml overnight culture to the 900 ml LB medium, followed by incubation at 30 °C for 1–2 h in a shaking incubator until the culture reaches an optical density (OD) of 0.8–1.2 at 600 nm (*see* **Note 21**).
8. Induce protein expression by addition of 0.2 mM IPTG (final concentration), and incubate the 1000 ml culture for 3–4 h at 30 °C while shaking (*see* **Note 22**).
9. Pellet the bacteria by centrifugation for 15 min at 4 °C at 2000 × *g* (*see* **Note 23**). In the meantime, pre-chill 50 ml tubes on ice.

10. Remove carefully the supernatant, and resuspend the bacteria pellet in a total volume of 50 ml of 200 mM column washing buffer (*see* **Notes 8** and **23**). Make sure to keep the sample on ice.
11. Lyse the bacteria by sonicating the bacteria suspension on ice (*see* **Note 24**).
12. Pellet the bacterial debris by centrifugation for 15 min at 4 °C at 2000 × *g*, followed by filtering the supernatant through a 40 μm cell strainer into fresh pre-chilled 50 ml tubes on ice. The easiest is to pipette the supernatant onto the strainer.
13. Add amylose resin to the lysate, and incubate the lysate-amylose resin mix overnight at 4 °C on a gentle shaker (*see* **Note 25**).
14. The next day, collect the protein-amylose resin mix in a pre-chilled glass column at 4 °C (*see* **Note 26**).
15. Discard the flow-through and wash the resin once with 20 ml of 200 mM column washing buffer.
16. Wash the resin once with 1 M column washing buffer, followed by one wash with 20 ml of 200 mM column washing buffer.
17. Pre-chill two 2 ml Eppendorf tubes on ice, and prepare freshly 10 ml of 200 mM column washing buffer supplemented with 10 mM maltose (hereafter termed "elution buffer"). We do not add protease inhibitors to this elution buffer.
18. Once the final wash is completed, add 2 ml of elution buffer to the resin and incubate the elution buffer-resin mix for 15 min at room temperature.
19. Collect the eluate in the pre-chilled 2 ml Eppendorf tube (*see* **Note 27**). Store on ice.
20. Prepare a dialysis tube in sterile ice-cold PBS. Separately, prepare at 4 °C 2 l of sterile ice-cold PBS in a pre-chilled glass beaker containing a stirring bar. Cover the beaker with a foil.
21. Transfer the 2 ml eluate into the dialysis tube, and perform the dialysis against 2 l of prepared PBS (*see* **step 20** above) overnight at 4 °C on a magnetic stirrer.
22. Prepare 2 l of 20 mM Tris–HCl pH 7.5 and store at 4 °C overnight in a beaker covered with foil.
23. The next day, transfer the dialysis tube containing the eluate into the prepared 20 mM Tris–HCl pH 7.5 buffer (*see* **step 22** above) and dialyze for 4–6 h at 4 °C on a magnetic stirrer.
24. Prepare 15 ml and 2 ml tubes on ice.
25. Collect the dialysis tube and carefully transfer the eluate into a 15 ml tube.

26. Briefly centrifuge the sample to collect the eluate at the bottom and transfer the eluate into a 2 ml Eppendorf tube. Store on ice.
27. To concentrate the eluate, transfer the dialyzed eluate into a centrifugal filter unit and process at 4 °C (*see* **Note 28**).
28. Add glycerol to the dialyzed and concentrated eluate to achieve a final concentration of 10% of glycerol. Store on ice.
29. Mix the protein eluate by gentle pipetting, and subsequently collect 1, 5, 10, and 20 μl samples for SDS-PAGE-based testing (*see* **step 30**). Aliquot the remaining protein sample into 50 μl portions and store aliquots at −80 °C.
30. Add 5 μl of 5× Laemmli SDS-PAGE loading buffer to each eluate sample, and boil them at 95 °C for 5 min, before allowing them to cool down to room temperature. In parallel also prepare your protein standard of choice (*see* **Note 29**). To define the quantity and purity of your recombinant protein preparation, separate by SDS-PAGE your samples of 1, 5, 10, and 20 μl alongside your protein standards, followed by Coomassie staining to visualize the proteins.

3.2 Recombinant Kinase Assays Followed by SDS-PAGE and Immunoblotting

1. Prepare kinase assay master mix (*see* **step 6** in Subheading 2.2).
2. Pre-chill the required amount of 1.5 ml Eppendorf tubes on ice (*see* **Note 30**).
3. In the pre-chilled 1.5 ml tubes, mix together 200 ng of your purified recombinant Mal-tagged protein (*see* Subheading 3.1) with 50 ng of recombinant GST-MST1 or GST-MST2 kinases. Using sterile ddH_2O fill up each reaction to a total volume of 15 μl. Store on ice.
4. Add 5 μl of kinase assay master mix per kinase reaction (total volume per reaction is 20 μl). Mix the samples by flicking the tubes, followed by brief centrifugation at 4 °C and immediate placement on ice.
5. Incubate the kinase reactions for 30 min at 30 °C on a thermomixer (*see* **Note 31**).
6. Stop the kinase reactions by adding 10 μl of 5× Laemmli SDS-PAGE loading buffer to each sample and boil them at 95 °C for 5 min.
7. Next allow the samples to cool down, before processing them for SDS-PAGE, followed by standard Western blotting using the materials outlined in Subheading 2.2 (*see also* **Notes 14–18**). Representative results using recombinant full-length Mal-MOB1 proteins as a MST1/2 substrate are shown in Fig. 1 (*see* **Note 32**).

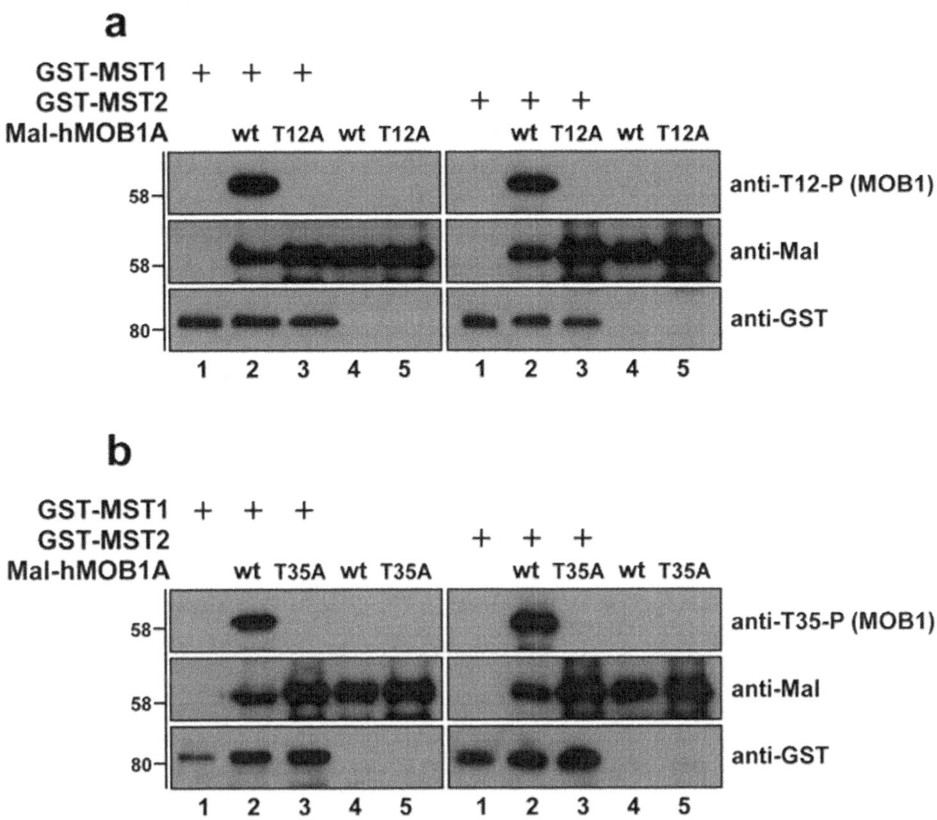

Fig. 1 Kinase assays using recombinant GST-MST1/2 and Mal-MOB1. Recombinant full-length GST-MST1 or GST-MST2 wild-type (wt) were incubated with full-length recombinant Mal-MOB1A(wt) or the indicated phospho-acceptor mutants of Mal-MOB1A (T12A and T35A). Following kinase reactions, samples were processed for SDS-PAGE, followed by Western blot analysis using indicated antibodies. Relative molecular weights are shown. Please note that the displayed figure has been adjusted from the Supplementary Information section of [14]

4 Notes

1. Our preferred vector is pMALc2 (New England Biolabs) (*see* also [14]). If possible, we subclone the full-length cDNA of interest into the *Bam*HI and *Sal*I restriction sites of pMALc2 and subsequently screen for inserts using *Bam*HI and *Hind*III. Make sure to confirm the cDNA insert using sequence analysis of the entire cDNA.
2. *E. coli* BL21 (DE) bacteria carry the gene for T7 RNA polymerase under control of the lacUV5 promoter. IPTG is required to maximally induce expression of the T7 RNA polymerase in order to express recombinant genes. It is suitable for expression from a T7 or T7-lac promoter recognized by the *E. coli* RNA polymerase. A transformation protocol of this

bacteria strain is readily available online at https://www.neb.com/protocols/0001/01/01/transformation-protocol-for-bl21-de3-competent-cells-c2527.

3. We prefer to concentrate our recombinant proteins using centrifugal filter units. While we indicated our product of choice, you can of course use alternative sources. However, please make sure that you use a unit with an appropriate molecular weight cutoff. We prefer a cutoff of 10 kDa, but you may want to vary your cutoff range dependent on the molecular weight of your recombinant protein of interest.
4. You can of course use alternative dialysis tubing systems, but please make sure that you match the molecular weight cutoff of the dialysis tubing and the centrifugal filter units (*see* **Note 3** above).
5. 10% SDS-PAGE gels work nicely for our purposes; however please make sure to adjust the gel percentage to your personal needs.
6. To prepare 1 l of LB medium, dissolve 10 g of peptone, 5 g of yeast extract, and 10 g of NaCl in 950 ml deionized water. Adjust the pH to 7.0 and the total volume to 1 l, before autoclaving. Then allow the solution to cool down to 40 °C, before adding antibiotics (i.e., ampicillin).
7. Prepare LB medium as described in **Note 6**, but add 15 g of agar per liter before autoclaving. Then allow the agar solution to cool down to 40 °C, before adding antibiotics (i.e., ampicillin) and finally pouring into plastic petri dishes.
8. To prepare 500 ml of 200 mM column washing buffer (20 mM Tris–HCl pH 7.5, 1 mM EDTA, and 200 mM NaCl), add to 350 ml of deionized water 10 ml of 1 M Tris–HCl pH 7.5, 1 ml of 0.5 M EDTA pH 8.0, and 20 ml of 5 M NaCl. Mix well and fill up to 500 ml with deionized water. Add protease inhibitors and DTT before use.
9. To prepare 200 ml of 1 M column washing buffer (20 mM Tris–HCl pH 7.5, 1 mM EDTA, and 1 M NaCl), add to 150 ml of deionized water 4 ml of 1 M Tris-HCl pH 7.5, 0.4 ml of 0.5 M EDTA pH 8.0, and 40 ml of 5 M NaCl. Mix well and fill up to 200 ml with deionized water. Add protease inhibitors and DTT before use.
10. To prepare 800 ml of Coomassie staining solution (25% methanol, 10% acetic acid, and 0.1% (w/v) Coomassie), add to 500 ml of deionized water 200 ml of methanol (99.5%) and 100 ml of acetic acid (80%). Mix well and dissolve two Coomassie tablets (#B4921, Merck).
11. To prepare 1 l of destaining solution (10% ethanol, 10% acetic acid), add to 780 ml of deionized water 100 ml ethanol

(≥99.5%) and 120 ml of acetic acid (80%). Mix well before storage.

12. To prepare 100 ml of 5× Laemmli SDS-PAGE loading buffer, add to 57.5 ml of deionized water 12.5 ml of 1 M Tris–HCl pH 6.8, 25 ml of 20% SDS, and 5 ml of beta-mercaptoethanol (99%). Mix gently and add bromophenol blue to a final concentration of 0.02% (w/v).
13. Our preferred sources of recombinant full-length wild-type GST-MST1 and GST-MST2 are #M9697 and #S6573 from Sigma (Merck). Good alternatives are also recombinant full-length wild-type GST-MST1 and GST-MST2 from Carna Biosciences (#07-116 and #07-117) or Invitrogen (#PV3854 and #PV4805).
14. Our SDS-PAGE and Western blotting equipment are from Bio-Rad (Mini-Protean Tetra Cell system), which we use together with Immobilon PVDF membranes from Millipore (#IPVH00010) and our self-made standard buffers (SDS running buffer, 25 mM Tris, 192 mM glycine, 0.1% SDS; Western blot transfer buffer, 25 mM Tris, 192 mM glycine, 20% methanol; TBS-T, 100 mM Tris-HCl pH 7.4, 300 mM NaCl, 0.5% (v/v) Tween 20). All buffers are stored at room temperature. Of course, you can also use your SDS-PAGE and immunoblotting system of choice, as long as you are certain that it is robust and reliable.
15. Our preferred molecular weight marker is the blue pre-stained protein standard with broad range from New England BioLabs (#P7706). However, at the end, any reliable molecular weight marker will do.
16. For the described protocol, we used the following antibodies: anti-T12-P (MOB1; #8843, used at 1:1000), anti-T35-P (MOB1; #8699, used at 1:1000), anti-Mal (Maltose-binding protein; #E8032, used at 1:10,000), and anti-GST (#ab6613, used at 1:5000) from Cell Signaling, New England BioLabs, and Abcam, respectively. It is essential that the primary antibodies are highly specific. If possible, try to use primary antibodies from different host species (like rabbit anti-T12-P, mouse anti-Mal, and goat anti-GST), so that you can swiftly re-probe the same membrane with all your antibodies of interest without the need of any membrane stripping.
17. Our secondary antibodies of choice are sheep donkey anti-rabbit HRP (#NA934), anti-mouse HRP (#NA931), and donkey anti-goat HRP (#sc-2020) from GE Healthcare and Santa Cruz Biotechnology, respectively. We recommend using these secondary antibodies in a range of 1:5000 to 1:10,000 dilutions in TBS-T with 5% milk.

18. A cheap and easy way to prepare your own ECL solutions is as follows. For "ECL solution A," mix, for a total volume of 50 ml, 5 ml of 1 M Tris–HCl pH 8.5, 44.28 ml of sterile ddH2O, 0.5 ml of 250 mM luminol (prepare in DMSO and store in aliquots at 4 °C), and 0.22 ml of 90 mM p-coumaric acid (prepare in DMSO and store in aliquots at 4 °C). For "ECL solution B," mix, for a total volume of 50 ml, 5 ml 1 M Tris–HCl pH 8.5, 45 ml of sterile ddH2O, and 31 μl of H_2O_2 (30%). To initiate the ECL reaction, mix equal amounts of "ECL solution A" and "ECL solution B" at room temperature. Store solutions at 4 °C until use.
19. Our bacteria transformation protocol is as follows: thaw competent bacteria on ice while allowing LB agar plates with the appropriate antibiotic to warm up at room temperature. Add about 10 ng of plasmid to competent cells, and incubate the plasmid/bacteria mixture for 30 min on ice, before heat shocking at 42 °C for 45 s, followed by placing the mixture back on ice for at least 2 min. Add 500 μl LB medium without antibiotic, and grow the plasmid/bacteria mixture for 30 min or longer at 37 °C in a shaking incubator. Then pellet the bacteria for 3 min at 3000 rpm in a standard tabletop centrifuge, remove most of the supernatant, and resuspend the bacteria in about 100 μl fresh LB medium, before plating the bacteria onto the agar plates. Incubate the plates overnight at 37 °C and finally place the plates with single bacterial colonies at 4 °C.
20. We check protein expression on a small scale as follows: inoculate 3 ml of LB medium with antibiotic with a single bacteria colony and incubate at 37 °C overnight in a shaking incubator. The next day, split the overnight culture into two 1 ml cultures and add 2 ml of fresh LB medium with antibiotic to each one, so that in each the final volume is 3 ml. Incubate both cultures for 1 h at 30 °C in a shaking incubator, before adding IPTG at a final concentration of 0.2 mM (dilute the 100 mM IPTG stock at 1/500) to one of the cultures (which we normally label as "plus"). Do not add any IPTG to the other culture (the control "minus" one). Incubate "plus" and "minus" for a further 3 h at 30 °C in a shaking incubator. Pellet 500 μl of each for 30 seconds at 11,000 rpm in a standard tabletop centrifuge. Remove the supernatant, and add 100 μl of 5× Laemmli SDS-PAGE loading buffer to each culture. Vortex for 30 s, boil the samples for 5 min at 95 °C, vortex once more, boil a second time, and then allow the sample to cool down to room temperature. Finally, analyze varying volumes of the "minus" and "plus" cultures by SDS-PAGE, followed by Coomassie staining (normally we analyze 10, 20, and 40 μl of each sample side by side). Check at the protein size of interest whether you can detect your Mal-tagged protein in the IPTG-treated sample (the

"plus" sample), while you should not be able to detect any expression of your Mal-tagged protein in the control ("minus" sample).

21. The important point is to achieve an OD of 0.8–1.2 at 600 nm; hence the incubation time at 30 °C may vary and needs to be adjusted accordingly.
22. Please note that the optimal IPTG induction times will vary depending on the protein of interest. In case you choose longer induction times, we recommend lowering the incubation temperature.
23. Although we prefer to complete the protein production and purification in 1 day, it is possible to freeze the bacterial pellets at this step at −20 °C for a few days and then to recommence by adding the 200 mM column washing buffer. In general, we prefer to split the 1000 ml culture into two 500 ml centrifugation tubes, followed by resuspension of each pellet in 25 ml of 200 mM column washing buffer (the total volume for 1000 ml culture remains at 50 ml of 200 mM column washing buffer).
24. We alternate between the sonication of each 25 ml sample. Normally, we perform six rounds of 15-second sonication per sample at 70% amplitude (model: VCx 130, Sonics Vibra-Cell). At the end any powerful sonicator of your choice should be able to do the job, as long as you make sure that you perform the sonication steps on ice to avoid temperature changes.
25. The binding capacity of the amylose resin is about 3 mg/ml, so a few milliliters are fully sufficient to purify several milligrams of your protein of interest. For the purification of protein from a 1 l bacterial culture, we normally use 2–3 ml of amylose resin.
26. We recommend that you perform **steps 14–16** in Subheading 3.1 in a cold room and make sure that all wash buffers are supplemented with protease inhibitors. Furthermore, ensure that the resin never runs dry during the washing steps.
27. We recommend collecting 2 × 1 ml of eluate in two separate tubes, which afterward can be pooled for the dialysis step. It makes the sample handling easier.
28. The centrifugation time and speed will depend on the manufacturer of your centrifugal filter units. In general, we aim to concentrate the 2 ml of eluate down to 300–500 μl.
29. Our protein standard of choice is BSA; hence, we prepare BSA standards of 100 ng, 500 ng, 1 μg, and 5 μg to run along our recombinant protein of interest for subsequent protein quantification.
30. In the described protocol, we are using recombinant Mal--MOB1 as a substrate for MST1/2. However, any recombinant protein could be tested as MST1/2 substrate using our

protocol. The important points when designing the experimental plan are as follows: (1) include a kinase reaction including the kinase without the substrate; (2) include a kinase reaction without the kinase, but with the substrate; and (3) if possible also include the testing of a phospho-acceptor substrate mutant.

31. We prefer to incubate our samples in a thermomixer at 600–700 rpm at 30 °C.

32. The representative results shown in Fig. 1 illustrate important points that need to be considered when designing recombinant kinase assays. First, the incubation of MST1/2 kinases alone (without any substrate) did not result in any detectable phosphorylation of recombinant Mal-MOB1A on Thr12 and Thr35. Second, the incubation of recombinant substrates alone (without any MST1/2 kinases) did not generate a detectable phosphorylation of Mal-MOB1A on Thr12 and Thr35. Third, the incubation of MST1/2 with recombinant Mal-MOB1A(T12A) or Mal-MOB1A(T35A) did not produce a detectable phosphorylation of Mal-MOB1A on Thr12 or Thr35, respectively. Thus, one can conclude that the anti-T12-P and anti-T35-P antibodies specifically detect the phosphorylation of MOB1A on Thr12 and Thr35, respectively, a phosphorylation process which is fully dependent on the presence of MST1/2 in our experimental settings.

Acknowledgments

The Hergovich laboratory was supported by the Wellcome Trust (090090/Z/09/Z), BBSRC (BB/I021248/1), Worldwide Cancer Research (AICR; 11-0634), UCL Cancer Research UK Centre and the National Institute for Health Research University College London Hospitals Biomedical Research Centre.

References

1. Johnson R, Halder G (2014) The two faces of Hippo: targeting the Hippo pathway for regenerative medicine and cancer treatment. Nat Rev Drug Discov 13(1):63–79. https://doi.org/10.1038/nrd4161
2. Yu FX, Zhao B, Guan KL (2015) Hippo pathway in organ size control, tissue homeostasis, and cancer. Cell 163(4):811–828. https://doi.org/10.1016/j.cell.2015.10.044
3. Meng Z, Moroishi T, Guan KL (2016) Mechanisms of Hippo pathway regulation. Genes Dev 30(1):1–17. https://doi.org/10.1101/gad.274027.115
4. Irvine KD, Harvey KF (2015) Control of organ growth by patterning and hippo signaling in Drosophila. Cold Spring Harb Perspect Biol 7(6):a019224
5. Sun S, Irvine KD (2016) Cellular organization and cytoskeletal regulation of the Hippo signaling network. Trends Cell Biol 26 (9):694–704
6. Pan D (2010) The hippo signaling pathway in development and cancer. Dev Cell 19 (4):491–505
7. Harvey KF, Zhang X, Thomas DM (2013) The Hippo pathway and human cancer. Nat Rev

Cancer 13(4):246–257. https://doi.org/10.1038/nrc3458
8. Hoa L, Kulaberoglu Y, Gundogdu R, Cook D, Mavis M, Gomez M, Gomez V, Hergovich A (2016) The characterisation of LATS2 kinase regulation in Hippo-YAP signalling. Cell Signal 28(5):488–497. https://doi.org/10.1016/j.cellsig.2016.02.012
9. Praskova M, Xia F, Avruch J (2008) MOBKL1A/MOBKL1B phosphorylation by MST1 and MST2 inhibits cell proliferation. Curr Biol 18(5):311–321. https://doi.org/10.1016/j.cub.2008.02.006
10. Ni L, Zheng Y, Hara M, Pan D, Luo X (2015) Structural basis for Mob1-dependent activation of the core Mst-Lats kinase cascade in Hippo signaling. Genes Dev 29 (13):1416–1431. https://doi.org/10.1101/gad.264929.115
11. Kim SY, Tachioka Y, Mori T, Hakoshima T (2016) Structural basis for autoinhibition and its relief of MOB1 in the Hippo pathway. Sci Rep 6:28488. https://doi.org/10.1038/srep28488
12. Couzens AL, Xiong S, Knight JD, Mao DY, Guettler S, Picaud S, Kurinov I, Filippakopoulos P, Sicheri F, Gingras A-C (2017) MOB1 mediated phospho-recognition in the core mammalian Hippo pathway. Mol Cell Proteomics 16(6):1098–1110
13. Xiong S, Couzens AL, Kean MJ, Mao DY, Guettler S, Kurinov I, Gingras A-C, Sicheri F (2017) Regulation of protein interactions by Mps One Binder (MOB1) phosphorylation. Mol Cell Proteomics 16(6):1111–1125
14. Kulaberoglu Y, Lin K, Holder M, Gai Z, Gomez M, Assefa Shifa B, Mavis M, Hoa L, Sharif AAD, Lujan C, Smith ESJ, Bjedov I, Tapon N, Wu G, Hergovich A (2017) Stable MOB1 interaction with Hippo/MST is not essential for development and tissue growth control. Nat Commun 8(1):695. https://doi.org/10.1038/s41467-017-00795-y
15. Vrabioiu AM, Struhl G (2015) Fat/Dachsous signaling promotes Drosophila wing growth by regulating the conformational state of the NDR kinase Warts. Dev Cell 35(6):737–749
16. Zanconato F, Cordenonsi M, Piccolo S (2016) YAP/TAZ at the roots of cancer. Cancer Cell 29(6):783–803
17. Hergovich A (2016) The roles of ndr protein kinases in hippo signalling. Genes 7(5):21
18. Zhang L, Tang F, Terracciano L, Hynx D, Kohler R, Bichet S, Hess D, Cron P, Hemmings BA, Hergovich A (2015) NDR functions as a physiological YAP1 kinase in the intestinal epithelium. Curr Biol 25 (3):296–305
19. Hergovich A (2011) MOB control: reviewing a conserved family of kinase regulators. Cell Signal 23(9):1433–1440. https://doi.org/10.1016/j.cellsig.2011.04.007
20. Sharif AA, Hergovich A (2017) The NDR/LATS protein kinases in immunology and cancer biology. In: Seminars in cancer biology. Elsevier.
21. Galan JA, Avruch J (2016) MST1/MST2 protein kinases: regulation and physiologic roles. Biochemistry 55(39):5507–5519
22. Qin F, Tian J, Zhou D, Chen L (2013) Mst1 and Mst2 kinases: regulations and diseases. Cell Biosci 3(1):31

Part IV

The Hippo Pathway and Mouse Models

Chapter 25

Visualizing HIPPO Signaling Components in Mouse Early Embryonic Development

Tristan Frum and Amy Ralston

Abstract

The HIPPO signaling pathway plays an early and essential role in mammalian embryogenesis. The earliest known roles for HIPPO signaling during mouse development include segregating fetal and extraembryonic lineages and establishing the pluripotent progenitors of embryonic stem (ES) cells. In the mouse early embryo, HIPPO signaling responds to multiple cell biological inputs, including cell polarization, cytoskeleton, and cell environment, to influence gene expression and the first cell fate decisions in development. Methods to monitor and manipulate HIPPO signaling in the mouse early embryo are fundamental to discovering mechanisms regulating pluripotency in vivo, but properties of the early embryo, such as small cell number and spherical architecture, pose unique challenges for signaling pathway analysis. Here, we share approaches for visualizing HIPPO signaling in mouse early embryos. In addition, these methods can be applied to visualize HIPPO signaling in other spherical or cystic structures comprised of relatively few cells, such as organoids, or for the examination of other signaling pathways in these contexts.

Key words Cell polarity, Stem cells, Infertility, Organoids, Single cell, Confocal

1 Introduction

The HIPPO signaling pathway performs important functions in development and tissue homeostasis [1–8]. The recent discovery that the HIPPO signaling pathway regulates the first cell fate decision in mammalian development has generated interest in investigating HIPPO signaling in the mammalian preimplantation embryo. Shortly after fertilization, when the embryo is a cluster of around 16 cells, the first cell fate decision establishes progenitors of the fetus and the placenta. This first cell fate decision is essential for normal development and for establishment of pluripotent progenitors of embryonic stem (ES) cells. Therefore, discovering the roles and regulation of HIPPO signaling in the early embryo is significant to stem cell, developmental, and reproductive biology.

The discovery that HIPPO signaling regulates the first cell fate decision was a major advance because it helped answer the long-

Alexander Hergovich (ed.), *The Hippo Pathway: Methods and Protocols*, Methods in Molecular Biology, vol. 1893,
https://doi.org/10.1007/978-1-4939-8910-2_25,

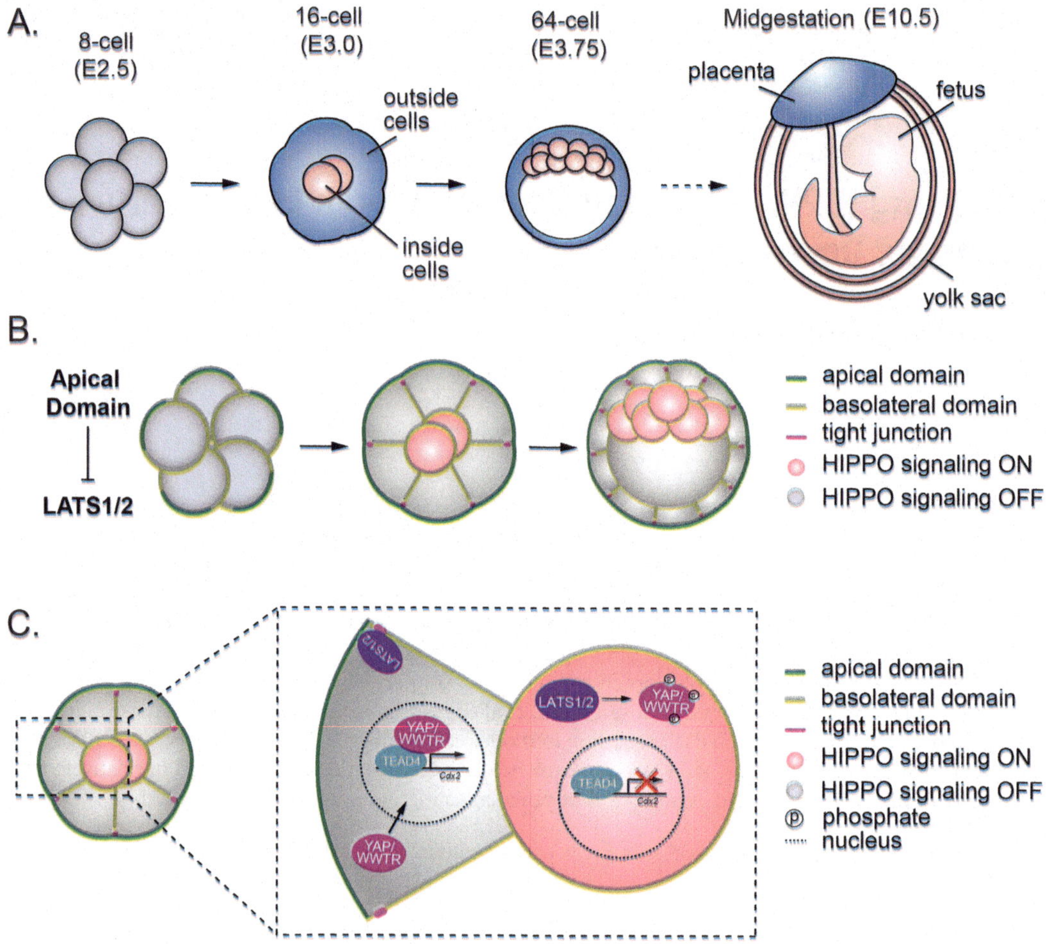

Fig. 1 Roles and regulation of HIPPO signaling in the mouse early embryo. (**a**) Fetal (pink) and placental (blue) lineages are specified in a position-dependent manner. (**b**) The apical domain is first established on the outside surface of embryonic cells at the eight-cell stage and is subsequently maintained in outside cells, which also assemble tight junctions and adherens junctions. Meanwhile, cells that move internally become depolarized. The pro-HIPPO kinases LATS1/LATS2 are sequestered and inactive in polarized cells. (**c**) In outside cells, LATS1/LATS2 are prevented from phosphorylating transcriptional activators YAP1 and WWTR1, which thus translocate to the nucleus where they partner with TEAD4 to induce expression of genes that reinforce trophectoderm cell fates. In inside cells, LATS1/LATS2 phosphorylate YAP1 and WWTR1, which therefore remain cytoplasmic, and TEAD4 is unable to promote gene expression

standing question as to how cell position influences the first cell fate decision. During preimplantation stages, cell fates are responsive to positional information, with outside cells acquiring trophectoderm fate (future placenta) and inside cells acquiring inner cell mass fate (future fetus) (Fig. 1a). Remarkably, changes in cell position are sufficient to induce changes in cell fate, indicating cell fates are highly plastic [9–13]. However, the mechanisms enabling embryonic cells to interpret positional information to regulate gene expression and cell fate were largely unknown. The HIPPO signaling pathway was shown to provide this missing link [14], with cells located inside the embryo exhibiting higher levels of HIPPO signaling pathway activity than cells on the surface of the embryo (Fig. 1b).

The mechanisms that regulate HIPPO signaling pathway activity in the preimplantation embryo are an active and exciting area of investigation. In this context, no ligand/receptor combination is known to kick-start HIPPO signaling, as in traditional cell signaling. Rather, apicobasal polarity proteins act to repress HIPPO signaling activity by spatially segregating components required to activate LATS1/LATS2 kinases [15–18]. Thus, outside cells, which are polarized, contain inactive LATS1/LATS2, while inside cells, which are non-polarized, contain active LATS1/LATS2 (Fig. 1c). More recently, additional studies have pointed to myosin-dependent regulation of the HIPPO signaling pathway in the preimplantation embryo, suggesting that mechanical force may also contribute to position-dependent regulation of HIPPO signaling activity [19]. Therefore, the preimplantation mouse embryo provides a useful model for investigating how cell polarity and mechanical forces regulate the HIPPO signaling pathway.

The preimplantation embryo is also a useful model for understanding how HIPPO signaling regulates gene expression to establish cell fates. HIPPO signaling regulates gene expression through the transcriptional coactivators YAP1 and WWTR1, which partner with TEAD4 to promote expression of *Cdx2* and *Gata3* and repress expression of *Sox2* in outside cells [14, 20–23]. In turn, this regulated gene expression promotes cell fate decisions, with CDX2 and GATA3 promoting differentiation of the placenta lineage and SOX2 promoting pluripotency in the fetal lineage [22, 24–26]. Curiously, although YAP1/TEAD4 repress expression of *Sox2* in the trophectoderm [23], YAP1/TEAD4 have been shown to promote expression of *Sox2* in pluripotent ES cells [27]. This striking difference highlights the context-dependent nature of the HIPPO signaling pathway and underscores the importance of discovering mechanisms regulating the roles and regulation of HIPPO signaling in vivo.

Here, we provide protocols and describe reagents for evaluating HIPPO signaling in vivo during preimplantation mouse development. These protocols preserve the three-dimensional structure of the preimplantation embryo, which enables single-cell analysis of protein expression and localization. These protocols can also be used for examining other signaling pathways in preimplantation mouse embryos or even embryos of other mammalian species. Additionally, these protocols can be adapted to analysis of non-embryo cystic structures, such as stem cell-derived organoids.

2 Materials

2.1 Timed Matings

Female and male mice, at least 6 and 8 weeks old, respectively.

2.2 Embryo Harvest

1. Stereomicroscope.
2. Flushing needle (blunted 32-gauge, ½ in. length).
3. 1 ml disposable syringe (e.g., Becton Dickson, 309659).
4. Fine forceps (e.g., Fine Scientific Tools, 11254-20).
5. Coarse forceps (e.g., Fine Scientific Tools, 11210-10).
6. Surgical scissors (e.g., Fine Scientific Tools, 14090-09).
7. Fine scissors (e.g., Fine Scientific Tools, 14090-11).
8. M2 medium (e.g., Millipore, MR-015-D).
9. 70% (v/v) ethanol.
10. 10 cm plastic petri dish.

2.3 Embryo Transfer Device

1. Aspirator mouthpiece (e.g., Fisher Scientific, NC9048719).
2. Latex tubing (e.g., VWR, 62996-350).
3. Pipet tip, with filter for 101–1000 μl (e.g., USA Scientific, 1126-7810).
4. Pasteur pipets (9 in., e.g., VWR 14672-380).
5. Bunsen burner.

2.4 Embryo Fixation and Staining

1. 4-Well dish for IVF (e.g., Thermo Fisher, 179830).
2. Primary antibodies for detecting HIPPO signaling pathway components (*see* Table 1).
3. Fluorophore-conjugated secondary antibodies (e.g., Jackson ImmunoResearch).
4. 1× Dulbecco's Phosphate-Buffered Saline with $MgCl_2$ and $CaCl_2$ (DPBS): 137 mM NaCl, 2.7 mM KCl, 8 mM Na_2HPO_4, 1.5 mM KH_2PO_4, 0.9 mM $CaCl_2$, 0.5 mM $MgCl_2$, pH 7.4 (e.g., Thermo Fisher 14040-133).
5. Fixation solution: 4% (v/v) EM-grade formaldehyde in DPBS (e.g., Polysciences, 0408-1).
6. Permeabilization solution: 0.5% (v/v) Triton X-100 in DPBS.
7. Blocking solution: 0.2% (v/v) Triton X-100 (e.g., Sigma, X100), 10% fetal bovine serum (e.g., HyClone, SH30396.03) in DPBS.
8. Nuclear stain diluted to working concentration in blocking solution (*see* Table 2).
9. RNAse A (e.g., Qiagen, 19101).

Table 1
Antibodies for immunolocalization of regulators and targets of HIPPO signaling during preimplantation mouse development

Target	Host species	Supplier	Catalog number	Exemplary image	Suggested working conc.	Usage
PARD6B	Rabbit	Novus Biologicals	NBP1-87337	(Alarcon, 2010)	1:100	Apical membrane component
aPKCz	Mouse	Santa Cruz Biotechnology	U3254	(Zhu et al., 2017)	1:100	Apical membrane component
pERM	Rabbit	Cell Signaling Technology	3149	(Kono et al., 2014)	1:1000	Apical membrane component
CDH1	Rat	Sigma Aldrich	U3254	(Thomas et al., 2004)	1:500	Basolateral membrane component
YAP1	Mouse	Santa Cruz Biotechnology	sc101199	(Nishioka et al., 2009)	1:200	Excluded from the nucleus by HIPPO signaling
WWTR1	Rabbit	Cell Signaling Technology	2149	(Nishioka et al., 2009)	1:500	Excluded from the nucleus by HIPPO signaling
TEAD4	Mouse	Abcam	ab58310	(Hirate et al., 2012)	1:1000	Activity repressed by HIPPO signaling
ACTIN	Phalloidin	Thermo Fisher	Various	(Zhu et al., 2017)	1:50 (add with secondary antibodies)	Stains actin
CDX2	Mouse	BioGenex	CDX2-88	(Strumpf et al., 2005)	1:200	Repressed by HIPPO signaling
NANOG	Rabbit	ReproTech	RCAB002P-F	(Dietrich and Hiiragi, 2007)	1:400	Repressed by HIPPO signaling
SOX2	Goat	Neuromics	GT15098	(Wicklow et al., 2014)	1:2000	Promoted by HIPPO signaling

Table 2
Nuclear stains permitting embryo staging

DNA stain	Suggested laser line wavelength (nm)	Peak emission wavelength (nm)	Suggested working conc.	Incubation time (min)	For mounting
DAPI	405	460	1 μg/mL	15	Blocking solution (no DAPI)
Draq5	633	696	1:400	5	Draq5 working solution
YOYO1	488	508	1:400 + 20 μg/mL RNAse A	20	YOYO1 working solution

2.5 Confocal Microscopy

1. Microscope Cover Glasses, No. 1, 24 × 60 mm (e.g., VWR, 16004-096).
2. Microscope Cover Glasses, No. 1, 24 × 40 mm (e.g., VWR, 091514-9).
3. Secure-Seal spacers, 0.12 mm depth (e.g., Thermo Fisher, S24735).
4. 10 cm petri dishes.
5. Laser scanning confocal microscope with water immersion, ×20–40 objective.

2.6 Recovery and Genotyping of Embryos by PCR

1. Tissue PCR Kit (e.g., Sigma, REDExtract-N-Amp XNAT).
2. 1× Dulbecco's Phosphate-Buffered Saline without $MgCl_2$ or $CaCl_2$: 137 mM NaCl, 2.7 mM KCl, 8 mM Na_2HPO_4, 1.5 mM KH_2PO_4, pH 7.4 (e.g., Thermo Fisher, 14190-144).
3. Thermocycler.
4. PCR tubes.
5. PCR genotyping primers to detect alleles of interest.

2.7 Culturing Embryos

1. 35 mm dish (e.g., VWR, 25382-064).
2. KSOM Embryo Culture Medium (e.g., Millipore, MR-015-D).
3. ES Cell Grade Mineral Oil (e.g., Millipore, ES-005-C).
4. Cell culture incubator with 5% CO_2.

3 Methods

3.1 Timed Natural Matings

1. Transfer a 6-week to 12-month-old female mouse to cage containing a singly housed, 2–9-month-old male stud mouse.

Checking females for visible signs of estrus can lead to improved prediction of time of mating [28]. Additionally, pairing the mice at a consistent time of day can lead to improved prediction of embryo stage in pregnant females (*see also* **Notes 1** and **2**).

2. The next day, check the vaginal opening of the female mouse for the presence of a copulation plug. To ensure copulation plug is not missed, mice should be checked within 9 h of the middle of the dark cycle (i.e., by 9 a.m. for colonies that maintain mice on a 7 a.m. to 5 p.m. dark cycle). For most strains, breeding should result in a plugged female within 4–5 days.
3. After detection of a copulation plug, the female mouse may remain paired with the male stud mouse or may be transferred to her own cage. At noon on the day that the plug is detected, the embryos are considered to have reached the E0.5 stage, since mating around midnight is defined as E0.0 (*see* Table 3).

3.2 Making an Embryo Transfer Device

1. Assemble mouthpiece, tubing, and p1000 pipet tip as in Fig. 2. Be careful to store the transfer device mouthpiece in a clean, food-safe, sealable bag when not in use (for alternative, hand-operated transfer device, *see* **Note 3**).
2. Make several finely pulled embryo transfer pipets by holding the thin portion of a Pasteur pipet in the inner cone of a 2 in. flame of a Bunsen burner for several seconds, while rotating the pipet (Fig. 2a). As soon as the glass becomes noticeably soft, remove from the flame and quickly pull the ends in opposite directions. Break the two pieces apart, and then graze the smaller piece along the pulled end of the larger piece. This will create an orthogonal break in the end of the pulled pipet. Many shapes and sizes of pulled pipet are useable, but a typical embryo transfer pipet will possess a finely pulled region that is 200–300 μm in diameter and 2–3 in. in length. If desired, flame polish the tip of the embryo transfer pipet by holding the tip near the Bunsen burner flame for a few seconds.
3. Insert the embryo transfer pipet into the open end of the embryo transfer device (Fig. 2b).

3.3 Harvesting Preimplantation Embryos

1. Euthanize pregnant female mouse on the appropriate day, using your institution-approved method.
2. Place the animal on its back and wet the abdomen with 70% ethanol. Using surgical scissors, make a small lateral incision 1–3 cm anterior to the vaginal opening (Fig. 3a). Grip both sides of the incision firmly with gloved hands, and tear the skin to completely expose the peritoneal membrane (Fig. 3b).

Table 3
Timing and expected cell numbers of developmental stages in preimplantation mouse embryos

Stage	E0.5	E1.5	E2.5	E2.75	E3.0	E3.5	E4.5
Timing (E0.0 = 12 a.m. day 0)	12 p.m. day 0	12 p.m. day 1	12 p.m. day 2	6 p.m. day 2	12 a.m. day 3	12 p.m. day 3	12 p.m. day 4
No. of cells	1	2	8	8	16	32–64	>100
Morphology	Zygote	Two cells	Eight cells	Eight cells compacted	Morula	Blastocyst	Implanted
Location	Ampulla	Oviduct				Uterus	

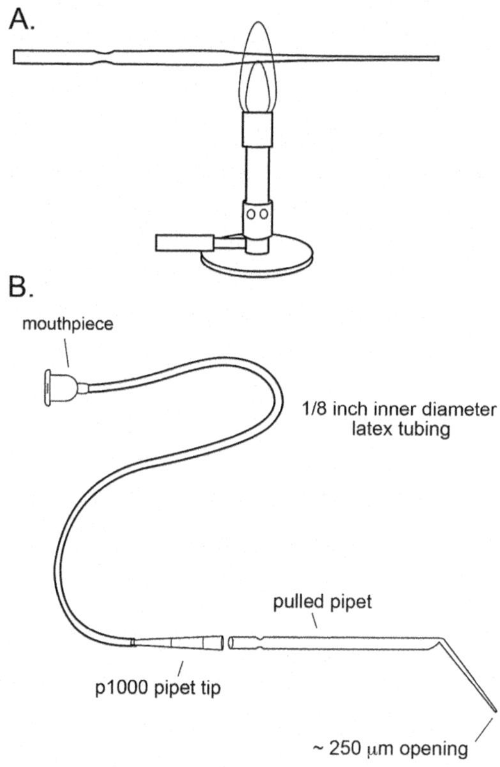

Fig. 2 Preparing embryo transfer device. (**a**) Pasteur pipet softened in the flame of a Bunsen burner and then pulled to a fine (250 μm diameter) tip. Pulling an angled tip, as shown in panel (**b**), can provide a more ergonomically designed transfer device. (**b**) Components of the embryo transfer device are assembled

3. Using coarse forceps and fine scissors, open the peritoneal cavity by cutting the membrane laterally, left and right to expose intestines and fat pads (Fig. 3c). Lift the intestines and fat pads to the side to expose the two uterine horns (Fig. 3c′).

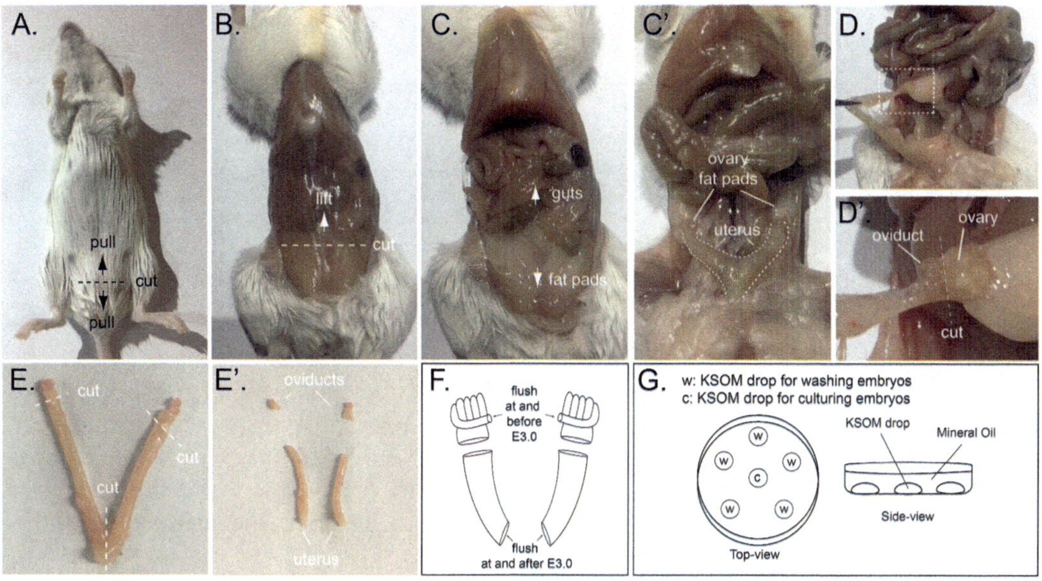

Fig. 3 Harvesting preimplantation embryos. (**a**) Sacrificed, pregnant mouse. (**b**) Fur is peeled back to expose peritoneum. (**c**) Peritoneal membrane is peeled back to expose organs. (**c′**) Organs are displaced to expose reproductive tract. (**d**) One uterine horn stretched to the side. (**d′**) Close-up of junction between the oviduct and ovary. (**e**) Isolated reproductive tract, with mesometrial membranes removed. (**e′**) Oviducts and uterine horns separated. (**f**) Cartoon depiction of (**e′**). (**g**) Sample layout of embryo culture dish. In a 35 mm dish, embryos collected in M2 are first washed through several drops of equilibrated KSOM (indicated by w) under light mineral oil, to dilute M2. Embryos are then pooled in the final drop of equilibrated KSOM (indicated by **c**) and then transferred to 37 °C incubator with 5% CO_2

Posteriorly, the two uterine horns meet at the cervix. Anteriorly, each uterine horn terminates with an oviduct and ovary, held together by a thin membrane called the bursa.

4. Remove the uterus and oviducts with three incisions. First, using coarse forceps, grip a single uterine horn midway between the cervix and ovary. Gently stretch the uterus until somewhat taut and the oviduct/ovary are well exposed (Fig. 3d). Use the blade of the scissors to slice the bursa, between oviduct and ovary (Fig. 3d′). Repeat on the other side. Make a final cut across the base of the uterus and transfer the two horns, with intact oviducts, to a 10 cm dish.
5. Separate each uterine horn by cutting vertically at the base of the uterus (Fig. 3e–e′). Cut each uterine horn laterally 0.2–0.3 cm below the oviduct to separate the oviduct from the uterus (Fig. 3e-e′).
6. Recover embryos by flushing M2 medium through the appropriate section of the reproductive tract (Fig. 3f). To recover embryos prior to E3.0, insert the blunted flushing needle into the opening at the end of the oviduct. Flush 0.5 mL of room temperature M2 medium through each oviduct. To recover

Table 4
Average embryo cell number in differing conditions

Condition	Culture start	Embryos fixed	[O_2]	Avg. no. cells	St dev. no. cells	Morphology
Flushed	n.a.	8 a.m., day 3	n.a.	34.7	7.2	Blastocyst
Cultured	12 p.m., day 1	8 a.m., day 3	20%	16.7	3.4	Morula
Cultured	12 p.m., day 1	8 a.m., day 3	5%	24.7	6.4	Blastocyst and Morula

embryos after E3.0, insert the blunted flushing needle into either end of the uterine horn, and flush each uterine horn with ~1 mL room temperature M2 medium. At E3.0 it is recommended to flush both the oviducts and uterine horns since embryos are transitioning from the oviduct to uterus (Table 3).

7. Make five 100 μl drops of M2 medium on a new 10 cm dish. Using appropriate embryo transfer device (Fig. 2 and **Note 3**), collect embryos from flushed M2 medium, and wash embryos by transferring the embryos through each drop of M2 medium. Embryos are now ready for fixation as described in Subheading 3.4. Alternatively, embryos may continue development in culture (*see* Subheading 3.7). However, freshly flushed embryos should be used whenever possible, since developmental timing is slowed in culture (Table 4 and **Note 4**).

3.4 Embryo Fixation and Immunostaining

1. Prepare a 4-well or multi-well dish by adding 500 μl fixation solution, 500 μl permeabilization solution, and 500 μl blocking solution to separate wells of the 4-well dish. Adjust the volume of buffers to cover the entire surface of the well if using format other than a 4-well dish. Let all buffers equilibrate to room temperature.
2. Transfer embryos from the final drop of M2 medium to the well containing fixation solution. Embryos may float initially but will eventually settle. Incubate embryos in fixation solution for 10 min at room temperature.
3. Transfer embryos from fixation solution to permeabilization solution with minimal carryover. Incubate embryos in permeabilization solution for 30 minutes at room temperature.
4. Transfer embryos from permeabilization solution to blocking solution with minimal buffer carryover. Incubate embryos in blocking solution for 1 h at room temperature. Embryos may be stored in blocking solution at 4 °C for up to 1 week with minimal loss in signal quality. Store embryos or proceed to primary antibody staining.

5. Dilute primary antibodies (Table 1) to working concentration in blocking solution, and transfer 200–500 μl to a new well of the multi-well dish. Transfer embryos from blocking solution to primary antibodies in blocking solution with minimal carry-over. Incubate embryos in primary antibodies in blocking solution overnight at 4 °C.
6. Prepare a new 4-well or multi-well dish for processing embryos by placing 500 μl of blocking solution in separate wells of the dish. To wash embryos, transfer embryos from primary antibodies in blocking solution to a well of freshly aliquoted blocking solution containing no primary antibody. Incubate embryos in blocking solution for 30 min at room temperature.
7. Dilute the secondary antibodies to working concentration with blocking solution, and aliquot 500 μl of secondary antibody in blocking solution to a new well of the multi-well dish. Transfer embryos from blocking solution to secondary antibody diluted in blocking solution. Incubate embryos in secondary antibodies in blocking solution for 1 h at room temperature in the dark.
8. Wash embryos by transferring from secondary antibodies in blocking solution to a second well of blocking solution as in **step 6**. Incubate embryos in blocking solution for 30 minutes at room temperature in the dark.
9. Transfer embryos to your choice of nuclear stain (Table 2 and **Note 5**), and incubate according to Table 2.

3.5 Mounting for Imaging by Confocal Microscopy

1. Using household scissors, cut the spacers into four equal sections, with two wells per section (Fig. 4a and *see* **Note 6**).
2. Peel the backing from the adhesive side of the spacer (Fig. 4b). The backing is opaque and very weakly adhesive on one side, while the spacer is clear and strongly adhesive on one side. Discard the backing.
3. Attach the adhesive side of the spacer to the large (24 mm × 60 mm) cover glass (Fig. 4c). Smooth the spacer down with the edge of the pipet tip. If the spacer is not smooth, then liquid will run out of the well when the top cover glass is placed.
4. Transfer the large cover glass with spacer to the lid of a 10 cm petri dish (Fig. 4d).
5. Add 18–20 μl appropriate mounting solution (Table 2) to each well within the spacer. Using the pipet tip, drag the solution to the edges to completely fill the well (Fig. 4e). If embryos are to be recovered for genotyping after imaging, number the cover glass near each well using a permanent marker.

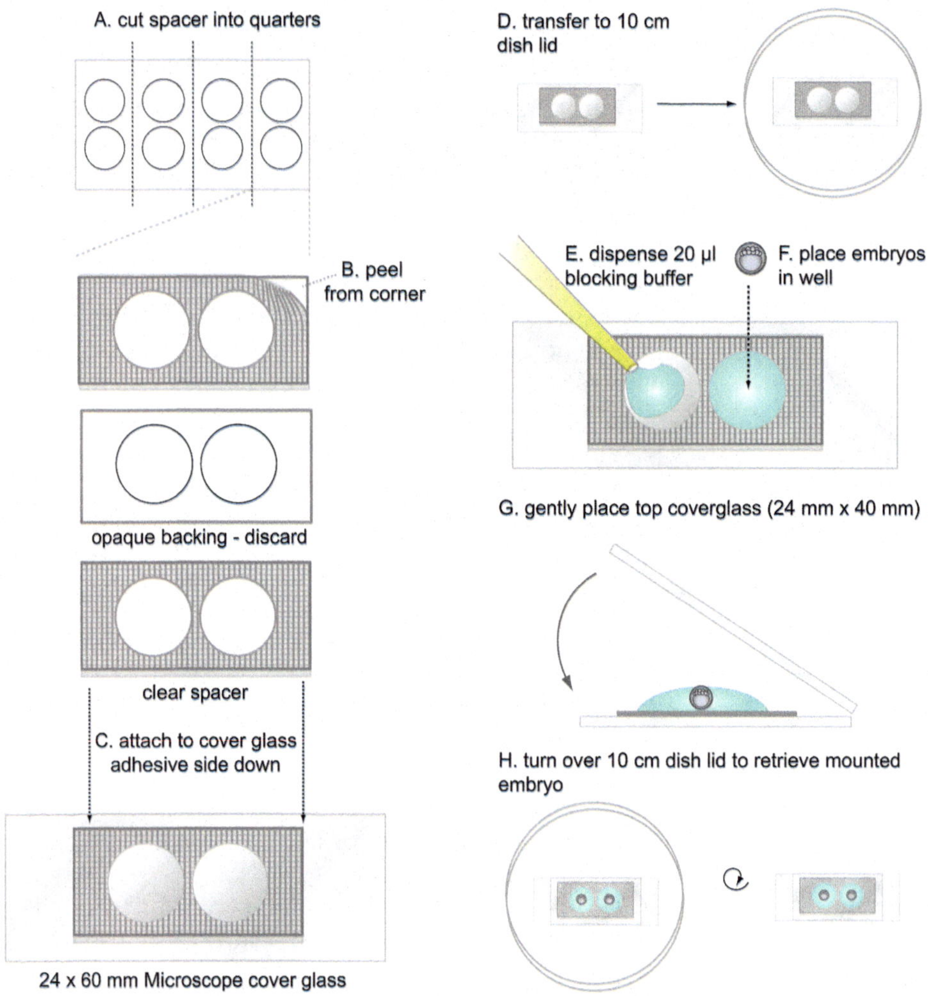

Fig. 4 Mounting embryos using Secure-Seal Spacers. (**a–h**) Cartoon summary of procedures described in Protocol 3.5

6. Transfer a single embryo in each well (Fig. 4f). If embryos do not need to be identified after imaging, multiple embryos can be transferred to each well (we recommend a maximum of five embryos per well).
7. Place the top cover glass by aligning the top cover glass parallel to the bottom cover glass and placing one long side in contact with the bottom cover glass. Gently lower upper cover glass until it makes contact with the spacer, avoiding bubbles (Fig. 4g). If an embryo becomes trapped in a bubble of air, slide the upper cover glass until the embryo enters the blocking solution, and then slide the upper cover glass back into place. Press gently on the cover glass sandwich to create a seal that is held by surface tension (*see* **Note** 7).

8. Collect the cover glass by turning over the 10 cm dish lid and catching the cover glass sandwich in your hand (Fig. 4h). Embryos should be imaged immediately by laser scanning confocal microscopy (*see* **Note 8**).

3.6 Recovery and Genotyping of Embryos by PCR

1. Prepare solutions for DNA extraction of embryos using the Sigma REDExtract-N-Amp Tissue PCR Kit (*see* **Note 9**). For each embryo, add 4.4 μl extraction solution and 1.1 μl tissue preparation solution to an Eppendorf tube to produce DNA extraction master mix. Next, aliquot 5.5 μl DNA extraction master mix to individual PCR tubes for each embryo.
2. Place a cover glass sandwich containing embryos to recover on the lid of a 10 cm petri dish, with the smaller cover glass facing up. To unseal the top cover glass, add 60–100 μl of DPBS without $MgCl_2$ or $CaCl_2$ to the perimeter of the cover glass sandwich so that the upper cover glass floats just slightly.
3. Slide the cover glass until the edge of one well becomes exposed, while keeping the other well sealed. Recover the embryo from the open well, and transfer in minimal solution to a PCR tube containing the DNA extraction master mix.
4. Slide the top cover glass to expose the edge of the second well. Recover the embryo and transfer to the next PCR tube, as before.
5. When all embryos have been recovered and transferred to PCR tubes containing the DNA extraction master mix, place the PCR tubes in thermocycler, and run the following program:
   (a) 56 °C for 30 min
   (b) 24 °C for 5 min
   (c) 95 °C for 5 min
   (d) 24 °C hold
6. To each PCR tube, add 4.4 μl neutralization solution and mix thoroughly. The extracted embryo DNA can now be stored at 4 °C until further use.
7. For PCR, follow the manufacture's recommendations for the REDExtract-N-Amp PCR Reaction Mix, and then add 1 μl DNA extract to a 10 μl PCR reaction with 200–500 nM allele-specific primers. Typically, 35–40 cycles are sufficient to produce enough amplified DNA for detection by gel electrophoresis.

3.7 Culturing Embryos

1. One day prior to collecting embryos, set up and equilibrate a plate for embryo culture. In a 35 mm dish, make 6× 20 μl drops of KSOM medium. Layer ES cell grade mineral oil on top of the drops of KSOM medium until they are completely submerged (Fig. 3g). Return the embryo culture plate to the incubator. Allow 12–18 h for KSOM to equilibrate.

2. Collect embryos and wash through drops of M2. Transfer embryos to the embryo culture plate and wash through all six drops of equilibrated KSOM. Leave embryos in final drop of KSOM and return culture dish to the incubator. Embryos will continue development in culture, although on a delayed schedule (Table 4).
3. Embryos may be observed under the microscope during culture, but care should be taken to minimize the time spent out of the incubator, as embryos are extremely sensitive to small perturbations in temperature.

4 Notes

1. *Best practices for in-house matings.* Because preimplantation development begins immediately after fertilization, it is optimal to set up timed matings in-house, rather than ordering pregnant females from a commercial vendor. Maintaining mice on a 12-h light cycle will result in peak mating activity during the sixth hour of the dark cycle. Male stud mice become sexually mature at approximately 6 weeks of age. It is ideal to allow studs to first practice mating for 5–7 days and then 2–3 days to rest before pairing with the experimental female. This initial mating should also be used to confirm that the stud is fertile. Depending on strain, male mice are typically fertile for up to 1 year but should be replaced every 6 months if possible for optimal mating performance.
2. *Natural mating versus superovulation.* Injecting female mice with hormones to trigger estrus (i.e., superovulation) can be convenient for controlling the timing of reproduction. However, embryos from superovulated females are often abnormal and fail to develop. To ensure embryo quality and reproducibility of results, using naturally mated females, rather than superovulated ones, is recommended.
3. *Approaches to handling embryos.* Preimplantation embryos, which are ~100 μm in diameter, are too small to transfer by pipetman because they can be easily lost. Accordingly, the standard in most transgenic facilities is to transfer embryos using the mouth-operated embryo transfer device we describe here. However, hand-operated embryo transfer devices also exist and provide enhanced compliance with biosafety standards [29].
4. *Culturing embryos, caveat emptor.* Culturing embryos is to be avoided whenever freshly flushed embryos can be used, because cultured embryos exhibit a developmental delay (Table 4), suggesting that culture methods do not completely recapitulate in vivo development. However, culturing embryos is necessary

when embryos are to be exposed to signaling pathway agonists or antagonists. For example, inhibitor of Rho-associated protein kinase (ROCK) has been used to interfere with cell polarization in cultured preimplantation mouse embryos [15]. In this case, treated embryos are to be compared with cultured (and possibly flushed) controls at equivalent cell number. Reducing the incubator oxygen concentration to 5% greatly improves the rate of cultured embryo development. This can be achieved with specialized hypoxia chamber or incubator.

5. *The value of nuclear stains.* Counterstains, such as a nuclear stain, are commonly used to identify locations of cells within a sample. For preimplantation analyses, nuclear stains provide an additional benefit, which is to permit the quantification of cell number and thereby infer the embryo's developmental stage. To achieve a complete count of embryo cell number, it is important to collect z-stacks of the entire embryo by confocal microscopy, with confocal cross sections collected every 5 μm.
6. *Preserving embryo structure and immunofluorescent signal.* The quality of the embryo images is greatly influenced by how they are mounted and imaged. To preserve the three-dimensional structure of the embryo, embryos are mounted in blocking solution or dilute DNA stain (Table 2), between two cover glasses separated by a Secure-Seal spacer. Mounting in more viscous solutions or without a spacer tends to crush the embryos and diminish immunofluorescent signal (compare Fig. 5a–c).
7. *Recovering embryos after imaging.* Cover glass sandwiches are held together only by the surface tension of the mounting medium. Therefore, this protocol is not compatible with long-term storage of embryo samples. However, the advantage of this approach is that embryos can be recovered for genotyping after image collection by sliding the upper cover glass aside after imaging. This is especially useful for genotyping embryos that may be carrying alleles of interest (*see* Subheading 3.5 and **Note 9**).
8. *Confocal best practices.* Signal is lost when imaging by wide-field fluorescence (Fig. 5d), rather than confocal, where a z-stack of optical sections can be collected. An additional advantage of collecting z-stacks is that every cell within each embryo is analyzed, thereby permitting phenotype analyses at the single-cell level.
9. *Recovering embryos for further analysis after imaging.* When knockout or other mutant alleles are used, embryo genotype can be determined after imaging, by carefully recovering embryos and performing single embryo genomic DNA extraction and PCR.

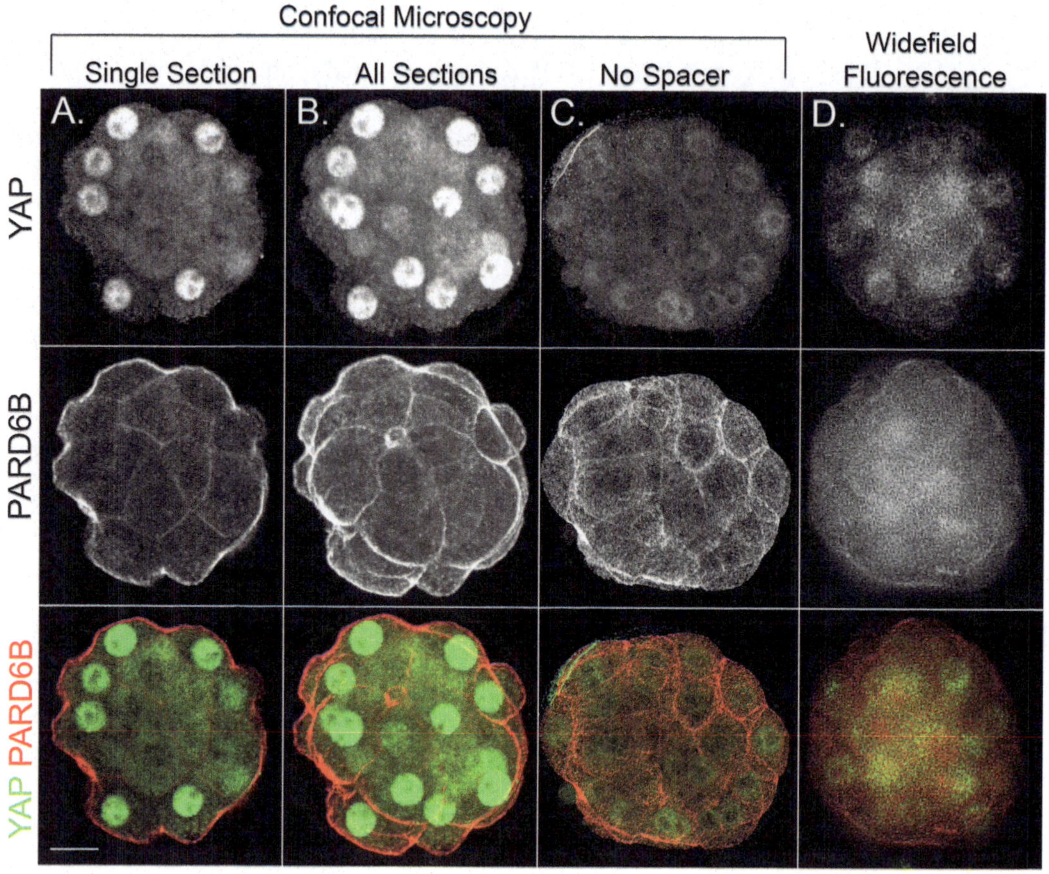

Fig. 5 Differing microscopy approaches influence image quality. (**a**) Typical images resulting from laser scanning confocal microscope analysis of a single section of a preimplantation mouse embryo stained with indicated antibodies. (**b**) Maximum intensity projection of a z-stack of sections collected every 5 μm for the embryo shown in panel **a**. (**c**) Images resulting when sample was imaged as in panel **a**, but mounted without a spacer. (**d**) Images resulting when sample was imaged using wide-field fluorescence and LED light source. Bar = 25 μm

Acknowledgments

Work in our lab is supported by National Institutes of Health grant R01 GM104009 and the James K. Billman, Jr., M.D. Endowment Fund.

References

1. Fu V, Plouffe SW, Guan KL (2018) The Hippo pathway in organ development, homeostasis, and regeneration. Curr Opin Cell Biol 49:99–107. https://doi.org/10.1016/j.ceb.2017.12.012
2. Yu FX, Zhao B, Guan KL (2015) Hippo pathway in organ size control, tissue homeostasis, and cancer. Cell 163(4):811–828. https://doi.org/10.1016/j.cell.2015.10.044
3. Harvey KF, Zhang X, Thomas DM (2013) The Hippo pathway and human cancer. Nat Rev Cancer 13(4):246–257. https://doi.org/10.1038/nrc3458

4. Johnson R, Halder G (2014) The two faces of Hippo: targeting the Hippo pathway for regenerative medicine and cancer treatment. Nat Rev Drug Discov 13(1):63–79. https://doi.org/10.1038/nrd4161
5. Pan D (2010) The hippo signaling pathway in development and cancer. Dev Cell 19 (4):491–505. https://doi.org/10.1016/j.devcel.2010.09.011
6. Sharif AAD, Hergovich A (2018) The NDR/LATS protein kinases in immunology and cancer biology. Semin Cancer Biol 48:104–114. https://doi.org/10.1016/j.semcancer.2017.04.010
7. Yimlamai D, Fowl BH, Camargo FD (2015) Emerging evidence on the role of the Hippo/YAP pathway in liver physiology and cancer. J Hepatol 63(6):1491–1501. https://doi.org/10.1016/j.jhep.2015.07.008
8. Zanconato F, Cordenonsi M, Piccolo S (2016) YAP/TAZ at the roots of cancer. Cancer Cell 29(6):783–803. https://doi.org/10.1016/j.ccell.2016.05.005
9. Rossant J, Lis WT (1979) Potential of isolated mouse inner cell masses to form trophectoderm derivatives in vivo. Dev Biol 70 (1):255–261
10. Suwinska A, Czolowska R, Ozdzenski W, Tarkowski AK (2008) Blastomeres of the mouse embryo lose totipotency after the fifth cleavage division: expression of Cdx2 and Oct4 and developmental potential of inner and outer blastomeres of 16- and 32-cell embryos. Dev Biol 322(1):133–144. https://doi.org/10.1016/j.ydbio.2008.07.019
11. Tarkowski AK, Suwinska A, Czolowska R, Ozdzenski W (2010) Individual blastomeres of 16- and 32-cell mouse embryos are able to develop into foetuses and mice. Dev Biol 348 (2):190–198. https://doi.org/10.1016/j.ydbio.2010.09.022
12. McDole K, Xiong Y, Iglesias PA, Zheng Y (2011) Lineage mapping the pre-implantation mouse embryo by two-photon microscopy, new insights into the segregation of cell fates. Dev Biol 355(2):239–249. https://doi.org/10.1016/j.ydbio.2011.04.024
13. Toyooka Y, Oka S, Fujimori T (2016) Early preimplantation cells expressing Cdx2 exhibit plasticity of specification to TE and ICM lineages through positional changes. Dev Biol 411(1):50–60. https://doi.org/10.1016/j.ydbio.2016.01.011
14. Nishioka N, Inoue K, Adachi K, Kiyonari H, Ota M, Ralston A, Yabuta N, Hirahara S, Stephenson RO, Ogonuki N, Makita R, Kurihara H, Morin-Kensicki EM, Nojima H, Rossant J, Nakao K, Niwa H, Sasaki H (2009) The Hippo signaling pathway components Lats and Yap pattern Tead4 activity to distinguish mouse trophectoderm from inner cell mass. Dev Cell 16(3):398–410. https://doi.org/10.1016/j.devcel.2009.02.003
15. Kono K, Tamashiro DA, Alarcon VB (2014) Inhibition of RHO-ROCK signaling enhances ICM and suppresses TE characteristics through activation of Hippo signaling in the mouse blastocyst. Dev Biol 394(1):142–155. https://doi.org/10.1016/j.ydbio.2014.06.023
16. Cockburn K, Biechele S, Garner J, Rossant J (2013) The Hippo pathway member Nf2 is required for inner cell mass specification. Curr Biol 23(13):1195–1201. https://doi.org/10.1016/j.cub.2013.05.044
17. Hirate Y, Hirahara S, Inoue K, Suzuki A, Alarcon VB, Akimoto K, Hirai T, Hara T, Adachi M, Chida K, Ohno S, Marikawa Y, Nakao K, Shimono A, Sasaki H (2013) Polarity-dependent distribution of angiomotin localizes Hippo signaling in preimplantation embryos. Curr Biol 23(13):1181–1194. https://doi.org/10.1016/j.cub.2013.05.014
18. Leung CY, Zernicka-Goetz M (2013) Angiomotin prevents pluripotent lineage differentiation in mouse embryos via Hippo pathway-dependent and -independent mechanisms. Nat Commun 4:2251. https://doi.org/10.1038/ncomms3251
19. Maitre JL, Turlier H, Illukkumbura R, Eismann B, Niwayama R, Nedelec F, Hiiragi T (2016) Asymmetric division of contractile domains couples cell positioning and fate specification. Nature 536(7616):344–348. https://doi.org/10.1038/nature18958
20. Ralston A, Cox BJ, Nishioka N, Sasaki H, Chea E, Rugg-Gunn P, Guo G, Robson P, Draper JS, Rossant J (2010) Gata3 regulates trophoblast development downstream of Tead4 and in parallel to Cdx2. Development 137(3):395–403. https://doi.org/10.1242/dev.038828
21. Nishioka N, Yamamoto S, Kiyonari H, Sato H, Sawada A, Ota M, Nakao K, Sasaki H (2008) Tead4 is required for specification of trophectoderm in pre-implantation mouse embryos. Mech Dev 125(3–4):270–283. https://doi.org/10.1016/j.mod.2007.11.002
22. Strumpf D, Mao CA, Yamanaka Y, Ralston A, Chawengsaksophak K, Beck F, Rossant J (2005) Cdx2 is required for correct cell fate specification and differentiation of trophectoderm in the mouse blastocyst. Development 132(9):2093–2102. https://doi.org/10.1242/dev.01801

23. Wicklow E, Blij S, Frum T, Hirate Y, Lang RA, Sasaki H, Ralston A (2014) HIPPO pathway members restrict SOX2 to the inner cell mass where it promotes ICM fates in the mouse blastocyst. PLoS Genet 10(10):e1004618. https://doi.org/10.1371/journal.pgen.1004618
24. Avilion AA, Nicolis SK, Pevny LH, Perez L, Vivian N, Lovell-Badge R (2003) Multipotent cell lineages in early mouse development depend on SOX2 function. Genes Dev 17(1):126–140. https://doi.org/10.1101/gad.224503
25. Home P, Ray S, Dutta D, Bronshteyn I, Larson M, Paul S (2009) GATA3 is selectively expressed in the trophectoderm of peri-implantation embryo and directly regulates Cdx2 gene expression. J Biol Chem 284(42):28729–28737. https://doi.org/10.1074/jbc.M109.016840
26. Ma GT, Roth ME, Groskopf JC, Tsai FY, Orkin SH, Grosveld F, Engel JD, Linzer DI (1997) GATA-2 and GATA-3 regulate trophoblast-specific gene expression in vivo. Development 124(4):907–914
27. Lian I, Kim J, Okazawa H, Zhao J, Zhao B, Yu J, Chinnaiyan A, Israel MA, Goldstein LS, Abujarour R, Ding S, Guan KL (2010) The role of YAP transcription coactivator in regulating stem cell self-renewal and differentiation. Genes Dev 24(11):1106–1118. https://doi.org/10.1101/gad.1903310
28. Behringer R, Gertsenstein M, Nagy KV, Nagy A (2016) Selecting female mice in estrus and checking plugs. Cold Spring Harb Protoc 2016(8). https://doi.org/10.1101/pdb.prot092387
29. Behringer R, Gertsenstein M, Nagy KV, Nagy A (2014) Manipulating the mouse embryo: a laboratory manual, 4th edn. Cold Spring Harbor Laboratory Press, Cold Spring Harbor, NY

Chapter 26

The Hippo Signaling Pathway in Regenerative Medicine

Lixin Hong, Yuxi Li, Qingxu Liu, Qinghua Chen, Lanfen Chen, and Dawang Zhou

Abstract

The major role of Hippo signaling is to inhibit their downstream effectors YAP/TAZ for organ size control during development and regeneration (Nat Rev Drug Discov 13(1):63–79, 2014; Dev Cell 19 (4):491–505, 2010; Cell 163(4):811–828, 2015). We and others have demonstrated that the genetic disruption of kinases Mst1 and Mst2 (Mst1/2), the core components of Hippo signaling, results in YAP activation and sustained liver growth, thereby leading to an eight- to tenfold increase in liver size within 3 months and occurrence of liver cancer within 5 months (Curr Biol 17(23):2054–2060, 2007; Cancer Cell 16(5):425–438, 2009; Cell 130(6):1120–1133, 2007; Cancer Cell 31(5):669–684 e667, 2017; Nat Commun 6:6239, 2015; Cell Rep 3(5):1663–1677, 2013). XMU-MP-1, an Mst1/2 inhibitor, is able to augment mouse liver and intestinal repair and regeneration in both acute and chronic injury mouse models (Sci Transl Med 8:352ra108, 2016).In addition, YAP-deficient mice show an impaired intestinal regenerative response after DSS treatment or gamma irradiation (Proc Natl Acad Sci U S A 108(49):E1312–1320, 2011; Nature 493(7430):106–110, 2013; Genes Dev 24(21):2383–2388, 2010; J Vis Exp (111), 2010). IBS008738, a TAZ activator, facilitates muscle repair after cardiotoxin-induced muscle injury (Mol Cell Biol. 2014;34(9):1607–21). Deletion of Salvador (Sav) in mouse hearts enhances cardiomyocyte regeneration with reduced fibrosis and recovery of pumping function after myocardial infarction (MI) or resection of mouse cardiac apex (Development 140(23):4683–4690, 2013; Sci Signal 8(375):ra41, 2015; Nature 550(7675):260–264, 2017). This chapter provides a detailed description of procedures and important considerations when performing the protocols for the respective assays used to determine the effects of Hippo signaling on tissue repair and regeneration.

Key words Hippo signaling, Tissue regeneration, Protocol

1 Introduction

1.1 Liver Regeneration

The liver's regenerative capacity has been known since mythological times and has been intensively studied by scientists since the early years of last century. Recent studies reveal that the Hippo pathway plays an important role in regulation of liver repair and

Lixin Hong and Yuxi Li contributed equally to this work.

Alexander Hergovich (ed.), *The Hippo Pathway: Methods and Protocols*, Methods in Molecular Biology, vol. 1893, https://doi.org/10.1007/978-1-4939-8910-2_26, © Springer Science+Business Media, LLC, part of Springer Nature 2019

regeneration [1, 3, 4, 9]. Pharmaceutical intervention of Hippo activity augments liver regeneration but also brings risks, as Hippo deficiency results in hepatomegaly and liver cancer[2, 5–8]. External stimulants induce liver injury and regeneration. Here, we will focus on one humanized chimeric murine model, one surgical model, and two chemical-induced liver injury models.

1.1.1 Repopulation of Human Hepatocytes in Fah$^{-/-}$/Rag2$^{-/-}$/Il2rg$^{-/-}$ *(FRG) Mice*

Cultured primary human hepatocytes have different gene expression and metabolic profile compared with hepatocytes in liver, and the cells are difficult to be substantially expanded in vitro [19]. Murine livers show different metabolic enzymes, which are just not suitable for all studies. Repopulating human primary hepatocytes in murine liver to produce a chimeric murine model overcomes the limitations mentioned above and provides broad potential application. The urokinase-type plasminogen activator (uPA) mouse model is the first reported to allow liver repopulation by xenogeneic hepatocyte. Transplantation of primary human hepatocytes in homozygous uPA-SCID mice resulted in a stable engraftment, achieving >90% replacement of the mouse tissue liver [20]. However, this model shows several disadvantages, including high mortality due to intestinal bleeding or liver failure, low breeding efficiency, narrow timeframe (5–12 days after birth) to perform hepatocyte transplantation, and unhealthy status with renal disease [21], which limits its application.

In 2007, two groups overcome several disadvantages of previous model and successfully transplanted human hepatocytes in immunodeficient (*Rag2*$^{-/-}$/*Il2rg*$^{-/-}$), fumarylacetoacetate hydrolase-deficient (*Fah*$^{-/-}$) mice [22, 23]. FAH is an enzyme which catalyzes the last step of tyrosine catabolism in hepatocytes, and its deletion leads to hepatotoxic tyrosine catabolite accumulation [24]. 2-(2-Nitro-4-trifluoro-methylbenzoyl)-1,3-cyclohexanedione (NTBC), a chemical compound, blocks the enzyme hydroxyphenylpyruvate dioxygenase, which is upstream of Fah and protects hepatocyte from the accumulation of toxic metabolites. Without NTBC, FRG mice will suffer gradual liver injury and eventually die within 1–2 months [25]. Two immunodeficient genetic backgrounds are used in this model: *Rag2*$^{-/-}$ mouse with none mature T and B lymphocytes and *Il2rg*$^{-/-}$mice with deficient T and B cell development and impaired NK cell development; therefore FRG mice have no T cells, B cells, and natural killer cells and are tolerant to hepatocyte transplantation [22, 23]. The liver damage level of FRG mice is well controllable via NTBC administration. FRG mice have highly engrafted (up to 90%) efficiency, and engrafted hepatocytes show serial transplantation capability. Moreover, the timeframe of surgery in FRG model is wide; both neonatal and adult mice can be used through different injection methods. Therefore, FRG mice have great application potential in regenerative studies.

1.1.2 Partial Hepatectomy

Two-thirds partial hepatectomy (PH) is the most widely used model in liver regenerative studies. Mouse liver has four lobes, including median lobe (39.9 ± 8.0%), left lateral lobe (30.8 ± 1.3%), right liver lobe (28.5 ± 1.7%), and caudate lobe (8.8 ± 1.4%) [26]. Two-thirds PH is usually to remove the median and left lateral lobes which are about 2/3 of liver mass. In response to the PH, the remnant liver enlarges until it restores normal mass and functions. Compared to chemical-induced liver damages, PH model induces less inflammation, and the original source of liver mass restoration is from remaining hepatocytes but not progenitor cells. The liver regeneration process after PH usually takes about 7 days which can be divided into three stages: priming, progression, and termination [27].

1.1.3 APAP-Induced Hepatotoxicity

Acetaminophen (APAP) is the leading worldwide cause of drug overdose and acute liver failure (ALF) [28]. The therapeutic dosage of APAP is metabolized to APAP-glucuronide or APAP-sulfate, which are nontoxic metabolites. However, in overdose situations, cytochrome P (CYP)-450 family member CYP2EI oxidizes APAP into hepatotoxic N-acetyl-p-benzoquinone imine (NAPQI). NAPQI neutralizes hepatic glutathione (GSH), which results in reactive oxygen species (ROS) production and inhibits mitochondrial biogenesis, leading to liver necrosis [29, 30]. In addition, APAP-induced hepatotoxicity partially results from the strong induction of hepatoprotoxicant cytokines including INF-ɣ, MIF, IL-1, and TNF-α [31, 32].Antioxidant N-acetylcysteine (NAC) is an efficient medication when treated within a short period of time post-APAP overdose [33]. We recently demonstrated that the survival rate of APAP overdose is significantly increased by Mstl/2 inhibitor: XMU-MP-1 alone or the inhibitor combined with NAC [10], which indicates that Hippo signaling may be a therapeutic target in drug-induced liver injury and regeneration studies.

1.1.4 CCl_4 Treatment-Induced Chronic Liver Injury

Carbon tetrachloride (CCl_4)-induced hepatic fibrosis and cirrhosis in rodents is a well-established and widely accepted experimental model for the study of liver fibrosis and cirrhosis [34, 35]. Chronic low dosage of CCl_4 induces liver cirrhosis [36]. CCl_4 administration results in oxidative stress and increases lipid peroxidation. Lipid peroxidation results in liver injury, necrosis, inflammation, and liver fibrosis. Similar to APAP-induced liver injury, CCl_4 administration also induces hepatoprotoxicant cytokine production, including TNFα, IL-10, and MCP-1 [31, 32, 37].Similarly, inhibition of Mstl/2 with small molecular inhibitor XMU-MP-1 ameliorates CCl_4-induced chronic liver injury [10].

1.2 Intestine Regeneration

Intestine epithelium replaces and repairs itself through the activation and expansion of stem cells. Yap is involved in the regulation of the balance between mature epithelial cells and stem cells in the intestine [4, 11–13]. Among the chemical-induced intestinal injury and colitis models, dextran sodium sulfate (DSS)-induced colitis model is a relatively simple and very widely used model of experimental colitis. DSS is a water-soluble, sulfated polysaccharide with variable molecular weight. DSS treatment results in mucin depletion, epithelial degeneration, and infiltrate of inflammatory cells. After the withdrawal of DSS, intestinal repair and regeneration occur [15, 38, 39]. We have used this protocol to address the role of Mst1/2 inhibitor in DSS-induced colitis in our previous work [10].

1.3 Heart Regeneration

Heart is a poorly regenerative organ and susceptible to organ failure, while recent researchs indicate that the Hippo pathway restricts regeneration of cardiomyocytes and deficiency of Hippo signaling enhances the regenerative potential of hearts [16–18]. In fact, cardiac tissues show complete structural regeneration after LAD ligation in newborn mice before postnatal day 7 (P7). However, the regeneration potential was quickly lost after P7.Two different models of neonatal heart injury are used for cardiac regeneration research. The first method is apical resection. The second method is the ligation of the left anterior descending artery (LAD). LAD ligation results in the loss of oxygen to and the ensuing necrosis of the heart muscle in that area [14, 40–43].

2 Materials

2.1 Liver Regeneration

1. 2-(2-Nitro-4-trifluoro-methylbenzoyl)-1,3-cyclohexanedione (NTBC) (e.g., Orfadin® (nitisinone), from Swedish Orphan Biovitrum AB).
2. 0.5% (w/v) sodium bicarbonate.
3. 0.22 μm filter.
4. Cryopreserved human hepatocytes (e.g., from XenoTech or Lonza).
5. 70% ethanol.
6. Dulbecco's modified essential medium.
7. Single-use syringe with 27-gauge needle.
8. Heparinized blood capillary.
9. Human Albumin ELISA Quantitation Kit.
10. Water bath.
11. $Fah^{-/-}/Rag2^{-/-}/Il2rg^{-/-}$ (FRG) mice.
12. Sterile surgical instruments: microsurgery scissors, microsurgery forceps, surgical suture needle (16 mm, 3/8 circle double, curved cutting or ½ circle, reverse cutting), surgical suture (4-0), 16G × 11/2″ needle, cotton swap, skin staple.

13. Eight- to ten-week-old C57BL/6 male mice.
14. Normal saline 0.9%: 0.9 g NaCl in 100 mL distilled water.
15. Anesthesia: 1% (w/v) pentobarbital sodium in normal 0.9% saline.
16. Disposable warming pad (e.g., from Kent Scientific).
17. PBS: 137 mM NaCl, 2.7 mM KCl, 10 mM Na_2HPO_4 and 2 mM KH_2PO_4, pH 7.4.
18. 100% xylene.
19. 10% neutral-buffered formalin (10% (v/v) formaldehyde, 37% in PBS).
20. Paraffin.
21. Tris-buffered saline (50 mM Tris and 150 mM NaCl, pH 7.6).
22. 3% (v/v) H_2O_2 in PBS.
23. Anti-Fah antibody (Cat#ab81047, Abcam).
24. Biotinylated anti-Rabbit IgG (Cat#BA-1100, Vector Laboratories).
25. Streptavidin-HRP conjugates (Cat# PK-7100, Vector Laboratories).
26. DAB solution (Cat# SK-4100, Vector Laboratories).
27. BSA.
28. Heparinized blood capillary (e.g., from Fisher Scientific).
29. Body temperature monitoring system.
30. 1 mL single-use syringe with a needle.
31. Tapes.
32. Spray bottle.
33. 35 mm or 60 mm cell culture dish.
34. Electric fur shaver.
35. Acetaminophen: 25 mg/ml normal 0.9% saline.
36. *N*-Acetylcysteine (NAC): 25 mg/ml in normal saline.
37. Mst1/2 inhibitor XMU-MP-1: 0.1 mg/ml in 20% (v/v) Solutol.
38. Stainless steel feeding (gavage) tubes, 22ga × 25 mm, straight, sterile.
39. Alanine Aminotransferase Assay Kit.
40. Aspartate Aminotransferase Assay Kit.
41. In Situ Apoptosis Detection Kit (*see* **Note 1**).
42. 20% (w/v) CCl_4 in corn oil.
43. 4% (v/v) formalin in normal saline 0.9%.
44. Solution A (Picrosirius Red) (*see* **Note 2**).
45. Solution B (acidified water) (*see* **Note 3**).

2.2 Intestine Regeneration

1. Dextran sulfate sodium (DSS) salt, Reagent Grade (molecular weight 36,000–50,000 kDa).
2. Male or female C57BL/6 J or BALB/c mice, preferably 8 weeks old.
3. Fecal occult blood (*see* **Note 4**).
4. RNAlater (Thermo Fisher Scientific).
5. Liquid nitrogen.
6. 10% neutral-buffered formalin: 10% (v/v) formaldehyde in PBS.

2.3 Heart Regeneration

1. Anesthesia: 1% (w/v) pentobarbital sodium in normal saline 0.9%.
2. Animal: C57BL/6 male mice.
3. Ventilator.
4. Suture.
5. Sterile surgical instruments: microsurgery scissors, microsurgery forceps, surgical suture needle.
6. Chest retractor.

3 Methods

3.1 Liver Regeneration

3.1.1 The FRG Mouse Model (See Notes 5 and 6)

1. Water with 8 μg/ml NTBC is used to maintain FRG homozygous mice.
2. Thaw cryopreserved human hepatocytes (*see* **Note 7**) rapidly (<1 min) in a 37 °C water bath.
3. Centrifuge cell at 200 × *g* for 5 min.
4. Disinfect the tube with 70% ethanol and carefully aspirate off the supernatant.
5. Resuspend the cells with Dulbecco's modified essential medium.
6. Count the cells and check cell viability with trypan blue staining.
7. Prepare 1–2 million viable hepatocytes in 50–100 μl per mouse and put the cell on ice (*see* **Note 8**).
8. Anesthetize the mice with pentobarbital sodium in normal 0.9% saline (80 mg/kg body weight) by intraperitoneal injection, and start the following procedure after the mouse is deeply anesthetized.
9. Skin preparation: shave the left side abdominal area of the mice.
10. Tape the anesthetized mouse on its back onto a warming pad.

11. Preoperative disinfection: use 70% ethanol to disinfect the skin.
12. Make a midline laparotomy via a 1–2-cm-long abdominal skin and muscle incision to expose the spleen with a scissor.
13. Inject the prepared 50–100 μl with1–2 million viable hepatocytes slowly into the lower pool of the spleen at a speed about 10 μl per second. Then take out the needle slowly to prevent bleeding (*see* **Note 9**).
14. Use one 10–12 cm suture to start the first stitch from the sternum, and leave around 3 cm suture at the sternum side, close abdominal muscle layer in S-shape, and make a knot at the end. Close skin layer by skin staples. Use antiseptic to wipe the skin surrounding the suture.
15. Take off the tapes and lay the mice on the warming pad to recover.
16. Next, you determine the NTBC cycling (*see* **Note 10**). Collect small amounts of blood once per week after surgery from the left saphenous vein with a heparinized blood capillary.
17. In the first cycle, you treat the animals as follows (*see* **Note 11**):

Days 0–2, water with 25% NTBC (2 μg/ml).

Days 3–4, water with 12% NTBC (0.96 μg/ml).

Days 5–6, water with 6% NTBC (0.48 μg/ml).

Days 7–14, withdrawn NTBC.

Days15–21, water with 100% NTBC (8 μg/ml).

18. In the second cycle you treat the animals as follows:

Days 22–23, water with 25% NTBC (2 μg/ml).

Days 25–25, water with 12% NTBC (0.96 μg/ml).

Days 26–27, water with 6% NTBC (0.48 μg/ml).

Day 28, withdrawn NTBC completely.

19. Determine the levels of human albumin.
20. Dilute serum 1000- or 10,000-fold with Tris-buffered saline.
21. Measure the human albumin concentration with a Human Albumin ELISA Quantitation Kit (Wantai, Beijing, China) according to the manufacturer's protocol.
22. Finally perform FAH staining as follows (*see* **Note 12**):
    - Collect liver samples 6–8 weeks after surgery.
    - Fix the liver tissue in 10% neutral-buffered formalin and then embed it in paraffin.
    - Section the tissue into 5 μm.
    - Deparaffinize and rehydrate paraffin sections according to standard protocol.

- Retrieve antigen by incubating sections in citrate buffer at 95 °C for 40 min.
- Wash the slides 3 × 5 min in PBS at room temperature with gentle agitation.
- Incubate the slides in 3% H_2O_2 in PBS for 10 min.
- Block the slides in 10% normal serum with 1% BSA in PBS for 30 min at room temperature.
- Incubate sections with anti-Fah antibody at 4 °C for overnight.
- Wash the slides 3 × 5 min by PBS with gentle agitation.
- Incubate with biotinylated anti-rabbit IgG for 1 h at room temperature.
- Wash the slides 3 × 5 min by PBS with gentle agitation.
- Add streptavidin-HRP conjugates and incubate for 45 min at room temperature.
- Wash the slides 3 × 5 min by PBS with gentle agitation.
- Add DAB solution, and check the slide under microscope; stop the stain when positive staining shows up.
- Counterstain the slides with hematoxylin for 10 s.
- Rinse the slides quickly with PBS, and wash in running tap water for 3 min.
- Dehydrate the slides through four times of alcohol (95%, 95%, 100%, and 100%), 5 min each.
- Clear the slides with xylene for three times and coverslip using mounting solution.

3.1.2 The Partial Hepatectomy Model (See ***Notes 13–15****)*

1. Anesthetize the mice with pentobarbital sodium in normal saline 0.9% (80 mg/kg body weight) by intraperitoneal injection, and start the following procedure after the mouse is deeply anesthetized (*see* **Note 16**).
2. Skin preparation: shave the abdominal area of the mice.
3. Tape the anesthetized mouse on its back onto a heating pad.
4. Preoperative disinfection: use 70% ethanol to disinfect the skin.
5. Make a midline laparotomy via a 2–3-cm-long abdominal skin and muscle incision to expose the xiphoid with a scissor.
6. Expose and cut off the adhesion of the liver to thoracic septum, and identify four liver lobes and cut off the adhesion between the left lateral lobe and caudate lobe (*see* **Note 17**).
7. Put one 5-0 suture under the left lateral lobe, and wind two circles of the free right end of the suture on the tip of microsurgery forceps, then clip the free left end of the suture, and

make a knot with the forceps. Resect the liver tissue distal to the knot. Cut off the suture to leave 0.3 cm on the knot.

8. Similarly, repeat **step 7** to remove median lobe of the liver.
9. Check and clean the intraperitoneal cavity and organ with warm saline and cotton swab.
10. Close abdominal muscle layer with a 5-0 suture. Close skin layer with 4-0 sutures.
11. (Optional) Intraperitoneally inject the mice with the regenerative drug or the vehicle control after surgery (*see* **Note 18**).
12. Take off the tapes, and lay the mice on the warming pad to recover (*see* **Note 19**).

*3.1.3 APAP-Induced Hepatotoxicity (See **Notes 20–25**)*

1. Fast the mice for 16 h.
2. Administrate 300–600 mg/kg APAP via oral gavage carefully to the mice.
3. 1.5–2.5 h later, intraperitoneally inject 250 mg/kg NAC and/or XMU-MP-1 or the vehicle control to the mice.
4. Observe the mice every 30 min and make records for surviving curve.
5. Collect blood via cardiac puncture with 1 ml single-use syringe for ALT/AST test according to the manufacturer's instructions.
6. Collect liver samples on proper time after APAP and NAC treatment, and perform TUNEL assays according to the manufacturer's instructions.

*3.1.4 CCl_4 Treatment-Induced Chronic Liver Injury (See **Notes 26–28**)*

1. Administrate 20% CCl_4 in corn oil or the corn oil control twice per week for 1 month by intraperitoneal injections.
2. After 1-month CCl_4 treatment, half of each group mice intraperitoneally received XMU-MP-1 (1 mg/kg) and the other mice received the vehicle control daily for 10 days.
3. Collect blood via cardiac puncture for ALT/AST test according to the manufacturer's instructions.
4. Collect liver samples for standard H&E and TUNEL staining according to the manufacturer's instructions and Sirius Red staining according to below protocol.
5. To perform a Sirius Red staining, do as follows:
   - Fix tissue in 10% neutral-buffered formalin embedded in paraffin.
   - Section the tissue into 5 μm.
   - Deparaffinize and rehydrate paraffin sections according to standard protocol.

- Stain in solution A (Picrosirius Red) for 1 h.
- Rinse the slides in solution B (acidified water) twice.
- Dry the slides by blotting with damp filter paper.
- Dehydrate the slides through three times of alcohol, 5 min each.
- Clear the slides with xylene for three times and coverslip using mounting solution.

3.2 Intestine Regeneration (See Notes 29–32)

1. Label and weight the mice.
2. Prepare an optimized concentration of DSS in autoclaved water until completely dissolved.
3. Fill the water bottle in the cages with DSS-containing water. Control mice are provided with autoclaved water without DSS.
4. Measure body weight fecal occult blood every day. Score the intestinal bleeding.
5. Usually, DSS is given for 5 or 7 days. After DSS treatment, mice were supplied with normal drinking water for another 7 days to repair the damage.
6. Four hours before sacrifice, mice should be fasted for food, administered with BrdU solution (to measure epithelial cell proliferation) and FITC-dextran tracer (to measure intestinal permeability) intraperitoneally.
7. Perform last bleeding and euthanize mice with CO_2.
8. Open the mouse by abdomen incision. Remove and weigh the spleen.
9. Lift the colon until the cecum is visible and separate the colon.
10. Take pictures and measure length of the colon.
11. Rinse the colon with cold PBS to remove feces and blood.
12. Cut colons into pieces depending on need: one for Western blotting (quick frozen in liquid nitrogen), one for RNA isolation (stored in RNA later), and one for histology analysis (fixed in 10% formalin), for example.

3.3 Heart Regeneration (See Notes 33–35)

1. Anesthetize the mice with 0.9% pentobarbital sodium in normal saline (80 mg/kg body weight) by intraperitoneal injection, and ensure the mouse reaches a deep plane of anesthesia.
2. Tape the anesthetized mouse on its back onto a heating pad.
3. Disinfect the skin with 70% ethanol.
4. Make a midline incision in the neck; retract the skin and muscle to expose the trachea.
5. Cut a small hole into the trachea, and insert the endotracheal tube of the ventilator (respiration rate (RR), 110 per minute).

6. Turn the mouse to its left side. Perform skin incision on the left chest and thoracotomy in the fourth intercostal space, and dissect the tissue and muscle carefully.
7. Drip normal saline into the cavity to protect the lungs.
8. Open the thorax carefully. Spread apart and fix the ribs with a chest retractor. Remove the pericardial sack covering the heart.
9. The LAD is located between the pulmonary artery and the left auricle. Ligate the LAD with one single suture below the artery through the mid-ventricle below the left auricle.
10. Remove the chest retractor. Use two sutures to close the ribs. Close the thoracic incision in muscle and skin layers using sutures.
11. Remove the endotracheal tube; sew up the trachea and skin in the neck area using running sutures.
12. Mice are injected with buprenorphine (0.05 mg/kg) to alleviate pain in the next 3 days.

4 Notes

1. The ApopTag Peroxidase In Situ Apoptosis Detection Kit (Millipore).
2. Solution A: dissolve 0.5 g Direct Red 80 (Sigma-Aldrich) in 500 ml saturated aqueous solution of picric acid.
3. Solution B: add 5 ml acetic acid to 1 L of distilled water and mix well.
4. Fecal occult blood: Hemoccult Dispensapak Plus from Beckman Coulter (Ref = 61130).
5. Genetic background: C57bl/6 and NOD are two different backgrounds, and there is no significant difference in human hepatocyte transplantation [44–46] . However, NOD background is recommended if human hematopoietic stem cell expansion is required [47].
6. Prepare NTBC stock (1000×):scale 8 g of NTBC, and dissolve in 1 L of 0.5% sodium bicarbonate at 65 °C with shaking intermittently every 5 min until the entire solid compound is dissolved. And filter the solution with a 0.22 μm filter. Finally allocate and store at −20 °C.
7. Hepatocytes sources: the age of the hepatocyte donor and the origin of isolation (cadaveric or resection) has no discernible impact on hepatocyte repopulation. Cryopreserved human hepatocytes are widely utilized. And hepatocytes, expanded in chimeric mice with serial transplantation, also repopulate well.

8. The amount of cell: one to five million cells were used in different reports. More cells may lead to faster repopulation; however, it is easy to cause bleeding if high-concentration and large-volume cells are used [48].
9. Alternative hepatocyte transplantation method: although different groups use distinct cell amount or NTBC cycling manner, majority of researchers manage the transplantation via intra-splenic injection into adult mice. However, according to the report, intrahepatic transplantation into neonatal mice is also efficient [22]. Hepatocytes in neonatal liver have a higher division rate and provide the proliferative environment, which may be helpful in regenerative medicine research.
10. NTBC cycling: liver damage degree is negatively correlated with the level of NTBC. NTBC concentration and cycling timing are manageable. Repopulation efficiency also depends on the quality of human hepatocytes. Therefore it is important to set up a proper cycling manner. NTBC is stable for 3 weeks at room temperature. (Optional) Administrate the mice with the regenerative drug or the vehicle control 10–14 days after NTBC withdrawal.
11. If the surgery is unsuccessful, the mice will die in 10 days.
12. FRG mice are FAH negative, and transplanted human hepatocyte can be distinguished from mouse hepatocyte by FAH staining.
13. Genetic background: though wild-type animals with different genetic background show similar recovery capacity, mutations under different background [49] or HBV infection or fibrosis level have interfered with the results [50, 51].
14. Gender: clinical studies demonstrate that female with HCC have a better prognosis than male after liver resection surgery [52]. Female mice show delayed regeneration rate after PH has been reported [53]. Thus, the same gender mice are recommended to be used in PH procedures.
15. Age: in general, liver regenerative capacity decreased with age in rodents and human. Old mice (older than 10 months) delay liver regeneration [54]. 8–14-week-old mice are recommended except age-dependent research is required.
16. Anesthesia: other injectable anesthetics such as the combination of ketamine (80–110 mg/kg) and xylazine (5–10 mg/kg) are also widely used. Due to its hepatotoxicity, Avertin is not advisable to be used in PH. It is reported that injectable drugs increase the risk of insufficient anesthetics and severe side effect including hepatotoxicity, and inhalant anesthetics such as isoflurane are highly recommended in some protocol [55].

17. During the procedure, saline-moistened cotton swabs can be used to lift or pull down the liver lobes or other organs gently.
18. When and how to administrate the regenerative drug depend on the purpose of the study and pharmacokinetic of the drug. Choose the proper method for each experiment.
19. Proper temperature: during disinfection, avoid too much 70% ethanol to prevent hypothermia. During the recovery stage after surgery, continuous warming is required until resume of physical activity.
20. We have used this protocol to address the role of Mst1/2 inhibitor XMU-MP-1 in APAP-induced hepatotoxicity.
21. Age: it is reported that neonatal mice (3–35 days old) are resistant to APAP-induced hepatotoxicity [56]. Moreover, no matter the intraperitoneal or oral APAP administration, younger mice show more resistance to APAP-induced hepatotoxicity than older ones [57]. It is required to use mice with the same age in this model.
22. Gender: male mice are more susceptible than females to APAP-derived hepatotoxicity [58]. Usually, it is recommended to use male mice in this model.
23. Mouse strains: the susceptibility to an APAP overdose is various in different mouse strains. According to the reports, B6C3F1/J is the most sensitive strain, and CAST/EiJ is the most resistant strain to APAP [59].
24. The nutritional status: most of the mice, which are used in APAP liver injury studies, need to be fasted for 12–20 h to reduce the hepatic GSH level [60]. However, fasting also reduces glycogen and ATP levels in hepatocytes and may prevent apoptosis but promote necrosis [61]. Furthermore, the mice with different nutritional status show varied susceptibility to APAP toxicity. Western-style diet with high fat and high carbohydrate induces liver steatosis, and the mice with hepatic steatosis are more resistant to APAP-induced liver injury [62].
25. Infections: bacterial and viral infections have an effect on the susceptibility to APAP hepatotoxicity by modulating cytochrome P450 enzyme expression and activities or immune cells recruitment and cytokine release [63–66]. Therefore, infection-free mice are required for APAP-induced liver injury studies.
26. We have used this protocol to address the role of Mst1/2 inhibitor XMU-MP-1 in CCl_4-induced hepatotoxicity. Design the experiments.
27. Age: several studies have shown that age is a vital parameter of tissue repair following CCl_4 exposure. Generally, newborn animals have the highest efficiency for damage repair and

regeneration [67, 68]. Mice with similar age are recommended except age-dependent research is required.

28. The nutritional status and disease condition: it should be noted that glucose loading inhibits CCl_4-induced centrilobular hepatotoxicity and tissue repair by suppressing tissue repair response [69]. In a noninsulin-dependent diabetic model, tissue repair after CCl_4-induced hepatotoxicity is also inhibited [70].
29. The colitogenic potential of DSS varies between diverse molecular weight. Administration of 40–50 kDa DSS results in the most severe colitis in mice.
30. It is recommended to purchasing DSS in quantity and determining the optimal dosage because the efficiency of DSS may be different between vendors and batches.
31. Mice with specific genetic modification or treatment may consume more or less water than the control group.
32. Change the DSS water every 2 days or if any growth is observed in the bottles.
33. Fur around the surgical area may be shaved to avoid infection.
34. The surgical area and instruments must be sterilized.
35. Neonatal ribs are very fragile. Operate carefully to avoid breaking it.

Acknowledgments

The authors thank Quan Yuan (Xiamen University, China) for critical reading of the protocol of FRG mice model. This work was supported by grants from National Key R&D Program of China 2017YFA0504502 to D.Z. and L.C., the National Natural Science Foundation of China (81790254, 31625010, and U1505224 to D.Z.; U1405225 and 81372617 to L.C.; and 81472229 to L.H.). Lixin Hong and Yuxi Li contributed equally to this work.

References

1. Johnson R, Halder G (2014) The two faces of Hippo: targeting the Hippo pathway for regenerative medicine and cancer treatment. Nat Rev Drug Discov 13(1):63–79. https://doi.org/10.1038/nrd4161
2. Pan D (2010) The hippo signaling pathway in development and cancer. Dev Cell 19 (4):491–505. https://doi.org/10.1016/j.devcel.2010.09.011
3. Yu FX, Zhao B, Guan KL (2015) Hippo pathway in organ size control, tissue homeostasis, and cancer. Cell 163(4):811–828. https://doi.org/10.1016/j.cell.2015.10.044
4. Camargo FD, Gokhale S, Johnnidis JB, Fu D, Bell GW, Jaenisch R, Brummelkamp TR (2007) YAP1 increases organ size and expands undifferentiated progenitor cells. Curr Biol 17

(23):2054–2060. https://doi.org/10.1016/j.cub.2007.10.039

5. Zhou D, Conrad C, Xia F, Park JS, Payer B, Yin Y, Lauwers GY, Thasler W, Lee JT, Avruch J, Bardeesy N (2009) Mst1 and Mst2 maintain hepatocyte quiescence and suppress hepatocellular carcinoma development through inactivation of the Yap1 oncogene. Cancer Cell 16(5):425–438. https://doi.org/10.1016/j.ccr.2009.09.026
6. Dong J, Feldmann G, Huang J, Wu S, Zhang N, Comerford SA, Gayyed MF, Anders RA, Maitra A, Pan D (2007) Elucidation of a universal size-control mechanism in Drosophila and mammals. Cell 130(6):1120–1133. https://doi.org/10.1016/j.cell.2007.07.019
7. Zhang S, Chen Q, Liu Q, Li Y, Sun X, Hong L, Ji S, Liu C, Geng J, Zhang W, Lu Z, Yin ZY, Zeng Y, Lin KH, Wu Q, Li Q, Nakayama K, Nakayama KI, Deng X, Johnson RL, Zhu L, Gao D, Chen L, Zhou D (2017) Hippo signaling suppresses cell ploidy and tumorigenesis through Skp2. Cancer Cell 31(5):669–684 e667. https://doi.org/10.1016/j.ccell.2017.04.004
8. Wu H, Wei L, Fan F, Ji S, Zhang S, Geng J, Hong L, Fan X, Chen Q, Tian J, Jiang M, Sun X, Jin C, Yin ZY, Liu Q, Zhang J, Qin F, Lin KH, Yu JS, Deng X, Wang HR, Zhao B, Johnson RL, Chen L, Zhou D (2015) Integration of Hippo signalling and the unfolded protein response to restrain liver overgrowth and tumorigenesis. Nat Commun 6:6239. https://doi.org/10.1038/ncomms7239
9. Wu H, Xiao Y, Zhang S, Ji S, Wei L, Fan F, Geng J, Tian J, Sun X, Qin F, Jin C, Lin J, Yin ZY, Zhang T, Luo L, Li Y, Song S, Lin SC, Deng X, Camargo F, Avruch J, Chen L, Zhou D (2013) The Ets transcription factor GABP is a component of the hippo pathway essential for growth and antioxidant defense. Cell Rep 3(5):1663–1677. https://doi.org/10.1016/j.celrep.2013.04.020
10. Fan F, He Z, Kong LL, Chen Q, Yuan Q, Zhang S, Ye J, Liu H, Sun X, Geng J, Yuan L, Hong L, Xiao C, Zhang W, Sun X, Li Y, Wang P, Huang L, Wu X, Ji Z, Wu Q, Xia NS, Gray NS, Chen L, Yun CH, Deng X, Zhou D (2016) Pharmacological targeting of kinases MST1 and MST2 augments tissue repair and regeneration. Sci Transl Med 8(352):352ra108. https://doi.org/10.1126/scitranslmed.aaf2304
11. Zhou D, Zhang Y, Wu H, Barry E, Yin Y, Lawrence E, Dawson D, Willis JE, Markowitz SD, Camargo FD, Avruch J (2011) Mst1 and Mst2 protein kinases restrain intestinal stem cell proliferation and colonic tumorigenesis by inhibition of Yes-associated protein (Yap) overabundance. Proc Natl Acad Sci U S A 108(49):E1312–E1320. https://doi.org/10.1073/pnas.1110428108
12. Barry ER, Morikawa T, Butler BL, Shrestha K, de la Rosa R, Yan KS, Fuchs CS, Magness ST, Smits R, Ogino S, Kuo CJ, Camargo FD (2013) Restriction of intestinal stem cell expansion and the regenerative response by YAP. Nature 493(7430):106–110. https://doi.org/10.1038/nature11693
13. Cai J, Zhang N, Zheng Y, de Wilde RF, Maitra A, Pan D (2010) The Hippo signaling pathway restricts the oncogenic potential of an intestinal regeneration program. Genes Dev 24(21):2383–2388. https://doi.org/10.1101/gad.1978810
14. Blom JN, Lu X, Arnold P, Feng Q (2016) Myocardial infarction in neonatal mice, a model of cardiac regeneration. J Vis Exp (111). https://doi.org/10.3791/54100
15. Chassaing B, Aitken JD, Malleshappa M, Vijay-Kumar M (2014) Dextran sulfate sodium (DSS)-induced colitis in mice. Curr Protoc Immunol 104.:Unit(15):25. https://doi.org/10.1002/0471142735.im1525s104
16. Heallen T, Morikawa Y, Leach J, Tao G, Willerson JT, Johnson RL, Martin JF (2013) Hippo signaling impedes adult heart regeneration. Development 140(23):4683–4690. https://doi.org/10.1242/dev.102798
17. Morikawa Y, Zhang M, Heallen T, Leach J, Tao G, Xiao Y, Bai Y, Li W, Willerson JT, Martin JF (2015) Actin cytoskeletal remodeling with protrusion formation is essential for heart regeneration in Hippo-deficient mice. Sci Signal 8(375):ra41. https://doi.org/10.1126/scisignal.2005781
18. Leach JP, Heallen T, Zhang M, Rahmani M, Morikawa Y, Hill MC, Segura A, Willerson JT, Martin JF (2017) Hippo pathway deficiency reverses systolic heart failure after infarction. Nature 550(7675):260–264. https://doi.org/10.1038/nature24045
19. Brandon EF, Raap CD, Meijerman I, Beijnen JH, Schellens JH (2003) An update on in vitro test methods in human hepatic drug biotransformation research: pros and cons. Toxicol Appl Pharmacol 189(3):233–246
20. Meuleman P, Libbrecht L, De Vos R, de Hemptinne B, Gevaert K, Vandekerckhove J, Roskams T, Leroux-Roels G (2005) Morphological and biochemical characterization of a human liver in a uPA-SCID mouse chimera.

Hepatology 41(4):847–856. https://doi.org/10.1002/hep.20657

21. Tateno C, Yoshizane Y, Saito N, Kataoka M, Utoh R, Yamasaki C, Tachibana A, Soeno Y, Asahina K, Hino H, Asahara T, Yokoi T, Furukawa T, Yoshizato K (2004) Near completely humanized liver in mice shows human-type metabolic responses to drugs. Am J Pathol 165(3):901–912. https://doi.org/10.1016/S0002-9440(10)63352-4
22. Bissig KD, Le TT, Woods NB, Verma IM (2007) Repopulation of adult and neonatal mice with human hepatocytes: a chimeric animal model. Proc Natl Acad Sci U S A 104 (51):20507–20511. https://doi.org/10.1073/pnas.0710528105
23. Azuma H, Paulk N, Ranade A, Dorrell C, Al-Dhalimy M, Ellis E, Strom S, Kay MA, Finegold M, Grompe M (2007) Robust expansion of human hepatocytes in Fah−/−/Rag2−/−/Il2rg−/− mice. Nat Biotechnol 25(8):903–910. https://doi.org/10.1038/nbt1326
24. Grompe M, al-Dhalimy M, Finegold M, Ou CN, Burlingame T, Kennaway NG, Soriano P (1993) Loss of fumarylacetoacetate hydrolase is responsible for the neonatal hepatic dysfunction phenotype of lethal albino mice. Genes Dev 7(12A):2298–2307
25. Overturf K, Al-Dhalimy M, Tanguay R, Brantly M, Ou CN, Finegold M, Grompe M (1996) Hepatocytes corrected by gene therapy are selected in vivo in a murine model of hereditary tyrosinaemia type I. Nat Genet 12 (3):266–273. https://doi.org/10.1038/ng0396-266
26. Inderbitzin D, Gass M, Beldi G, Ayouni E, Nordin A, Sidler D, Gloor B, Candinas D, Stoupis C (2004) Magnetic resonance imaging provides accurate and precise volume determination of the regenerating mouse liver. J Gastrointest Surg 8(7):806–811. https://doi.org/10.1016/j.gassur.2004.07.013
27. Fausto N, Campbell JS, Riehle KJ (2012) Liver regeneration. J Hepatol 57(3):692–694. https://doi.org/10.1016/j.jhep.2012.04.016
28. Ostapowicz G, Fontana RJ, Schiodt FV, Larson A, Davern TJ, Han SH, McCashland TM, Shakil AO, Hay JE, Hynan L, Crippin JS, Blei AT, Samuel G, Reisch J, Lee WM, Group USALFS (2002) Results of a prospective study of acute liver failure at 17 tertiary care centers in the United States. Ann Intern Med 137(12):947–954
29. Hanawa N, Shinohara M, Saberi B, Gaarde WA, Han D, Kaplowitz N (2008) Role of JNK translocation to mitochondria leading to inhibition of mitochondria bioenergetics in acetaminophen-induced liver injury. J Biol Chem 283(20):13565–13577. https://doi.org/10.1074/jbc.M708916200
30. Mitchell SJ, Kane AE, Hilmer SN (2011) Age-related changes in the hepatic pharmacology and toxicology of paracetamol. Curr Gerontol Geriatr Res 2011:624156. https://doi.org/10.1155/2011/624156
31. Blazka ME, Wilmer JL, Holladay SD, Wilson RE, Luster MI (1995) Role of proinflammatory cytokines in acetaminophen hepatotoxicity. Toxicol Appl Pharmacol 133(1):43–52. https://doi.org/10.1006/taap.1995.1125
32. Kellokumpu-Lehtinen P, Iisalo E, Nordman E (1989) Hepatotoxicity of paracetamol in combination with interferon and vinblastine. Lancet 1(8647):1143
33. Whyte IM, Francis B, Dawson AH (2007) Safety and efficacy of intravenous N-acetylcysteine for acetaminophen overdose: analysis of the Hunter Area Toxicology Service (HATS) database. Curr Med Res Opin 23 (10):2359–2368. https://doi.org/10.1185/030079907X219715
34. Soni MG, Mehendale HM (1991) Protection from chlordecone-amplified carbon tetrachloride toxicity by cyanidanol: regeneration studies. Toxicol Appl Pharmacol 108(1):58–66
35. Kodavanti PR, Kodavanti UP, Faroon OM, Mehendale HM (1992) Pivotal role of hepatocellular regeneration in the ultimate hepatotoxicity of CCl_4 in chlordecone-, mirex-, or phenobarbital-pretreated rats. Toxicol Pathol 20(4):556–569. https://doi.org/10.1177/019262339202000402
36. Slater TF, Cheeseman KH, Ingold KU (1985) Carbon tetrachloride toxicity as a model for studying free-radical mediated liver injury. Philos Trans R Soc Lond Ser B Biol Sci 311 (1152):633–645
37. Bourdi M, Reilly TP, Elkahloun AG, George JW, Pohl LR (2002) Macrophage migration inhibitory factor in drug-induced liver injury: a role in susceptibility and stress responsiveness. Biochem Biophys Res Commun 294 (2):225–230. https://doi.org/10.1016/S0006-291X(02)00466-7
38. Perse M, Cerar A (2012) Dextran sodium sulphate colitis mouse model: traps and tricks. J Biomed Biotechnol 2012:718617. https://doi.org/10.1155/2012/718617
39. Das S, Batra SK, Rachagani S (2017) Mouse model of Dextran Sodium Sulfate (DSS)-induced colitis. Bio-Protocol 7(16):e2515. https://doi.org/10.21769/BioProtoc.2515
40. Kolk MV, Meyberg D, Deuse T, Tang-Quan KR, Robbins RC, Reichenspurner H, Schrepfer

S (2009) LAD-ligation: a murine model of myocardial infarction. J Vis Exp (32). https://doi.org/10.3791/1438
41. Muthuramu I, Lox M, Jacobs F, De Geest B (2014) Permanent ligation of the left anterior descending coronary artery in mice: a model of post-myocardial infarction remodelling and heart failure. J Vis Exp (94). https://doi.org/10.3791/52206
42. Haubner BJ, Schuetz T, Penninger JM (2016) A reproducible protocol for neonatal ischemic injury and cardiac regeneration in neonatal mice. Basic Res Cardiol 111(6):64. https://doi.org/10.1007/s00395-016-0580-3
43. Haubner BJ, Adamowicz-Brice M, Khadayate S, Tiefenthaler V, Metzler B, Aitman T, Penninger JM (2012) Complete cardiac regeneration in a mouse model of myocardial infarction. Aging 4(12):966–977. https://doi.org/10.18632/aging.100526
44. Vaughan AM, Mikolajczak SA, Wilson EM, Grompe M, Kaushansky A, Camargo N, Bial J, Ploss A, Kappe SH (2012) Complete Plasmodium falciparum liver-stage development in liver-chimeric mice. J Clin Invest 122 (10):3618–3628. https://doi.org/10.1172/JCI62684
45. Li F, Cowley DO, Banner D, Holle E, Zhang L, Su L (2014) Efficient genetic manipulation of the NOD-Rag1−/-IL2RgammaC-null mouse by combining in vitro fertilization and CRISPR/Cas9 technology. Sci Rep 4:5290. https://doi.org/10.1038/srep05290
46. Bissig KD, Wieland SF, Tran P, Isogawa M, Le TT, Chisari FV, Verma IM (2010) Human liver chimeric mice provide a model for hepatitis B and C virus infection and treatment. J Clin Invest 120(3):924–930. https://doi.org/10.1172/JCI40094
47. Takenaka K, Prasolava TK, Wang JC, Mortin-Toth SM, Khalouei S, Gan OI, Dick JE, Danska JS (2007) Polymorphism in Sirpa modulates engraftment of human hematopoietic stem cells. Nat Immunol 8(12):1313–1323. https://doi.org/10.1038/ni1527
48. Li F, Nio K, Yasui F, Murphy CM, Su L (2017) Studying HBV infection and therapy in immune-deficient NOD-Rag1−/-IL2RgammaC-null (NRG) fumarylacetoacetate hydrolase (Fah) knockout mice transplanted with human hepatocytes. Methods Mol Biol 1540:267–276. https://doi.org/10.1007/978-1-4939-6700-1_23
49. Mutant mice and neuroscience: recommendations concerning genetic background (1997) In: Banbury conference on genetic background in mice. Neuron 19(4):755–759
50. Churin Y, Roderfeld M, Stiefel J, Wurger T, Schroder D, Matono T, Mollenkopf HJ, Montalbano R, Pompaiah M, Reifenberg K, Zahner D, Ocker M, Gerlich W, Glebe D, Roeb E (2014) Pathological impact of hepatitis B virus surface proteins on the liver is associated with the host genetic background. PLoS One 9(3):e90608. https://doi.org/10.1371/journal.pone.0090608
51. Shi Z, Wakil AE, Rockey DC (1997) Strain-specific differences in mouse hepatic wound healing are mediated by divergent T helper cytokine responses. Proc Natl Acad Sci U S A 94(20):10663–10668
52. Ng IO, Ng M, Fan ST (1997) Better survival in women with resected hepatocellular carcinoma is not related to tumor proliferation or expression of hormone receptors. Am J Gastroenterol 92(8):1355–1358
53. Wang Y, Ye F, Ke Q, Wu Q, Yang R, Bu H (2013) Gender-dependent histone deacetylases injury may contribute to differences in liver recovery rates of male and female mice. Transplant Proc 45(2):463–473. https://doi.org/10.1016/j.transproceed.2012.06.063
54. Iakova P, Awad SS, Timchenko NA (2003) Aging reduces proliferative capacities of liver by switching pathways of C/EBPalpha growth arrest. Cell 113(4):495–506
55. Mitchell C, Willenbring H (2008) A reproducible and well-tolerated method for 2/3 partial hepatectomy in mice. Nat Protoc 3 (7):1167–1170. https://doi.org/10.1038/nprot.2008.80
56. Hart JG, Timbrell JA (1979) The effect of age on paracetamol hepatotoxicity in mice. Biochem Pharmacol 28(19):3015–3017
57. Taguchi K, Tokuno M, Yamasaki K, Kadowaki D, Seo H, Otagiri M (2015) Establishment of a model of acetaminophen-induced hepatotoxicity in different weekly-aged ICR mice. Lab Anim 49(4):294–301. https://doi.org/10.1177/0023677215573041
58. Mohar I, Stamper BD, Rademacher PM, White CC, Nelson SD, Kavanagh TJ (2014) Acetaminophen-induced liver damage in mice is associated with gender-specific adduction of peroxiredoxin-6. Redox Biol 2:377–387. https://doi.org/10.1016/j.redox.2014.01.008
59. Harrill AH, Watkins PB, Su S, Ross PK, Harbourt DE, Stylianou IM, Boorman GA, Russo MW, Sackler RS, Harris SC, Smith PC, Tennant R, Bogue M, Paigen K, Harris C, Contractor T, Wiltshire T, Rusyn I, Threadgill DW (2009) Mouse population-guided resequencing reveals that variants in CD44 contribute to acetaminophen-induced liver injury

in humans. Genome Res 19(9):1507–1515. https://doi.org/10.1101/gr.090241.108

60. Jaeschke H, Wendel A (1985) Diurnal fluctuation and pharmacological alteration of mouse organ glutathione content. Biochem Pharmacol 34(7):1029–1033
61. Antoine DJ, Williams DP, Kipar A, Laverty H, Park BK (2010) Diet restriction inhibits apoptosis and HMGB1 oxidation and promotes inflammatory cell recruitment during acetaminophen hepatotoxicity. Mol Med 16 (11–12):479–490. https://doi.org/10.2119/molmed.2010.00126
62. Ito Y, Abril ER, Bethea NW, McCuskey MK, McCuskey RS (2006) Dietary steatotic liver attenuates acetaminophen hepatotoxicity in mice. Microcirculation 13(1):19–27. https://doi.org/10.1080/10739680500383423
63. Liu J, Sendelbach LE, Parkinson A, Klaassen CD (2000) Endotoxin pretreatment protects against the hepatotoxicity of acetaminophen and carbon tetrachloride: role of cytochrome P450 suppression. Toxicology 147 (3):167–176
64. Chang KC, Bell TD, Lauer BA, Chai H (1978) Altered theophylline pharmacokinetics during acute respiratory viral illness. Lancet 1 (8074):1132–1133
65. Nguyen GC, Sam J, Thuluvath PJ (2008) Hepatitis C is a predictor of acute liver injury among hospitalizations for acetaminophen overdose in the United States: a nationwide analysis. Hepatology 48(4):1336–1341. https://doi.org/10.1002/hep.22536
66. Pichlmair A, Reis e Sousa C (2007) Innate recognition of viruses. Immunity 27 (3):370–383. https://doi.org/10.1016/j.immuni.2007.08.012
67. Dalu A, Mehendale HM (1996) Efficient tissue repair underlies the resiliency of postnatally developing rats to chlordecone + CCl_4 hepatotoxicity. Toxicology 111(1–3):29–42
68. Dalu A, Cronin GM, Lyn-Cook BD, Mehendale HM (1995) Age-related differences in TGF-alpha and proto-oncogenes expression in rat liver after a low dose of carbon tetrachloride. J Biochem Toxicol 10(5):259–264
69. Chanda S, Mangipudy RS, Warbritton A, Bucci TJ, Mehendale HM (1995) Stimulated hepatic tissue repair underlies heteroprotection by thioacetamide against acetaminophen-induced lethality. Hepatology 21(2):477–486
70. Sawant SP, Dnyanmote AV, Shankar K, Limaye PB, Latendresse JR, Mehendale HM (2004) Potentiation of carbon tetrachloride hepatotoxicity and lethality in type 2 diabetic rats. J Pharmacol Exp Ther 308(2):694–704. https://doi.org/10.1124/jpet.103.058834

INDEX

Alexander Hergovich (ed.), *The Hippo Pathway: Methods and Protocols*, Methods in Molecular Biology, vol. 1893,
https://doi.org/10.1007/978-1-4939-8910-2, © Springer Science+Business Media, LLC, part of Springer Nature 2019

Printed by Printforce, the Netherlands